THE NEW TECHNOLOGY

-a survival guide to new materials, supercomputers & global communications for the 1990s

Dimitris N. Chorafas

SIGMA PRESS – Wilmslow, United Kingdom

First published in 1990 by

Sigma Press, 1 South Oak Lane, Wilmslow, Cheshire SK9 6AR, England.

British Library Cataloguing in Publication Data

A CIP catalogue record for this book is available from the British Library.

ISBN: 1-85058-134-7

Typesetting and design by

Sigma Hi-Tech Services Ltd

Printed in Malta by
Interprint Ltd.

Distributed by

John Wiley & Sons Ltd., Baffins Lane, Chichester, West Sussex, England.

Foreword

This comprehensive work by Professor Chorafas comes at a time when burgeoning technology developments and related issues threaten to engulf us all. In one epic work Dimitris Chorafas has managed to describe, in his characteristic down to earth style, many of the most significant high technology developments and related issues which should be of concern to a wide spectrum of managers and decision makers. But it is not just managers responsible for the well being and development of enterprises operating in an increasingly competitive and global business environment who will benefit from this work. Anyone in, or aspiring to, a responsible position in science and research, in government and regulation, in education, in medicine, in politics, who is concerned with environmental issues or who is involved with social, welfare, ethical or moral progress will find much in this work to interest them.

As someone whose job is to track and seek to apply high tech developments in one of the fastest changing and demanding sectors (financial services), I find few key subjects or issues Professor Chorafas has not addressed. I might only point to the emerging science of Chaos, Catastrophe Theory, and new concepts in production which are coming forward as ''Nanotechnology'' to complete the picture in my eyes. I have no doubt that it will not be long before the prolific Professor Chorafas covers these subjects also.

The new technologies are a powerful force for change. For instance they can:

❖ facilitate massive increases in productivity in all sorts of processes

❖ create opportunities for new cheaper and better types of product and service

❖ break down barriers of geography, time, understanding and class

❖ allow the control of increasingly complex processes

❖ maintain and increase corporate and state power, whilst paradoxically increasing the power of individuals in the know

❖ aid greatly the distribution, retention and harbouring of knowledge, and so on

If used intelligently, the new technologies can improve our lives and well being. However, many are coming to see in the new technologies a Pandora's box of threats. This is a recurrent theme of this work. The new technologies provide burgeoning opportunities for commercial, military and political interests to increase their power and influence. We are only now coming to terms with the largely unbridled exploitation of technologies which have arrived at an increasing rate over the last one hundred years. The exploitation of these technologies has been driven largely by commercial self interest or military imperatives. The increasing numbers of world scale catastrophes which have arisen in recent times, which are directly attributable to failures to control potent technologies, are a cause for increasing public concern. I need only mention Minimata, Seveso, Three Mile Island, Bhopal, Chernobyl, the Exxon Valdez spill, the recent Soviet gas pipeline explosion, incidents involving computer viruses, increasing numbers of commercial aircraft incidents, including two mistakenly shot down by the military, to illustrate the prevarious control we have over high technology. In this litany, I would not exclude the industry in which I am involved. Some would hold that the recent global market crash (or ''break'' as it is euphemistically called) was attributable in no small way to the increasing use of global information and communications systems. These episodes along with a realisation that our entire ecosystem is under threat through unbridled exploitation and use of new technologies service to highlight the need for increased public concern, awareness and involvement in the march of progress. The public has a real interest in seeing that the exploitation and application of high technology does not get out of hand. We need to ensure that the necessary public investments are made in programmes designed to research, to properly inform and educate the public as well as business and world leaders, and that timely and sensible regulation is brought forward to steer the development and application of the potent new technologies.

This need to inform ourselves was brought home to me recently by a friend and colleague, Peter Schwartz, who has created the Global Business Network of world class consultants and achievers. This is helping business and industry to develop strategies for navigating the future, to manage risks and to capitalise on opportunities. He said that we are moving from a time when power which has historicaly rested with the ''haves'' verses the ''have nots'' will move to the ''knows'' versus the ''know nots''.

Professor Chorafas is undoubtably one of the ''knows'', a technologoy enthusiast and an optimist, and this shines through in his work. However, in the midst of his optimism he does not shrink from providing us with the appropriate health warnings. It is in all of our interests to educate ourselves about the opportunities and threats presented by high technology. I am sure that many will find in this work the ideal starting point for this journey of discovery.

Peter Bennet
Executive Director, Strategic Research
The International Stock Exchange
London

Contents

1

Research, the Spearhead of Technology

Introduction

Modern science is much more than a list of impressive, but isolated, research results. It has for its goal *discovery*, which is a continuous, uninterrupted, and dedicated process. The key ingredients of discovery are brainpower, capital investments and competent exploitation of the knowledge acquired so far.

Research is done at the frontiers of knowledge. It is a search into what is still unknown to us. But, the frontiers of knowledge can only be extended by those who have absorbed the knowledge that is currently available and know how to penetrate the unknown.

Successful research is based on revolutionary ideas. It requires long term goals, and needs to be well funded. To attract scientists, laboratories must have a first class image, invest in their infrastructure, employ mature scientists to help the young researchers, be fundamental (avoiding surface quality), provide an environment where scientists can grow, help them emotionally when necessary, and welcome free communication.

If knowhow, experience, dedication, imagination and hard work are the ingredients which make up the personality of the researcher, why has the opening paragraph made reference to capital investments? Are they really important? The answer is ''yes'', for *big-science*. Today big-science is the mighty engine which rolls back the frontiers of the unknown.

During World War II, the Manhattan Project built an atomic bomb and ushered in the era of big science. Big-science projects can cost billions of dollars and marshal the talent of hundreds or thousands of people, as well as a very impressive array of facilities, from laboratories to supercomputers.

The 120 or so big-science research centres and vehicles built in the 1945 to 1989 timeframe include nuclear engineering accelerators, space missions, oceanographic efforts, and advanced observatories. These projects had an estimated total cost of about $35 billion. But one project alone, the space station, may cost nearly as much as that. These are the capital investments which could permit the next wave of breakthroughs in *discovery.*

The projected space station, for instance, will be an array of platforms and modules in which astronauts will run experiments, as well as perform astronomical and earth observations. This investment is just starting. NASA was granted $425 million for the station in 1988, and hopes to complete the first building phase in 1996.

Another big-science project is the Superconducting Super Collider (SSC). Projected to be the most powerful accelerator ever, this machine would be housed in a tunnel 53 miles in circumference. It will cost between $4-6 billion, its mission being to probe the structure of matter and shed light on how the universe was born.

Capital investment is only part of the story. For any big-science project, the operating expenses, too, are large. It is estimated that running the SSC will require more than $250 million a year. Besides, megaprojects tend to acquire a life of their own. Once a nation has spent billions on a space station, it is hard to shut it down regardless of its future value. And who is to say what the future value will be?

While exceptions always exist, the large majority of scientists do not do research because of the money that is involved. Lust and greed often characterises the private or institutional investor. Not the scientist. However, the scientist is not alone in the world. To get results, he must be supported by other professions.

Though at first sight it may sound illogical, the banker, the trader, the manufacturer, the farmer and the other productive sectors of the economy, are as important to sustaining scientific efforts as the researcher. They are providing the surplus an economy needs to invest in research and development (R&D) therefore fostering its technological future. However, what constitutes the economic infrastructure is not a subject of the present book.

Our emphasis is on the researcher and on the mental process which is his hallmark. One of the strongest motives that lead men to science is to escape from everyday life into the abstract and invisible world of human imagination and speculation. The outcome of such escape is an art and, as experience shows, the greater the artist, the greater the art.

When man speculates in the scientific domain he often attempts to arrive at the description (theory) of a given natural process, including life. He does so by means of *deduction* and *induction*, which are the basic rules of thought.

Deduction starts with facts, states and values, hence with data, and tries to establish underlying rules. It rains; therefore it must be cloudy. It snows; the temperature should be below zero. *Induction* is generalisation. We start from established laws and

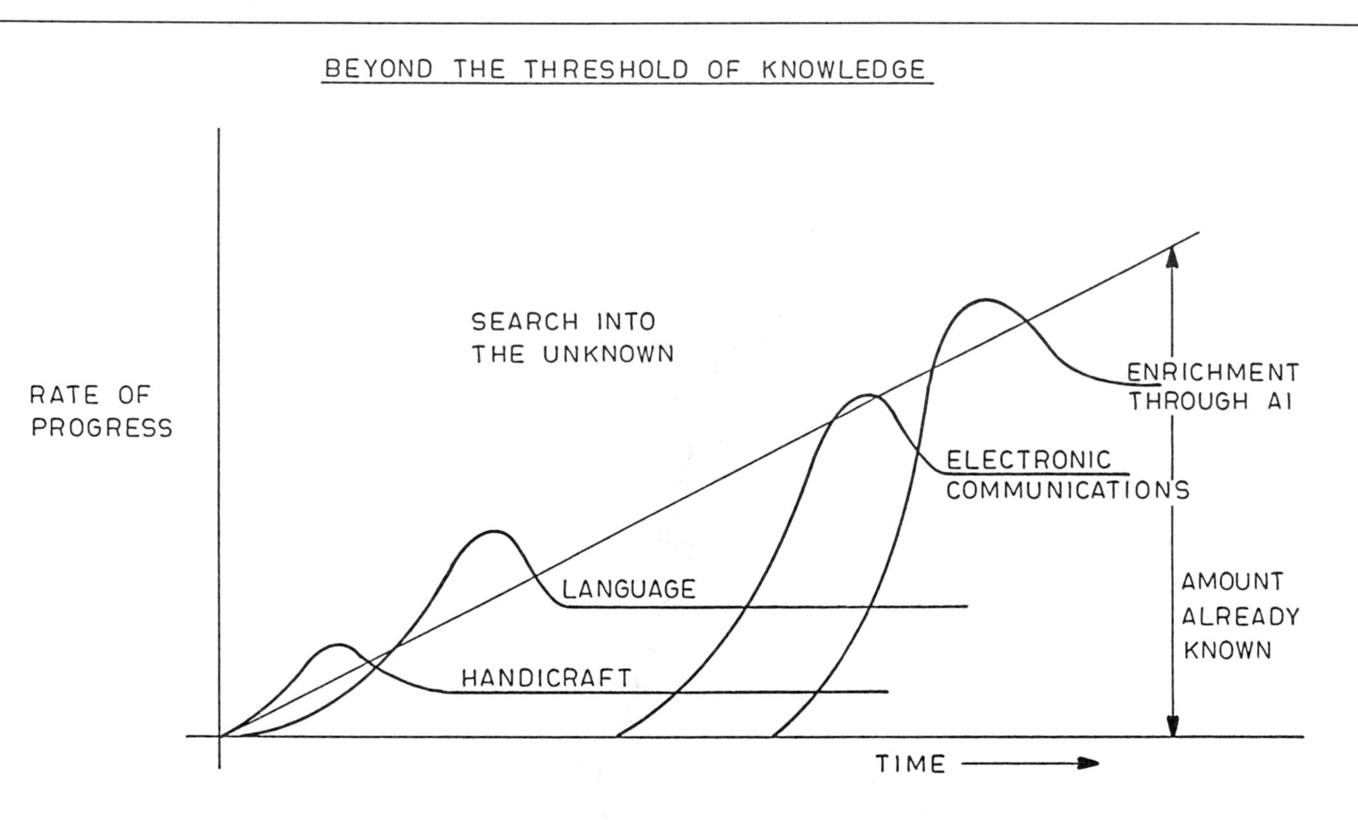
BEYOND THE THRESHOLD OF KNOWLEDGE
SEARCH INTO
THE UNKNOWN
RATE OF
PROGRESS
ENRICHMENT
THROUGH AI
ELECTRONIC
COMMUNICATIONS
LANGUAGE
HANDICRAFT
AMOUNT
ALREADY
KNOWN
TIME
DISCOVERY TYPICALLY HAPPENS AT THE PERIPHERY
OF KNOWLEDGE

Figure 1.1

expand. "If you do not believe in induction, you do not believe in anything", Bertrand Russell once said.

Induction and deduction have always been the distinguishing mental processes of great thinkers, and of philosophers. Induction and deduction tempt our appetite with a variety of delicacies from the table of wisdom. As Socrates underlined, *knowledge does not reside in impressions, but in our reflection upon them.* Induction and deduction are the tools for such reflection.

Discovery typically happens at the periphery of knowledge. Our tools make it feasible for us to successively cross new thresholds. In the material sciences this has been the case with handicraft. In laws of thought, with language and logic. During the last two centuries, electrical, and then electronic, communications provided the next quantum jump. Enrichment through artificial intelligence (AI) is today the frontier we are crossing. (Figure 1.1)

The language which we use shapes our thoughts, and the process of thinking permits us to have reflections. Perception is knowledge, but not all knowledge is perception. An *idea* is something we conceive in our own mind, but we do not necessarily perceive its details and implications. It is precisely in the field of objects, known, conceived, and perceived, that judgment turns and twists about, and proves true or false.

"*Thinking* is a discourse one engages with himself," suggested Plato. Alan Turing had a different definition calling *thinking those mental processes we do not understand.* Be that as it may, thinking distinguishes man from other animals, though there is also the finding that some other animals, whales for instances, have a bigger brain than our own. Nothing is wasted in nature.

As far as we know, among human beings thought is the process which assures us that the outside world exists in our power of formulating ideas. To a large measure, the concept of this word, of our environment, is our production, our image of reality, reflected or transformed by our thoughts.

Thinking with meaning can be applied to both theoretical and practical problems. A practical problem reflects a mental situation in which we plan to modify reality. This often requires introducing something (concept of fact) which does not yet exist, but which would be more convenient to our scientific work if it did.

Thought provides us with our knowledge of the universe. Without thought, the universe would look to us as an enormous, amorphous, or monolithic structure able to conceal until infinity what is the most important to us. But, we wish to conquer this concept of the universe through our thought. Therefore, we engage in *research* effort.

Research Laboratories

We have been talking of big-science, which also means *organised* scientific efforts. In the past, for over 100 centuries, there have been individual thinkers and researchers, but organised approaches were minimal. It has only been during the 20th century that the human potential we dispose has been structured into research laboratories, which often produce impressive results.

Most scientific disciplines of the 20th Century have been characterised by a team concept, and the team itself is not much different in its composition from an athletic team. There are many players, but the star players are very few. There may only be one or two in the team. These are people who help make the star teams, which are themselves rare within an athletic establishment. Yet they are those which really get results.

The notion of a research laboratory dates back to 1905. It was conceived and constructed by General Electric at Schenectady, NY. The originator was Dr. Charles Proteus Steinmetz, who used some 19th century precedents as his organisational background.

According to available evidence, in 1872 at the Siemens company in Berlin, Ing. Hefner-Alteneck introduced the idea that an industrial firm should allocate well-trained scientific personnel to science research activities. Hefner-Alteneck was the first person on record with a university degree to be hired by an industrial firm, in the first place. Just think of the millions of graduates who, a century later, concentrate their brains and their efforts on forcing open the doors of the unknown.

Another input to Steinmetz' imaginative development was the *concept of a research project*. This was also a 19th Century innovation, originated by Thomas Alva Edison. It concerned systematic organisation, starting with a clear definition of:

- Project objectives, *and*
- The successive phases the work should go through.

It takes both the research concept and project organisation to establish the foundations of a research laboratory, as well as to reach results.

One of Steinmetz' own contributions was to put into the same work group persons of different specialisations and levels of competence. This ran contrary to the then prevailing notion of *specialisation* - but it has been a fundamental ingredient of research work ever since. Furthermore, it was Steinmetz who defined the proportion to be observed between science and technology, this leading to the distinction between:

- Pure research;
- Applied research, *and*
- Development.

We are still working on the basis of these fundamental concepts. In fact, without them it would have been impossible to structure the powerful research endeavours which we have seen throughout this century.

Twentieth century research results owe a great deal to Socrates, Plato, Russel, and Turing as thinkers; and Edison, Hefner-Alteneck and Steinmetz as organisers. It is the depth and validity of our thoughts, together with the organisational skills that we possess, which provide the foundations for the following major step which needs to be taken. At no time has this been more true than today, as the world of the unknown we currently confront is characterised by an increasing complexity.

The emerging *science of complexity* will most likely provide the necessary foundation of knowledge for the next wave of the information-based technological revolution. Societies that:

- Master the sciences of complexity, *and*

- Convert their knowledge into products

become the new cultural and economic centres of power. New subjects, highly interdisciplinary in traditional terms, are emerging. In many cases, they represent the frontiers of current and foreseeable research. As particular subfields are joined together to make a new subject, the number of subjects treated by laboratory departments expands, making necessary a field of integration and coordination.

An example on the interdisciplinary character of the science of complexity is the attempt to create and study computer-based life forms. Research effort along this line is inspired by many other fields besides biology. Fields which interact with one another create a fitness landscape of interpersonal exchange of background and experience necessary to get results.

The parallel neural architecture of the brain has inspired a field that joins the computer with the cognitive sciences and neurosciences. The whole field of artificial intelligence, and particularly so its most advanced sectors, is an example of the implementation of the science of complexity. The Strategic Defense Initiative (SDI) would not have been possible without this background.

Research and development outlays made by the military find their way into private and public R&D laboratories, promoting intensive projects from basic to applied research. Such outlays typically come in waves which, more recently, tend to follow a decade or so of neglect. Figure 1.2 tells this story. It also suggests that each wave is higher than those which preceded it (in constant dollars), because megaprojects need megabucks for their funding. We will return to this issue.

"The consequence of the more or less efficient utilization of the state's productive resources ... relative to the other leading nations in the decades preceding the actual conflict, make the difference between success and failure of the bid for hegemony", suggests Paul Kennedy.* "The relative strengths of the leading nations in world affairs never remain constant, principally because of the uneven rate of growth among different societies and of the technological and organizational breakthroughs which bring a greater advantage to one society than to another". One of the strongest motors behind this uneven national development is scientific research.

The survival of the fittest dictates the fate of human tribes (and their states) as it does that of other animal species. As 19th Century intellectuals and politicians talked and wrote, it is a Darwinistic world of struggle, success and failure, growth and decline. "Those who do not advance go backwards and those who go back go under", said D'arcy, a 19th Century Frenchman. In 1920, L.S. Amery, of the British Empire

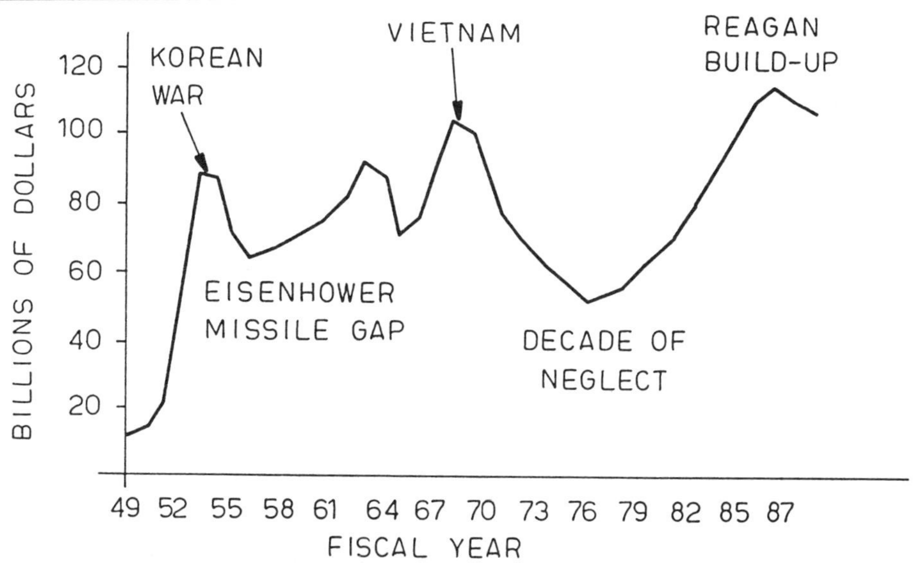

Figure 1.2 Research, Development & Procurement Outlays (in constant 1987 dollars).

Group advised to focus on:

1. Men

2. Money, and

3. Markets

Seventy years later we must add:

4. Means

Technology highly influences the other 3Ms, but also has a major impact of its own. Leo Amery was aware of this when he suggested: "The successful powers will be

* "The Rise and Fall of the Great Powers," Random House, New York, 1987

those who have the greatest industrial base. Those people who have the industrial power and the power of invention and science will be able to defeat all others''.

Ample financial resources, human capital and project complexity are also distinguished characteristics of *macroengineering*: large dams, global communications networks, satellite cities (among others). Such endeavours have far reaching economic, sociological, cultural, and environmental impact. They can be assimilated to large scale R&D.

The aftermaths of macroengineering projects extend well beyond the perspective of their sponsor, and include other organisations: Bankers, export credit agencies, insurers, suppliers, shippers, marketing outlets, computers and communications companies, software firms and consultants. But also communities, nations, governmental ministries, universities and public interest groups.

While much will still be contributed through the cottage industries, the interdisplinary nature of many research projects is now exceeding the walls of even the largest research laboratories. To provide solutions in an environment of complexity we require focused brain power and significant computer power. Such projects call for lavish funding. To obtain it, they have to maintain significant visibility.

Perspectives in Scientific Research

Today the Western countries invest in scientific research and technological development *ten times more* in share of Gross National Product (GNP) than the underdeveloped and lesser developed countries. The difference is not just of one order of magnitude. It is three orders of magnitude if we consider how much larger is the GNP of the developed countries.

Not all industries invest the same percentage of their business in research and development, some being more research-oriented than others. Figure 1.3 presents a correlation of R&D expenditures and growth rate of different industries, most of the data based on statistics by the National Science Foundation. Not surprisingly, the growth industries invest the most in research. It takes technology - not just information -to make a profit with new products.

Management needs technical expertise to nurture these products and bring them into play. But, there should first be management. The scarcity of financial resources left aside, the lack of high-calibre management is one of the problems cursing the Third World - and any slow-moving industry for that matter.

Superficially, the great lag in research looks like being one of the violations of the United Nations Charter. One of the rights underlined by the Universal Declaration of Human Rights of 1948 is that of ''participating in scientific progress and in the benefits which result from it''. Forty years later, this is one of the subjects which has never taken place - and the way things are going it never will.

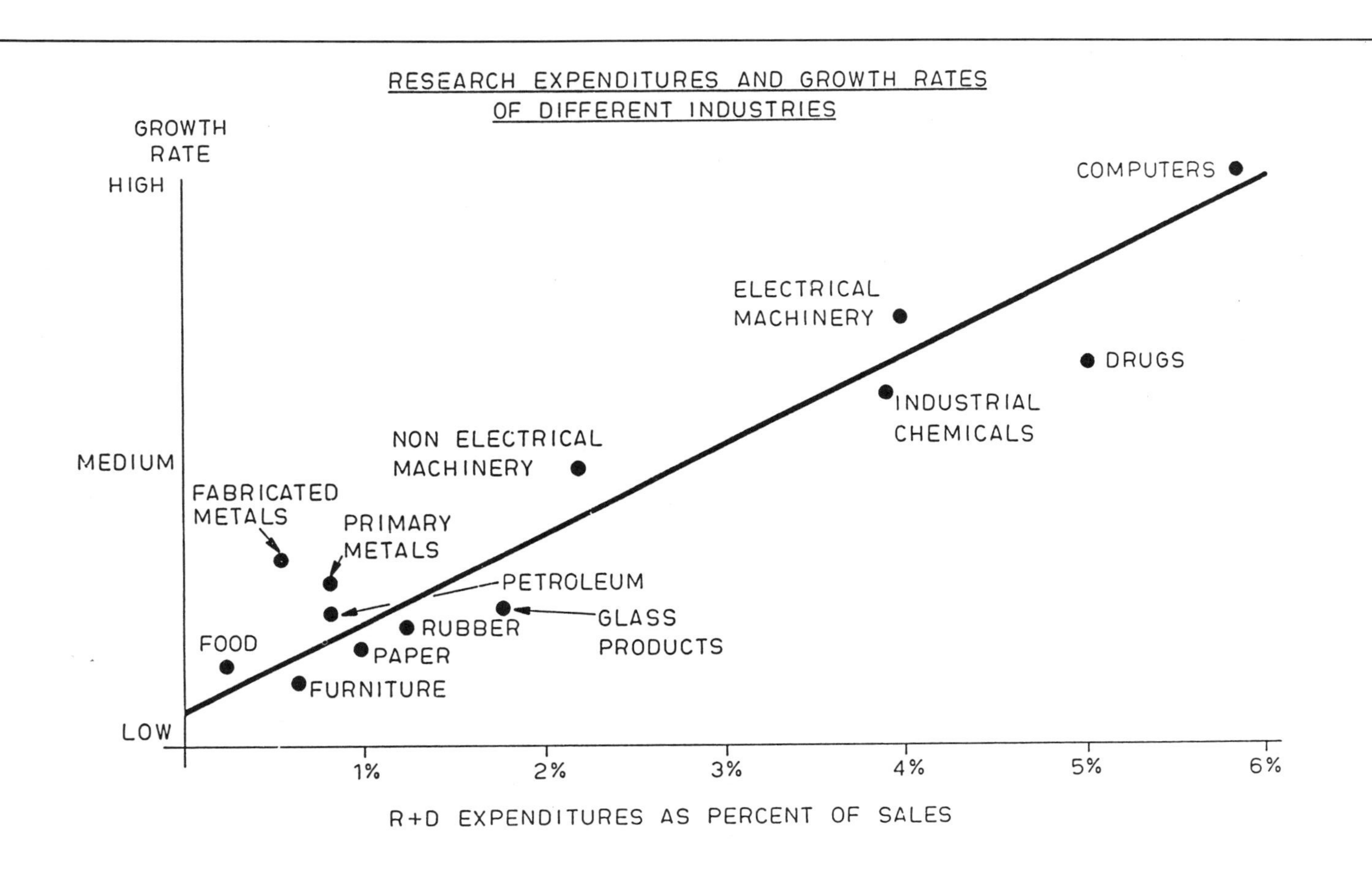
RESEARCH EXPENDITURES AND GROWTH RATES
OF DIFFERENT INDUSTRIES
GROWTH
RATE
HIGH
MEDIUM
LOW
COMPUTERS
ELECTRICAL
MACHINERY
DRUGS
INDUSTRIAL
CHEMICALS
NON ELECTRICAL
MACHINERY
FABRICATED
METALS
PRIMARY
METALS
PETROLEUM
GLASS
PRODUCTS
RUBBER
FOOD
PAPER
FURNITURE
1%
2%
3%
4%
5%
6%
R+D EXPENDITURES AS PERCENT OF SALES

Figure 1.3

Statistics are crushing. Twenty percent of the circa 160 nations on earth produce 94 percent of the scientific and technological knowhow. Four decades of experience have demonstrated that *research brings results only when an aggregate of political, social, cultural and economic preconditions exist.* Such preconditions:

- enhance,
- control, *and*
- modulate

the development and implementation of new knowhow. Not only are most nations not tuned to this wavelength, but also many companies are falling behind because of their own fault. Management which only cares about buying technology is not very interested in knowing the tedious details of making this technology work properly.

Since the arms market is thriving worldwide, it comes as no surprise that large sums of money are devoted to military R&D. An estimated 50 percent of all the scientific and technological researchers in the world work on weapon systems. An educated guess is that the cost of armaments on a worldwide basis exceeds $800 billion per year, *which is roughly equivalent to $1.5 million per minute.*

Nations spend more on military research, development, and procurement in times of crisis. In fact, Figure 1.2 reflected American expenditures associated to different build-ups. Military R&D money is not necessarily sterile. Money invested in government- sponsored R&D can have a significant impact on the economy beyond the old slogan that military build-up is the fly-wheel of the economy.

It is a general enough belief that every dollar spent on space has $4 fall-out on earth in terms of economics. One of the reasons is that, being government-financed, military R&D can at times become more fundamental than industry would risk if left alone. Fundamental R&D is one of the prerequisites to a steady flow of discoveries, particularly at a time when science and technology move very fast.

In America, the Strategic Defense Initiative, for instance, is largely based on photonics and artificial intelligence. The advent of *photonics* suggests that today's research is tomorrow's population engine. Lasers are likely to be part of the next generation's everyday equipment. Science does not come to a halt for the convenience of politicians. The power to create nuclear fusion, satellites, transistors, AI constructs and microprocessors was not science's last word.

Within a steadily evolving high technology environment, the new thrust is *opportunity analysis* - finding ways to steer global events favourably, as opposed to the traditional *risk analysis* or damage-control approach. Politically, that means keeping tabs on advanced R&D projects in other countries, in the hope of gaining information, or at least an early warning, of what is to come. Industrial spying is age old, coffee and rubber being examples. But it never reached the spotlight it has today.

Just because of the preeminence of research in our society and in our future, industrial and economic intelligence is widespread among the leading nations. As a general was to suggest in a meeting, top industrial managers, and the better-known researchers, merit as close watching as political leaders.

One of the often-discussed industrial spying approaches is *reverse engineering*. Reverse engineering is in itself a difficult task that does not necessarily contribute to the pool of basic knowledge, let alone to the factory floor. Says MIT's Stephen Meyern, a Soviet military policy specialist: ''You cannot reverse-engineer an entire industrial base, and that's what is important''.*

Although he acknowledges that the acquisition of such products as supercomputers or milling machines can help the Russians in specific areas, Meyern adds: ''They can steal all the microcomputers they want. It is not going to improve the level of their own indigenous technology''. And he is right. It is not enough to reverse-engineer the product. Even more important is to know the process necessary to make it in large numbers. Such processes are of increasingly high precision - and in the last analysis product and process development go hand in hand.

Planning and Financing High Technology

The successful use of technology requires people who can work with machines and with each other. As we begin to implement new methods we have to guard against too much automation, and too little training. We often need to redesign some of our products and processes - as well as to broaden our training perspectives.

A people possessing a far higher level of primary and technical education, an unrivalled university system, and a creative scientific establishment, has a significant advantage over other nations. Integral parts of this establishment are the physics and chemical laboratories, and the fine mechanics, electrical engineering and electronics institutes.

When it comes to the evaluation of the necessary effort for supremacy in high technology we must admit that, so far, only the Japanese companies created in significant numbers a manufacturing infrastructure that can respond with speed to market demands and changing opportunities. Products are designed from scratch:

- for ease of assembly;
- with emphasis on simplicity;
- with as few parts as possible;
- as well as for easy modification.

---

* Time Magazine, November 23, 1987

Unlike American and European operations, in Japan there is no so-called work-in-progress waiting for a drilling machine, then moving into another queue until it can be processed on a grinding machine. The whole production is orchestrated with such precision that it runs with virtually no inventories of purchased parts and materials or of partly finished products - all of which tie up money. Zero inventories characterise well-managed plants.

The Japanese also put the craft back in manufacturing by making quality the responsibility of each worker, not that of after-the-fact inspectors. Jobs that need mainly only hands are delegated to robots. This approach yields products that are both higher-quality and less expensive. We will return to these notions when we talk about the reasons behind Japan's competitiveness.

The results of better organisation, and more qualified management assisted by the proper computers and communications tools, are seen in the electronics market. In the semiconductor field, in 1982 the USA enjoyed a 49.1 percent world market shared, while Japan had 26.9 percent. In 1987, Japan was the No. 1 producer, with 45.5 percent of the world market, followed by the USA with 44 percent.

A study recently released by Integrated Circuit Engineering Corporation, of Scottsdale, Arizona, finds that three of the top five producers of semicustom chips are Japanese - including the No. 1 spot by a wide margin. And they use high technology to support their marketing lead.

Fujitsu America Inc., for instance, has a satellite link for sending chip designs to its waiting production line in Japan. As a result, Fujitsu can deliver semicustom chips to North America in less than eight weeks - no more than it takes most USA chipmakers.

Another statistic is telling what superior organisation and the conquest of the world's advanced markets means in terms of the economy and, by consequence, of employment. In 1980, the USA ran a record $27 billion trade surplus in advanced electronics products. In 1987, the American high technology balance was in a deficit -which started for the first time in 1986.

It is, however, no less true that the sprawling R&D expenditures tax both the economy and the human resources available to conduct research. In the years to come, even the mightiest nations will find difficulty to responding to all developmental requirements through their own knowhow. Hence the beginning policy toward multinational, cooperative research.

The cooperative effort started in some nations with cross-industry research projects. For instance, the USA plans to launch an experimental aircraft, in 1993, called the X-30. It is thought to be the forerunner of a series of 5,000-mph passenger jets that come into orbit from ordinary runways. But, to become reality, the industry must first find strong, new, lightweight materials that can withstand blistering temperatures of up to 3,000 degrees Fahrenheit.

To reach this goal, the USA government formed three new consortia of industry, public administration, and universities to study everything from the basic chemistry of materials to the tools and processes needed to make large sections of a fuselage. The research teams, *which may involve as many as 100 companies,* will be led by the three aerospace giants competing to build the X-30:

- Rockwell International will explore titanium-aluminum alloys.
- McDonnell Douglas will build composites of those two metals strengthened by fibres.
- General Dynamics will look at carbon-carbon composites, materials made by embedding carbon fibers in a matrix of carbonised phenolic plastic.

The Defense Advanced Research Projects Agency (DARPA) is said to have budgeted $100 million for this project, which is a joint effort of the Defense Department and the National Aeronautics & Space Administration.

If this case of cross-industry research works in one country, the now developing *nuclear teamwork* is a trans-national example. The four world heavyweights in fusion power research: Soviet Union, United States, European Community, and Japan, have agreed to design together what would be the world's *first nuclear fusion reactor*. The project, called the International Thermonuclear Experimental Reactor (ITER), would be the forerunner of an actual demonstration power plant. It would prove the engineering feasibility of fusion as an electric power source.

This promises to be a very interesting project indeed. The design team will be based at the Max Planck Institute of Physics, at Garching near Munich. The participating countries have agreed that the basic form of the machine will be a kind of magnetic ring, as first proposed in 1951 by Nobel prizewinner Andrei D. Sakharov. This design has been adopted the world over and has been the most successful fusion research machine to date. But the greatest challenge is that of multicultural, multilingual collaboration. We shall see.

Project *Esprit,* by the European Common Market, also had transnational goals and met them with reasonable success. The European Strategic Project for Research on Information Technology was built on the premise that the EEC countries needed a directed research programme focusing both on methodology and on technology. The goal is a programme able to put people together to solve the information systems problems of the years to come.

Partially overlapping with, and partly extending, the concept of Esprit, *Eureka* is another transnational research project. It is virtually free of government control, even though it gets half its money from its members' governments. In an effort to pool technological talents, a project must originate with an industrial firm and also involve at least one other firm in another country. Universities and government laboratories can be brought in, but industry is the initiator and the home base.

Race is still another transnational project of the EEC focusing on telecommunications. It rests on two pillars. The first is the perceived need for high quality, reliable telecommunications services. Knowledge societies began investing heavily in expanding public networks in the mid-1970s after waking up to the link between telecommunications and economic prosperity. But the pace quickened by the late 1980s.

Today, the 24 OECD member nations - that is, the industrialised countries: Western Europe, America, Canada, and Japan - have installed 330 million telephone main lines, or 78 percent of the world total. Their investments in telecommunications networks total over $50 billion per year and are increasing.

Some of the lesser developed countries are hurrying to lay the electronic pipeline that will fuel their own economic growth. Understandably, they are eager to catch up with the rest of the world, but it is not easy to avoid falling behind. Western governments have already upgraded networks to handle the digital information generated by communicating computers, while in the underdeveloped countries the lines are still analog, with very poor bit error rates.

The other pillar on which rests the thrust for a transactional collaboration in telecommunications networks - hence, the Race project - lies in the fact that research and development costs are skyrocketing. This is particularly true of software.

Seventy-five percent of the development cost of Siemens' Hicom, one of the newest private branch exchanges in the market, has gone into software. It is not that the hardware R&D got any cheaper. In fact, its cost also increased. But, it is software which got the lion share of the budget.

Such expenditures are justified only in the sense that the market potential could be enormous. Once in place, an advanced network stimulates demand by business users and, by consequence, the growth of markets for private communications equipment and services, such as PBX. If photonics is the electronics of the 21st century, communications is going to be the biggest market.

Companies, once content to limit their information processing to computers located on their own premises, are now linking up their systems with far-flung affiliates and other business partners. In some large companies, communications needs are estimated to be growing by up to 40 percent per year. Financial institutions, multinational corporations and airlines are among the most demanding users.

With effective telecommunications now recognised as being critical to post-industrial success, business users have become a driving force in the industry. Private organisations are pressing for political and industrial policies that protect their growing investments while enhancing their communications capability. Foremost among their concerns are the definition and implementation of industry-wide standards, and the liberalisation of markets.

In more than one way, business imperatives of the international marketplace mean that technological and political frontiers must be broken down. But, while telecommunications authorities in the United States, Japan and Britain have been privatised or deregulated, government ministries in other industrialised nations continue to maintain a monopolistic domination over the provision of services and the procurement of equipment. Still, the winds of change can be seen and continental Europe might be finally on the verge of market liberalisation. If so, research and technology would have played a major part in the change of hearts.

There is also a project in which Western Europe is still undecided as to whether or not to invest in a common effort. This is space research. The French and Germans have joined forces. The British go their own way. But neither party's budget is so impressive. Yet, politically as well as financially, Europe will count only if it becomes a leader in space. Otherwise, it will not be of any significance.

To count in space research and technology means to put up an all-out effort. This goes well beyond current approaches, which are largely arranged in retreat, to save face. In order to become a leader Western Europe should move ahead full speed, solving the space research problem head-on. Significant human and financial resources must be allocated to the project. This requires setting the proper priorities.

Man-Made Universe. An Example of Far-Out Research Goals

The projects we considered as examples of a transnational cooperation have, inherently, a rather short timespan, and rest on well-established technological fields. While the four-nation effort on new forms of nuclear power production is an exception to this reference, this project too does not involve the number of unknowns that we experienced with the original Manhattan Project. But there are other fields where scientific hypotheses are nearer to pure speculation.

Scientists often probe the workings of nature on a cosmic scale. One of the latest research goals is seeking a mechanism by which man might create a new universe from scratch. Dr. Alan H. Guth, professor at the Massachusetts Institute of Technology, is serious about his investigation: ''Ten years ago, we could not even have posed the question of whether a man-made universe would be possible''. However, physics has progressed a long way since then, and today we can ask this, and related questions, in the real hope of finding scientifically-testable answers. We are working in a new and exciting environment.

Just a few years ago astronomers, physicists and biologists thought they had figured out genesis. The universe began, they concluded, in a *Big Bang*. That is the explosion of an infinitely dense and hot point of space-time thought to have taken place 10 to 15 billion years ago. Now the time seems to be ripe for some earth-shaking discoveries, such as the one made by Galileo Galilei (1564-1642). Cosmologists do not know much about the universe's shape. They argue about its size, and they dispute its fate, but they do agree on a fairly basic point: It is the only universe around. Now even that dogma is being questioned.

Physicist Andrei Linde of Moscow's Lebedev Physical Institute pushed the idea that the universe actually consists of countless, separate universes whose very laws of nature may differ radically from ours. The earth, its galaxy and everything else astronomers can see would inhabit just one of them. Says Dr. Linde: "We should not insist we live in the only possible world. We live in the one in which life of our type can exist. In other universes, other types of life may exist".

For precisely the same reason, bothered by the flaws of the Big Bang model, Dr. Alan Guth suggested that an important aspect of this new research climate is a sense of optimism that even the most opaque of cosmic mysteries: the nature of universal creation itself, may be penetrated by scientific minds. Since 1979, when he began work on his theory of an *inflationary universe,* Guth has shaken the scientific community with what so far seem to be unorthodox ideas.

The mathematical consistency of his proposals, and their success in describing conditions that actually exist in the present-day universe have won Dr. Guth an increasing number of believers. Sheldon L. Glashow of Harvard University, a winner of the 1979 Nobel Prize in Physics, called Dr. Guth's work "exciting".

The new theory likens the universe in which we live to the two-dimensional surface of a sphere which, because of its immense size, appears to us to be almost perfectly flat. There are circumstances, Dr. Guth says, in which an *aneurysm* could develop on this surface, a region in which space and time bulge like a tumour, eventually pinching itself off into a new universe.

To a hypothetical observer inside the bulge, conditions may initially resemble those of the Big Bang explosion from which our own universe is thought to have arisen. But, to the observer in our universe, the aneurysm would resemble a *Black Hole* - a supermassive object whose immense gravity prevents the escape even of light. (In Chapter 2, Section 2, Black Holes will be discussed under a different light in connection with the theory of Dr. Stephen W. Hawking).

According to this theory, after a certain amount of time the Black Hole would evaporate, leaving little trace of space/time where a new universe had been born. Of the half-dozen objects which astronomers believe may be Black Holes, some might be the scars of aneurysms left by the births of new universes. Scientists would be unable to tell the difference between such ex-Black Holes and the supermassive objects thought to be the more common kind.

Dubbed *cosmic inflation,* this theory basically accepts the Big Bang as the moment of creation, in which it coincides with the originating bang theory after the first quadrillionth of a quadrillionth of a second. But in between, during that blink of the cosmic eye, cosmic inflation and the Big Bang diverge in a way that literally makes all the difference in the universe.

According to the theory's advocates, in this fraction of a second the universe quickly ballooned to a size 1050 times greater than when it started. Only afterwards did the cosmos settle into the more sedate rate of expansion seen today. The virtue of the cosmic inflation theory is that, unlike the pure big bang, it explains puzzling facets of the universe. For instance:

- Why each piece of sky resembles every other, and
- Why the cosmos is so perfectly balanced between endlessly expanding and eventually collapsing.

According to Dr. Linde, in the beginning separate regions of the universe inflated differently. For all its decidedly unproved qualities, this theory might provide answers to questions with deep roots, such as: Why is the world the way it is? Einstein once wrote: ''What I am really interested in is whether God could have made the world in a different way''.

There is no known reason, for instance, why the strength of gravity should be what it is, or even why the world should have four dimensions: three of space and one of time. In fact, some theorists calculate that nature allows as many as 26 dimensions. The myriad possibilities of nature may find expression in other universes which, researchers suggest, have different kinds of physics and different kinds of space-time.

Does the new theory mean that our universe could have been created as the conscious act of human beings in some other universe with which we no longer have any contact? ''That is the possibility we are exploring'', says Dr. Guth. ''Nothing in our calculations so far has ruled out such a case''.

How could two or more universes coexist and go on expanding without colliding? ''You just have to think of them as existing in different places at different times'', Dr. Guth suggests. ''Once their common space-time is pinched off, the two universes can never again have any point of contact or any channel of mutual communication''.

Always, according to this theory, the risk conditions for creating the aneurysm of an embryonic universe could come about by chance as well as by an intelligent agency. Chances are that the random creation of a universe would be extremely rare. Still, if a new universe could spontaneously leap into existence from time to time, infinite numbers of universes must have been created, and an infinite number more are still to come.

Following the theory advanced by these scientists, an essential point is that quantum physics may permit the spontaneous creation of something from nothing. The test of this hypothesis may not be far away. New, powerful laboratory facilities, e.g.: the CERN collider in Europe and the SSC project in America, make feasible a momentous leap forward in the exploration of matter and energy. The appearance of exotic new particles: higgs bosons, quarks and leptons, can open new vistas of inner space for scientists to study and on which to test their theories.

Sounds exciting? It is so. Do you think it is incredible? It might be, but do not rush to conclusions. Who would have thought 40 years ago that genetics would be creating new life forms never before seen on this planet. Bioengineering has been the most chilling act of science since the splitting of the atom. The next chill may come through man-made cosmic inflation.

Technology Negatives. The Bioengineering Puzzle

The first USA patent law, passed in 1790, protected the invention of "any useful art, manufacture, engine, machine or device, or any improvement thereon not before known or used". Almost two centuries later, in May 1987, the USA Patent and Trademark Office announced that it "considers non-naturally occurring nonhuman multicellular living organisms, including animals, to be patentable subject matter".

The new policy might enable biotechnology companies to take control of the livestock industry, though the really alarming facts are the far-reaching aftermaths of *patenting human beings*. Some scientists are afraid that the patenting of *genetically altered human beings* might be the next glamour issue, even if the Patent Office statement specified "nonhuman" life. "My fear is that we will begin valuing human beings as no different from other animals", said J. Robert Nelson, of the Texas Medical Center in Houston.*

Admittedly, there have been many "pluses" with bioengineering. It has brought about various health improvements, particularly through research into the procedures concerned with understanding how the behavior of proteins and other organisms is governed by the body's genetic code. That is, the sequence of chemicals in deoxyribonucleic acid (DNA) a vital constituent of every living cell.

We badly need new methods in medicine in order to cope with an aging population. We also need new drugs, as the viciousness of some illnesses proliferates due to pollution, and new, so far unknown, illnesses show up. Statistics indicate that for every 20 patients who check into a hospital, one will contract a serious infection while there - and some of these people will die. Often, toxins produced by related bacteria, such as those causing toxic shock syndrome or boils, have turned up as well.

Biotechnology does not necessarily correct these statistics, which discredit medical care, but it does make its contribution felt down to the queue level. Scientists today target genes for repair and change. They are closing in on the long-sought goal of changing, and even repairing, individual genes within living cells.

---

* Time Magazine, May 4, 1987

This concept, known as *gene targeting,* aims at the creation of a specific genetic change exactly when and where the scientist wants it. Ultimately, the targeting might lead to effective therapy for incurable genetic diseases. Both the opportunity and the risk lie in this simple sentence as, for the future, scientists see targeting as a potent way of studying human and animal development, and the functions of many recently discovered genes.

Targeting, for instance, is expected to be important in helping scientists to produce laboratory animals that have the same genetic defects as those that cause hereditary diseases in humans - as well as superior laboratory animals which can be unwilling donors to humans. ''In the long run it might be possible to replace a defective gene with a good copy'', said a well-known medical doctor. But who is to decide what is ''good'' and what is ''bad''?

Nobody would dispute the need to develop new solutions and vastly improve over current methodology. The great puzzle is where exactly the dividing line between *do* and *do not* should be drawn. The answer will not be found in the Ten Commandments. This problem was not actual 2,000 years ago, but it is quite vital today.

The new patent controversy, for instance, is just the latest in a series of ethical battles over biotechnology, and man's efforts to manipulate the genetic code. Even if today's goals of most genetic engineers are rather modest: leaner pigs, dairy cows to produce more milk, chickens that are resistant to infection and which thus can be raised with fewer antibiotics, ethicists are worried about what they see as a slippery slope that could lead to strange experiments on people. It happened in the past, and not only in the Nazi era.

On the one side, biotechnology laboratories promise to produce techniques to determine which of the myriad of compounds secreted naturally in the human body can be used, perhaps in higher concentrations than occur naturally, to attack illnesses. From this may come processes to turn out chemicals in large enough quantities to be useful as drugs.

Another important aspect of biotechnology concerns not the products which it may make available, but the greater understanding that it promises to give to researchers working on conventional drugs. That is, those which are based on chemicals, and not produced naturally. ''Biotechnology should be looked at as a toolkit for picking apart disease at a molecular level'', says Mr. Tom Kiley, vice president for corporate development at Genentech.

The products themselves are a third reference. They include human insulin, useful for diabetics; interferon, which may prove valuable in fighting cancers; and growth hormone, a therapeutic agent for people suffering from restricted growth. These are largely materials that scientists have known about for years, but which can be turned out to some degree more easily using biotechnology methods.

But, in the other side of the coin is the lack of ethical standards able to tell what should be *doable*. In coming years, the patent controversy will join what one scientist calls the *storm over biology* - the public's concern about the effects of genetic engineering on humans. The scientific methods that are used to create patentable animals are increasingly well worked out. The still unsolved questions that need answering are mainly ethical:

- How will we care for these bizarre farm animals?

- How fast will we be creating them, and why?

- Who would draw the line when it comes to manipulating human genes?

- Where should the line be drawn in the first place?

The USA Patent Office can be faulted for haste. Its policy should have been set only after a *thorough public debate* on this issue. A debate which is so fundamental that it could change our *moral code*. We have to admit that, when it comes to ethics, we are not doing our homework as we should.

As it is to be expected, opinions are divided, yet neither side has yet formulated its thesis in a factual and documented manner. While animal welfare advocates concede that conventional animal breeding has produced sickly misfits, they fear that genetic engineering will inflict greater suffering and disability. The evidence to prove or disprove such arguments is lacking.

''Researchers are creating new disease complexes that we certainly could not treat'', leading veterinarians were to suggest. Some object to another area of genetic engineering: the development of animals that suffer from human diseases, like muscular dystrophy, even if such creatures would be invaluable in testing new drugs for humans.

To animal welfare activists, a new government policy permitting scientists to patent genetically engineered animals raises the spectre of a brave new world of unforeseen horrors. They see a world where living things are accorded all the status of a manufactured toaster. ''It is really outrageous for a handful of officials to say all members of the animal kingdom are reducible to the same manufacturing process as an automobile or toaster even'', said Jeremy Rifkin, director of the Washington-based Foundation on Economic Trends.

The ethical issues are in the background of such reactions. Using computers as research instruments, scientists and system engineers are opening frontiers that alter our idea of reality about life and, possibly, death. This is particularly true of the study of complex systems: the brain, the cell, large molecules, DNA. Man has reached the gates of his own inheritance.

The human *Genome* project in America is an attempt to determine the sequence of the f the f the f the f the 3 billion chemicals in human DNA that spell out our genetic endowment. This project could cost some $3 to $5 billion, and take 10 to 20 years to complete. Knowing the genetic sequences may enable biologists to diagnose, and perhaps treat, both inherited conditions and diseases such as arteriosclerosis and cancer - but also to alter the DNA code.

Biotechnology's bid to build a better mouse (or man) is based on four far reaching steps:

- Incorporate a gene for producing a human protein, such as tissue plasminogen activator, into the mouse genes that control milk production.

- Insert the resulting hybrid genes into a fertilised mouse egg and implant those eggs in a surrogate mother mouse.

- Mate the offspring that carry the hybrid gene to build up a new generation, then

- Milk the females that have the gene and extract the protein.

The "logical" next step seems to be that of extending the experience to include cows, goats, or sheep. And why not homo sapiens?

The more disturbing part is what cannot be seen in these developments. Research results are only the tip of the iceberg. The seven-eighths are not seen. If we do not look inside the Troian Horse of gene manipulation well in advance the day we wake up to disaster may be too late.

It is true that biotechnological processes have played an important role since very early times. Alcoholic fermentation, cheese-making and lactic fermentation for the preservation of food, have been discovered empirically and used since the early ages. But, in a massive industrial sense, biotechnology is a product of the 20th century.

Microbial cell matter for animal feed has been the first significant thrust. Work with molasses, ammonium salts and sulphite waste liquor was carried out in Germany in the 1920s and 1930s. The development of stirred and ventilated reactors enabled microorganisms to be grown submerged in liquid nutrient solutions, and thereby opened the way to industrial microbial mass production.

Technically efficient fermentations for the production of butanol, acetone and ethanol have also been done in the timeframe between the two world wars. But, the true potential of genetic engineering is a matter of the 1970s and 1980s.

Because of the dangers involved, scientific circles begun to study the risks which may arise from the application of the new knowledge. Convened by leading researchers in the field, the Asimolar Conference of February 1975 led to a broad discussion on the safety aspects of the new science. A paper was drafted culminating in a series of guidelines which eventually were to be discarded by everybody.

The possibilities of engineering changes of unknown consequences in the genetic material of microorganisms evidently caused concern. It is still doing so. Though there may be misgivings among the general public, and also among the scientists, the truth is that we know very little, or nothing, about existing dangers and -most importantly - how to handle them once they develop.

Many scientific colleagues complain about the schizophrenic character of modern society which constantly presents more exacting demands, plus a *fast growing population to feed and shelter* - and, at the same time, a decreasing readiness to assume risks. But this argument truly misses the point.

The steadily-growing-expectations is a fake planted by the politicians. The Third World economic crises of the 1980s documents how ill-founded such arguments can be. By contrast, the population explosion is real, dramatic and of untold consequences for the future of humanity. Science does provide the means for birth control, but *cowardice by religious and other authorities,* coupled with ignorance by the general public, have brought our world into a position of drift.

The attitude of concern toward molecular biology, and particularly towards genetic engineering as an applied discipline, in no way reflects avoidance of *reasonable risk.* But the *megarisk* is another matter. Whatever we do in this society of ours these days, has a *megatrend pattern.* Now we are trying to correct one big mistake, *overpopulation,* with another big mistake, *genetic code manipulation.* Two errors do not make one thing that is right, as the late Ben Gurion wisely said. We will return to this subject in Chapter 3.

Richard Forsyth and C. Naylor have brought under perspective a very interesting and alarming concept. Humanity has opened two Pandora's boxes at the same time they suggest:

- One labelled *genetic engineering,*

- The other known as *knowledge engineering.*

According to Forsyth and Naylor, what we have let out with these two advanced science fields is not entirely clear... "But it is reasonable to hazard a guess that they contain the seeds of our successors. The question is not whether *Intelligence* will supercede *Life* - but how fast?"

2

The New Physics

Introduction

When we know what knowledge is, in the most complete meaning of the term, can judge whether or not the fragments of knowledge that we possess fulfill t meaning, or simply approximate it. Only when we can perceive the practi implementations of our research, and of our knowledge, can we appreciate contributions the knowhow we possess could make to our wellbeing.

First, the acquisition and practical implementation of knowledge is the reason w over the centuries, physics accumulated prestige. Out of it came both chemistry : medicine, but also the machine, as well as the foundations of the technological a Therefore, of industrialisation.

Through the knowledge which he acquired in physics man intervened in nature : bent it to his own advantage. He used what he learned through physics to promote material interests, but also those of society as a whole. Some authors talk of *imperialism of physics*. With a few exceptions, it has been a benevolent imperialisr

Physics is more than the discovery of basic laws and, through them, of the way t nature works. The reasoning mechanism is most important, and with it the string thoughts leading to the investigation of problems, the querying attitude, the disbel the tentative statements expressed as hypotheses, the heuristic approach of researcher. The same is true of the always tentative proof awaiting to be up tomorrow or after centuries, by another more fundamental discovery which may nearer to the "truth".

For the physicist, a problem is that which, in principle, can be solved. But, in a w the solution is anterior to the problem. With few exceptions the physicist ' consider as a physical problem only one for which he has metrics and which he submit to measurements. A priori, a physical problem accepts methodolog treatment, while a philosophical problem does not. Problems to which metrics can be applied do not belong to physics.

In the introduction to Chapter 1 we have spoken of the process of thinking. Both science and philosophy are based on thinking. Both start with conception and perception, but their basic goals, as well as the rationality of each, strike a different cord.

In science, when we reject a hypothesis we do so because of the evidence which we have. When we accept, our statement is made in the most tentative manner. All we say is that we have not found a reason for the rejection of a hypothesis, but this is not necessarily synonymous to the etimological meaning of the word "accept".

In his research efforts, man exhibits variations in thinking. Such variations are changes in his orientation. They lead him to see other truths that are different to those of yesterday. It is *we*, not the truths themselves, that change, and as *we* change we go on examining a widening series of hypotheses.

Since we never learned the details of how our brain works, it is difficult to define what goes on in human thinking. The differences between one man and another can be enormous. At the same time, the broader the concept on which we work, the more profound and accurate our work is.

In principle, the scientific man is a rational animal. Educated men are living beings who often think with meaning, and whose thinking we are often able to understand. Understanding is an integral part of literacy.

Physicists are scientists particularly concerned with definitions which they can reach through thought and the instruments available at their disposition. They appreciate that both thought and the results of their experiments are prerequisites to *cognition*. But, even if scientists are able to perceive things which are transparent to others, they will not be able to recognise them without further thought.

Science does not exist except for he who seeks it eagerly. We say that we have discovered a truth when we have found a thought satisfying our intellectual need. If we do not feel in need for that thought, it will not be a truth to us.

Scientific truth is what creates an anxiety in our intelligence. That is why men seeking an unknown truth are constantly renewing, correcting and restructuring science. These are curious men of excellence. They are *men who care* - and, therefore, careful men. What they do, they do with attention. They neither hide nor neglect whatever occupies them.

Men of science work both in quantitative and in qualitative terms. *Symbolic,* rather than numerical, operations characterise cognitive activities such as problem solving, forecasting and planning, induction, deduction, extrapolation. For problem solving, general approaches must be supplemented by knowhow specific to a given domain - and the same is true when we search for meaning.

The scientific approach contradicts phenomenology and existentialism, which brought to an abrupt end the issue of meaning: "What you do at this moment is what you are". But, if we do not have *meaning* we are lost in life. We do not have continuity and, therefore, we are not fit to survive. Meaning is Plato's *idealism*. Leaving it out of our thoughts, we empty the sense of our life.

This great search is the life of the physicist. To him, as to any other man, seeking meaning gives a sense in life. It also provides continuity. To the true scientist, his whole life is his work.

As a matter of fact, even with ordinary people lack of meaning eventually leads to serious depressions. We have a sense of meaning in our life if we have an identity and know who we are and what we want, as well as what we need. The sense of *identity* rests on these simple words.

Identity acts like a fixed point to a surveyor. It is a fixed point of departure and of return. Everything else will go. What we have left is ourselves and our work.

The sense of immortality comes from this search for meaning. It is part of most of us, particularly of those who wish to escape steadily meaningless activity. To the scientist, money for money's sake is *not* a meaningful act, but rather resembles the work of Sisyphos. What purpose can be served by rolling a rock uphill?

Sisyphos was punished, mythology tells us, because he angered the Gods. He did not know his place. His sufferings became a message to others; a message of obedience. But the true scientist has different priorities than routine and obedience would dictate. To him, the challenge is how to justify what he is doing to *himself* - not to others, but to his own conscience. In the bottom line, he is the only person to which he is answerable every day and in the longer term.

A World of Physics

Whether he deals with arcane or modern concepts, the work of the physicist is part of a spiritual tradition older than recorded history. He seeks to know how the *hidden parts of the universe* work and fit together. Physics is a vital part of the cultural and intellectual mainstream of humanity, though not always appreciated as such.

Research in the physical sciences demands an exceptional ability to concentrate, to remember, to make connections between facts and ideas. This is the more challenging, as our insights about gravity and matter are changing the way we look at the universe and, at the same time, we seek to unify the theoretical breakthroughs in 20th-Century physics.

The elite of modern physicists today is actively seeking to establish whether there is one bigger law from which all the other laws can be derived. This is a reversal of trends which characterised the larger part of the 20th century and, as well, the thousands of years in which human thought crystalised.

The domain of science has been subdivided, since the beginning of this century, into a large number of narrow specialities. Twenty years ago, no one would have thought of a rapid *unification.* However, the most remarkable characteristic of present-day science and technology is exactly this tendency towards unification of the different disciplines. We have finally come to realise that some of the most diverse scientific endeavours are based on essentially the same problems and principles.

Niels Bohr postulated that the gravitational and electro-magnetic forces were different aspects of the same phenomenon. This is the basis of the unification theory. Physicists have now begun to understand basic forces, and call them GUT, for Grand Unified Theories.

Central to this framework is the existence of new particles, tiny fragments of matter or energy, since the two are interchangeable. At the frontiers of theoretical physics what is needed is not precision but key overview ideas. That is, great organisational principles, from which the details can follow.

Just as important is working out the details, ultimately to compare them with experiment, with reality. The underlining design, and its details, are the two pillars on which science rests. The whole history of human thought has been to try to understand what the universe is like.

Physics has made great strides towards a unified theory that may eventually relate:

- *Cosmology,* the science of the ultimately large, to
- *Particle physics,* the science of the ultimately small.

But, basic scientific discovery, including the line leading to unification, seldom comes easily. It demands energy and persistence; intelligence and open-mindedness. These are the high marks of the trade in any profession - hence of scientific discovery which is, by excellence, a domain of applied intelligence.

Experiments lead to theories, and theories suggest a cause-and-effect relationship valid until newer theories show up to upset it. Research into so far unknown properties calls for better experiments, and more powerful machines to conduct them, while each type of search is likely to produce, as a spin-off, its own unique discoveries.

The major advances of science are, almost by definition those that are not foreseen, and this is just as true of the great leaps forward that technology has achieved. A case in point are the developments which took place with *semiconductors* during the late 1940s to late 1980s and now the *photonics* science.

Technological knowhow has its roots in scientific discovery, but there is a significant lead time between the establishment of the physical principle and its technological application. Revolutionary discoveries in physics are interspersed by years of steady, often slow, progress towards practical applications.

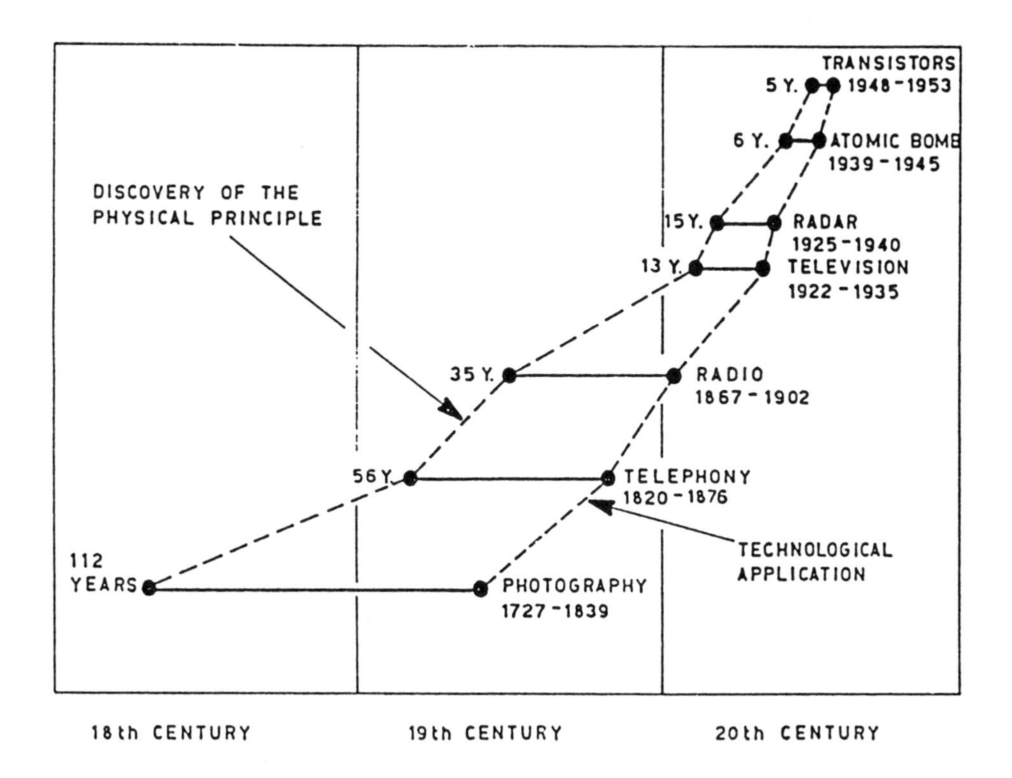

Figure 2.1

Another important observation is the accelerated pace of discovery and of development, and for good reason. Statistics aptly document that half the engineers and scientists that ever lived on earth are alive and working today. Massive, steady increases in brainpower are due to have some mighty fallout - and the same is true of the advanced tools that we currently have available.

At the same time, the products and processes that technology has brought about - such as the new physical sciences and microelectronics - can radically change the way we are looking at the world. Work in advanced science can also accelerate the time scales, bring up new technological breakthroughs and alter our values. This is the message to the reader from this chapter on new physics.

If we are going to attempt creative thinking, we need the broadest possible set of facts in front of us, so that we can sift them back and forth to see what will work. If we do not have the facts, we cannot eliminate the ones that are not worth our time; while a lack of facts severely limits our visibility. These are basic reasons why we will be well advised to take a look at the frontiers of knowledge. From there will come the new means, from tools to materials, to enhance the power of the human mind.

Scientific and technical progress has reached the point where it becomes inevitable. A civilisation that stops inquiring about the physical universe is giving up its future. To help themselves, and other scientists, with the understanding of the laws which govern the universe, physicists develop theories:

- *General relativity*, deals with predictable events and huge objects, such as stars and planets.
- *Quantum mechanics,* addresses itself to minute details inside the atom, an area where we have not yet succeeded is predicting events precisely.

While astronomers are on the verge of proving, by observation, the existence of black holes in space, physicists have drawn an increasingly detailed picture of what black holes ought to be like. They are not simply cold and dead collections of invisible matter, but they have temperature, and some can be extremely active, bright and hot.

Dr. Stephen W. Hawking, one of the most brilliant new physicists, has conceived startling new theories about black holes and the tumultuous events that immediately followed the Big Bang from which the universe sprang, according to the prevailing hypothesis. More recently, he unsettled both theologians and other physicists by suggesting that the universe:

- Has no boundaries;
- Was not created; *and*
- Will not be destroyed.

Having examined the possible characteristics of black holes, Hawking demonstrated that the forces of the Big Bang would have generated mini-black holes, each with a mass about that of a terrestrial mountain, but no larger than the subatomic proton. Applying the quantum theory, which accurately describes the random, uncertain subatomic world instead of general relativity, he made the hypothesis that the mini-black holes must emit particles and radiation.

A black hole's gravity swallows everything in its vicinity. The best chance of detecting a black hole is to watch for X-rays emitted by material falling into the hole from a companion star that it is orbiting. An object called Cygnus X-1, for example, seems to be a normal star with a black-hole companion.

Even more remarkable, these little black holes could gradually evaporate. The hypothesis has been formulated that, in 10 billion years or so after their creation, they will explode with the energy of millions of H-bombs. This is another tentative statement waiting to be proved or disproved.

The world of physics is changing. Our ancestors needed only their powerful intellect and imagination to postulate atoms as the basic building blocks of matter. Today, further exploration requires complex machines such as Tevatron of the Fermi

National Accelerator Laboratory, the accelerator of CERN* known as LEP (large electron-position storage ring) and the SSC mentioned in the introduction to Chapter 1.

While brilliant minds are the creators of new theories, experimentation has to confirm or dismantle them, and this increasingly means big science. That is how future progress will be achieved. By bombarding the nucleus with protons, or other subatomic particles, physicists tear apart the fabric of matter, and thus they gain insight, and show further foresight, through hypotheses.

Every higher level of sophistication requires increasingly powerful machines.

- The smaller the objects,

- The bigger the tools we must use to see them.

CERN's significant contribution is the discovery of a group of new particles dividing matter into two basic types:

1. *Quarks*, which are the building blocks of protons, neutrons and other "heavy" components of the atomic nucleus, *and*

2. *Leptons*, exemplified by light particles like the electron.

One major achievement was the demonstration that the heavier of the newly discovered particles (and, as well, the old familiar proton and neutron) was a compound structure, a combination of still more elementary components than earlier knowledge revealed. It is exactly these components which were given the name *quarks*, the prevailing hypothesis being that they are themselves similar in some respects to the lighter-weight particles, such as electrons.

Another fundamental physical principle relates to strong and weak forces. Other particles, not subject to the strong force, are themselves indivisible. The electron is the most familiar of these particles, which are called *leptons*. Quarks are thought to come in both matter and antimatter form: antimatter is like ordinary matter except that certain properties are reversed. A *positron* is the positively charged antimatter form of the electron.

Quarks are structureless and point-like. Pairs or triples of them are bound to each other by *gluons;* such conglomerates make up most particles. They are all influenced by the strong nuclear force that binds the nuclei of atoms. Protons and neutrons are the commonest quark-gluon combinations. New effects have been predicted by this theory, e.g. that the proton is slightly radioactive and, therefore, all matter is ultimately unstable and will decay.

---

* European Organization for Nuclear Research, Geneva, Switzerland.. A well known international laboratory for particle physics.

The structureless reference is very critical. Prior to the new theory, protons, neutrons and similar particles were supposed to have an inner structure. Now physicists postulate that they are made up of quarks; a finding based on human ingenuity. The detectors that surround the target areas of particle accelerators show no direct traces of any such particles. Only when the accelerators smash protons or electrons do they sometimes produce the kinds of debris expected if the quark concept is true.

Towards a Unification Theory

As Dr. Stephen W. Hawkins aptly suggests, the search for the beginning will not be complete until we are able to understand the boundary conditions. That is, what preceded the beginning:

- Matter,
- Space, *or*
- Time.

Had time a beginning? If yes, what was the universe like at the beginning? If time had no beginning, what does determines the condition of matter in the universe?

The search for boundary conditions is a leap forward from Einstein's work. The springboard for that leap is quantum theory, describing the behavior of elementary particles, i.e. the component parts of the atom.

Dr. Albert Einstein made major contributions to quantum mechanics, but he did not really like the theory. He viewed it as a step towards some better theory, because it depends on the *uncertainty principle* which says that an observer cannot precisely predict, at any moment, both the location and the speed of a particle within an atom. This introduces an element of chance for the observer.

But quantum theory works in situations where classical physics does not. Besides, acceptance of uncertainty may prove necessary to a complete understanding of the way the universe operates. Uncertainty and vagueness are at the roots of the new *possibility* theory, which is broader than the better-known probability theory, the latter being a special case of the former.

Physicists have long known that the photon (light particle) was the carrier of *electromagnetism*. In 1979, in Hamburg, they discovered the gluon, which conveys the strong forces. Four years later (in 1983) CERN scored its crowning achievement by confirming the existence of three particles, the W+, W- and Zł, known collectively as intermediate vector bosons, the agency of the *weak force*.

The leader of the successful CERN experiment is Nobel Prize winner Dr. Carlo Rubbia, who is also a faculty member at Harvard. Ironically, he had originally proposed this research to Fermilab, which decided to concentrate on the new machine instead.

Connecting links have been proven between quantum theory and every known physical field of force except gravity. Consistency requires that general relativity theory be brought in under the tent of quantum theory. This is known as *quantizing gravity,* one of the knottiest problems in physics today.

These are the reasons why elementary particle physics is a place where a truly creative genius can show his brains. It is also a domain where research efforts show their excellence. Dr. George Keyworth likens the building of big new accelerators to the Apollo program: ''Apollo wasn't just learning about the geology of the moon, it was also about leading the US into an era of high technology'' - which, incidently, is the reason for SDI.

The more physicists advance in their knowledge of the universe, the greater the impact science will feel, but also the faster the system of discovery accelerates. One of the latest theories goes beyond the four known forces of nature: gravity, electromagnetism, the strong force that binds atomic nuclei, and the weak force that governs certain types of radioactive decay. In August 1988, researchers at the Los Alamos National Laboratory announced that they may have found the best evidence yet for a hypothetical, but still elusive, *fifth force.*

The fifth force, if that is what it is, has been a source of debate among physicists since its possible existence was suggested in 1981 by Australian mineshaft experiments. Five years later, Purdue University physics professor Ephraim Fischbach measured a weak force he called *hypercharge,* and theorised that it caused objects of different composition to fall at different rates. Since Fischbach's finding as many as 45 experiments have sprung up in search of the mystery force, but so far each has served only to confound, rather than to clarify, the issues surrounding it.

We said that a Nobel prize was awarded to Dr. Carlo Rubbia in recognition of his theory regarding the unification of certain fundamental forces in the physical sciences. Modelled by some of the best minds of our time, such achievements promise to match the unification theory that Maxwell demonstrated in the early 19th century, between electricity and magnetism.

Noticing what was then an inconsistency between the equations for magnetism and those for electricity, James Clerk Maxwell put his insight to work, adding an extra term to the existing algorithms. This enabled electricity and magnetism to be treated as different aspects of the electromagnetic force, and paved the way for the invention of the *telephone,* the *radio, television* and, in general, of *electronics.*

Surrounded by the achievements of nuclear physics, and the insight they brought during the last fifteen years to properties thus far unknown, today scientists are on the threshold of new major breakthroughs. As Section 2 underlined, a different chapter is opening in particle physics, and this belief rests on two conceptual developments of the late 1960s and early to mid 1970s:

1. The growing evidence that virtually all of the hundreds of kinds of subatomic particles are composed of three basic units: *quarks, leptons* and *gluons.*

These seem to be the ultimate building blocks of matter, put together at work by the evidence which is unfolding in regard to the fundamental forces of nature.

2. The fact that scientists now think in terms of four basic forces: *gravity, electro-magnetism,* the *strong,* and the *weak* force -as well as the stated, elusive, fifth force.

A ''strong'' coupling holds atomic nuclei together. The weak force, a billion times weaker than the strong, can be found in many radioactive decay processes. With the electroweak theory only partially tested, physicists are already attempting a further unification combining the electroweak and the strong forces into GUT. In fact, the real scientific interest of weak coupling is the switch which it makes with the strong force.

Dr. David Politzer developed this theory in 1973, and it is unlike any that physics and physicists had heard of at an earlier time. A coupling force is weak when things are

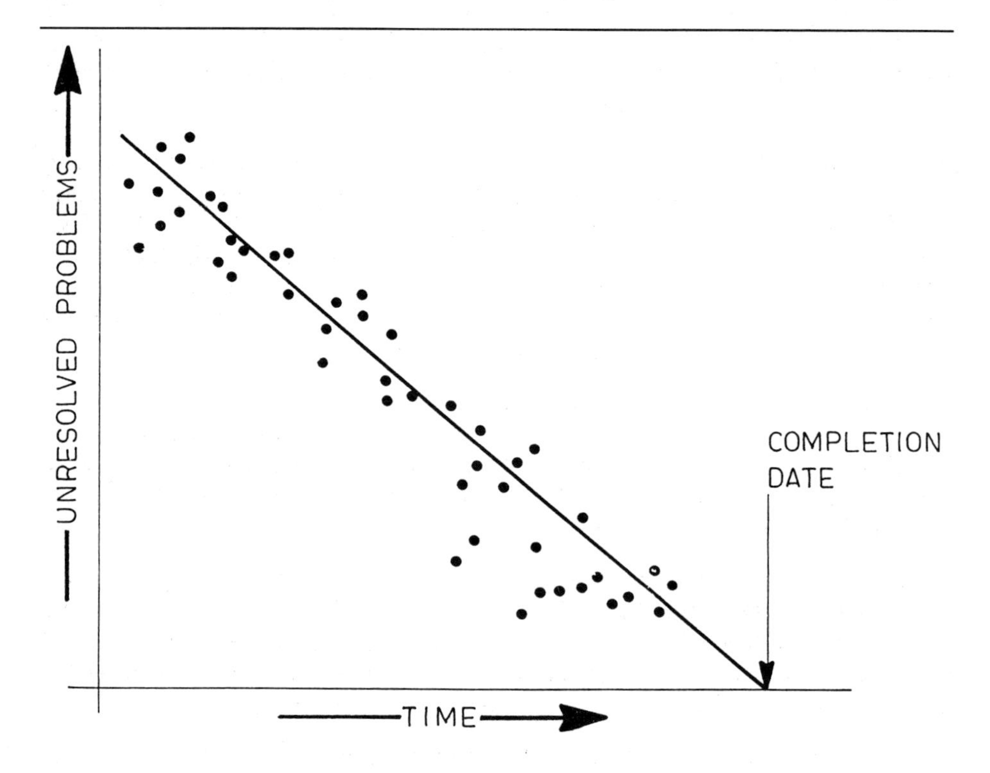
UNRESOLVED PROBLEMS
COMPLETION DATE
TIME
THE TIME CAN BE SHORT OR LONG -
DEPENDING ON HOW THE PROJECT IS MANAGED.

Figure 2.2 Project Status Chart Plotting unresolved problems vs Time

close together; it gets stronger as we pull them apart. Experiments have pretty well proved the accuracy of the quark hypothesis, paving the way to the larger theory "of quantum chromodynamics", or QCD, which predicts the behavior of quarks and their forces.

Quarks come in different varieties. One of the ways they differ is in a property that determines how different quarks will combine, and which is analoguous to electrical charge. In Section 2 we said that quarks are structureless and point-like. Politzer's theory about a quark force was an asnwer to a challenge which developed in the late 1960s. At that time, a similar feat to that of Maxwell was achieved by Steven Weinberg and Abdus Salam. They succeeded in providing a unified mathematical description of the electromagnetic and the weak forces.

Like all good theories, the blueprint laid down by Salam and Weinberg made precise quantitative predictions about subatomic matter. Experiments were needed to test the theory, the most decisive being the confirmation of the prediction that, at high energies, certain specific particles that have not been seen before will be created. These are the W+, W-, and Zł particles, to which we made brief reference in connection to research at CERN.

Hawking, Rubbia, Politzer, Salam, and Weinberg have been dedicated researchers. They also knew how to lead teams of physicists and other scientists able to generate significant results. Plotting unresolved problems versus time, and based on actual statistics, Figure 2.2 suggests that needed time can be short or long depending on how the project is managed - and on the *brains* on which it runs.

Another requirement for scientific experiments are machines measuring up to the effort. The more advanced these experiments are, the more powerful the required engines; and this is just as true of super cyclotrons as of supercomputers. *Intelligent machines are the lens of our mind's eye.*

Unlocking Nature's Secrets

Since the 1930s, the power of machines for accelerating subatomic particles has increased tenfold every seven years. We have spoken of the investments necessary for big science in Chapter 1. Parallels can be drawn between particle physics and the space program in terms of capital budgeting. Parallels can also be drawn between research in physics performed decades ago and the physical breakthroughs we experience today.

In 1911, the Dutch physicist Kammerlingh Onners discovered that, at close to absolute zero, some metals lose all their resistance to electrical current and become superconductors. Seventy years later, commercial applications of this have been developed in:

- Computer circuitry,

- Fusion reactors, *and*

- Other fields.

From the machinery that the physicists are using to unearth the laws of the universe will come the, admittedly rare, revolutionary discoveries to restructure the natural laws and eventually, offer humanity a new big jump in standards of living. The microprocessor has been one of these unforeseen, but mighty, results. More will be coming, provided that we are able to steadily renew the theoretical infrastructure on which further developments are based.

The age of the 1920s, 1930s, and the early 1940s was built on the conceptual revolution of quantum physics and the theory of relativity. But with the mammoth (for its time) nuclear programme of the 1940s and other scientific endeavours, including missile technology and space research, postwar physics faced a different challenge:

- At one end were the space probes, and the new telescopes, to stretch man's horizon to the farthest reaches of the universe.

- At the other end, particle accelerators became a vital tool in the search to identify the tiniest, indivisible constituents of matter.

In this race towards the unknown there was, and continues to be, no foreseeable end. The world's largest telescope was opened in 1980 in New Mexico: a 21-mile wide structure that cost $70 million; while the Voyager spacecraft reached Saturn, at a cost of some $340 million. A year later, in 1981, CERN completed a $120 million project to smash antimatter into matter -immediately thereafter asking for the budget to build a $550 million machine for crashing electrons into positrons; while Brookhaven built a $275 million piece of equipment to accelerate protons.

As we know from other experiences, matter and antimatter annihilate one another when they meet, transforming their mass into radiant energy. Conversely, pairs of matter-antimatter particles can form out of pure energy. These processes of creation and transformation produce the distinctive patterns of debris which physicists are looking for in accelerator experiments.

Such approaches are being followed to achieve the enormously powerful collisions that are necessary for new particles to be created. Some involve the use of superconducting magnets, as the power behind machines able to smash such protons into each other quite so energetically. Others call for smashing matter into antimatter as, when they collide, they are releasing far more energy than collisions between ordinary matter. Scientists would also like to crash electrons into positrons, with collisions probably resulting in a shower of new particles, including the "W" and "Z" we have mentioned.

What some of the advanced physical sciences are looking for, especially in electron-positron collisions, are patterns in which debris shoots out in jets. As science prepares for its next big leap forward, physicists speak of six flavours of quarks, each paired with its antiquarks: "up", "down", "strange", "charmed", "bottom" or "beauty", and "top" or "truth" - as with any truth, the latter is not yet experimentally confirmed.

Events involving gluons should produce two opposite directed jets that become increasingly thin as the collision energy increases. Double jet events observed over the past five years do support predictions: not only electromagnetism but the strong/weak force, too, derives from common origins on the quark scale. Eventually, gravitation and QCD might be contained into a Grand Unification Theory that would encompass every domain from elementary particles to the universe at large.

Quantum chromodynamics is, incidentally, one of the theories the knowledgeable man will be well advised to watch. Starting with the hypothesis that quarks stick together with gluons, QCD is based on a postulate analoguous to the proved quantum electrodynamics which describe the interaction of particles through electromagnetic force. Interestingly enough, running out of names for all the new properties that they have found in subatomic matter, physicists have assigned colours to the various kinds of quark charges - hence chromodynamics.

Will these developments lead to a wave of new processes and products similar to the inventions that followed Maxwell's electromagnetic unification theory? It is difficult to be assertive at this stage - but, as Dr. Robert Noyce (one of the inventors of the integrated circuit) was to suggest: "As an act of faith, I would say yes!".

It is a conditional "yes", betting both on brains and on equipment. As physics moves forward, the range of equipment necessary to serve the new physics widens. Proton machines are needed to study quarks and gluons; electro-positron systems are required for testing the electroweak theory; and other equipment is necessary to satisfy man's understandable curiosity about the universe.

Significant discoveries will almost certainly be made from these scientific investments. While no spectacular economic impact is likely to emerge until the next century, the wisdom of financing research in the new physics has become a major topic - and an expensive, but rewarding to commit, use of financial, technological and human resources.

Just as microprocessors changed many aspects of our everyday life, the discovery of new elementary particles has been changing the world of physics. A short twenty years ago, experimental work with high energy accelerators transformed the physicists' view of matter as something made up of a few elementary particles, electrons, protons and neutrons, into the realisation that they had to make room in their theories for literally hundreds of different particles, none more elementary than one another. Furthermore, the behaviour of these particles seemed to be governed by the distinct forces which we mentioned earlier. In astronomy, too, the pace of development moves faster. Some two decades ago, the sky was considered to be essentially unchanging except for planets, comets and very rare events such as novas and supernovas. Today, when looking at the sky with new types of telescopes, we see that it presents an ever-changing picture of fluctuating intensities.

If other types of radiation were translated into visible light, the sky would present a dramatic flaring up and fading away of stars, clusters and nebulae - the latter being clouds of gas and dust, in some of which new stars are forming. By contrast, novas and supernovas are stars in their death throes.

In a single explosive burst, a supernova can brighten to many billions of times its former intensity. The remnant star left at the centre of a supernova can be bizarre in nature. One possibility is a star which is so compressed that its atoms are crushed to neutrons. A neutron star with the mass of the sun would be only six miles in radius, and the effect of gravity would be enormously intense. Called pulsars, such stars rotate many times a second and this rotation is seen in radio telescopes as rapidly flickering radio signals.

These are some of the profusion of phenomena discovered in the sky by deploying powerful, accurate telescopes andalso telescopes that detect many types of radiation which prewar astronomers could not observe. Gamma rays, X-rays, ultraviolet, infrared, microwaves and radio waves are examples of this acquired knowhow. At the same time, they are only a small part of what we think we could learn.

By continually improving the definition of their instruments astronomers are able to probe deeper into the universe, and discover more about the range of cosmic phenomena. It is only since the late 1970s that scientists have possessed an X-ray telescope with good definition. Previously, they could not pinpoint X-ray sources: they could "see" the nearby Andromeda galaxy, about 2 million light-years away, but not its parts. Now 80 individual objects emitting X-rays have been observed in the spiral arms of the galaxy, where very massive stars appear to be forming.

For radio astronomers, a similar improvement in resolution has been achieved by the telescope opened in New Mexico in 1980. Called the very-large-array (VLA) telescope, it has 27 radio dishes that can be moved along a Y-shaped track covering nearly 38 miles in length.

Another way of improving the resolution of radio telescopes is to link several separate radio dishes, situated in different parts of the earth, and to coordinate the incoming signals. The next step is to place one or more of the telescopes linked into such a multi-telescope system in space - first on a space station, then on the moon and, later on, still further out.

Different media are necessary in particle physics. By the same token, an ever-increasing range of media are necessary in modern astrophysical research. A star that shines dimly in an optical telescope may be very bright in another type of telescope, and unveil astonishing qualities when examined by a space telescope. Very impressive research results are needed if we are to learn about the fiery birth and evolution of the cosmos.

Since the nuclear activity of the primeval universe has left imprints that can be detected even now from earth, astronomy has become a possible testing ground for ideas in subatomic physics. This points to a further sense of the unification of the physical sciences, whose outcome is going to be of tremendous importance.

Here exactly astronomy and particle physics *intersect.* The Big Bang which many scientists think marked the birth of the universe some 15 billion years ago, was a particle accelerator more powerful than man will ever build. In it quarks, leptons and gluons were created.

What Could We Expect from Space Exploration?

While astrophysics is instrumental in helping molecular physics to make a great step forward, this is not the only reason why investments are being made in this branch of big-science. Astrophysics has opened an avenue of development in space exploration, and space exploration at the scientific level has soon been followed by projects designed to open up the universe surrounding our earth for industrial, commercial and military uses.

Communications satellites were the first practical implementation of space exploration. They have been followed by applications such as weather detection systems, killer satellites, and reusable space shuttles. The latter have already made flights on a purely commercial basis a reality.

In an age in which the limits of economic growth on planet earth are becoming increasingly evident, the universe offers a field of development whose limits are at present unknown. So are the aftermaths that future generations may have to pay, but this is not the preoccupation of the present political and religious "leaderships".

While we talk of scientific endeavours and future progress, we should not lose sight of the fact that it is basically the same men who allocate large funds to big-science and who provide the greatest impedence. Power plays, egos, politics and sheer career making are behind the impedence reference, while national security, prestige, "me too" attitudes, pressure groups and, again, career making are the motor power behind the allocation of large financial resources.

Perhaps no better example illustrates how an established bureaucracy can blur scientific endeavour than the Jupiter and Vanguard rockets in America, in the late 1950s. The all-important officialdom was more concerned with the prestige of its baby than with the solutions which could swiftly provide valuable results to the nation (and its taxpayers) financing the rockets.

October 4, 1957, ushered a new page of living space. Plans for launching artificial satellites into orbit during the International Geophysical Year, 1957/58, were announced at the time by both the United States and Russia. Working for the US

Army, Dr. Wernher von Braun joined forces with scientists of the US Navy on Project Orbiter. The technological goal was that of launching an earth-orbiting satellite.

Capitalising on Germany's World War II experience with missile systems, at the beginning work progressed well. But, with the construction of the launching site at Cape Canaveral, the Pentagon reacted negatively: against further support for the Orbiter program. Pressure groups saw to it that it was decided to scrap Orbiter in favour of the Navy's Vanguard project. As a result, the von Braun group in Huntsville, AL., was directed to concentrate on ground-to-ground missiles. Not to work on satellites.

But, even at the Pentagon, an order does not necessarily slip down the organisation, much less get executed. Von Braun and his scientists set out to build the four-stage Jupiter version of the Redstone. On August 20, 1956, as a Jupiter C stood on the launching pad at Cape Canaveral, General John B. Medaris was hastily dispatched to Florida. A day later, Jupiter C flew further and faster than any rocket before it. Only its disabled nose cone - which Medaris assured his Pentagon superiors was a fake - prevented it from placing an American satellite in orbit a year ahead of Sputnik 1.

But Jupiter did not receive the official blessing of the Pentagon brass. The less promising Vanguard was the favoured carrier. Bogged down with problems, Vanguard increasingly seemed a failure, making von Braun realise that his hour had come. "The Navy project is not going to make it - give us just 60 days support and we will put a satellite into orbit", he is reported to have said to the generals. It took Sputnik 2, whose weight was stated as 500 kg, to convince officialdom in Washington that he was right. The Army Arsenal in Huntsville was given the green light on November 8, 1957. On December 6 that same year, the small rocket, Vanguard 1, exploded on its launch pad. But, by January 31, 1958, America had its very own "Explorer 1", just as von Braun had promised - through the Jupiter rocket.

The race to the moon was by then well launched. On April 12, 1961, Gagarin was the first man to survive a space flight, landing after one orbit round the earth. By then the Americans had taken up the challenge of the 1960s. A ten-stage preparation programme saw the light of day during 1965 to 1966. Its goal was a moon landing with the two-man Gemini capsule.

- This mission was carried out with a three-man craft.

- It landed on the moon in July 1969.

- The command ship had completed its maiden flight not earlier than October 1968.

- The complex lunar module was not put into service until 1969. In March, on an earth orbit; and in May, on a lunar orbit.

This speaks volumes about what a well-financed, dedicated programme with knowledgeable, hard-working people can achieve. The Jupiter/Vanguard story is a true happening, but at the same time it is an allegory in the world of physics. The European and American particle physics programmes are today in a head-on race to win first place. The same is true of significant efforts dedicated to astrophysics.

Can we talk of tangible results? First, as stated in the introductory paragraphs, there have been the communications satellites. This was the birth of a dynamic growth industry, but also of great communications capabilities, as Figure 2.3 documents.

Tax revenues alone from this new broadband communications industry have surpassed the original investment. And, though very important, communications are only one example. Other examples of by-products of space research include:

- New types of plastics;
- Fibre optics;
- Solar cells;
- Cardiac pacemakers; *and*
- Medical monitoring techniques.

Patents registered by the American National Aeronautics and Space Administration (NASA) over the years document the watershed of new products and processes. At the same time, companies that were able to gather innovative momentum during the ascendacy of space technology in the 1960s and 1970s are among those with the fastest growth rates today, provided that they exploited the opportunities available to them.

Still, these are only some of the examples and they focus strictly on the past or the present.

- An American study has shown that $1 invested in space research adds another $4 to Gross National Product (GNP) over ten years, by means of new products and services brought to the market, as has already been underlined.
- If successful, every kind of process, or type of machine, is likely to produce, as a spin-off, its own unique discoveries.

Past experience suggests at least one out-of-the-blue discovery with each new generation of new machines. And major advances in science are, almost by definition, those that are not foreseen.

The space station Skylab, whose aggregate cost was $2.4 billion, identified mineral resources in America totalling some $15 billion, which can be mined over a period of 15 years. Oil tankers save a fortune by using satellite pictures of ocean currents to steer energy-saving courses. Satellite navigation has brought further savings by considerably reducing the number of collisions at sea, including those caused by icebergs, due to its greater precision. Telephone satellites saw to it that the number of

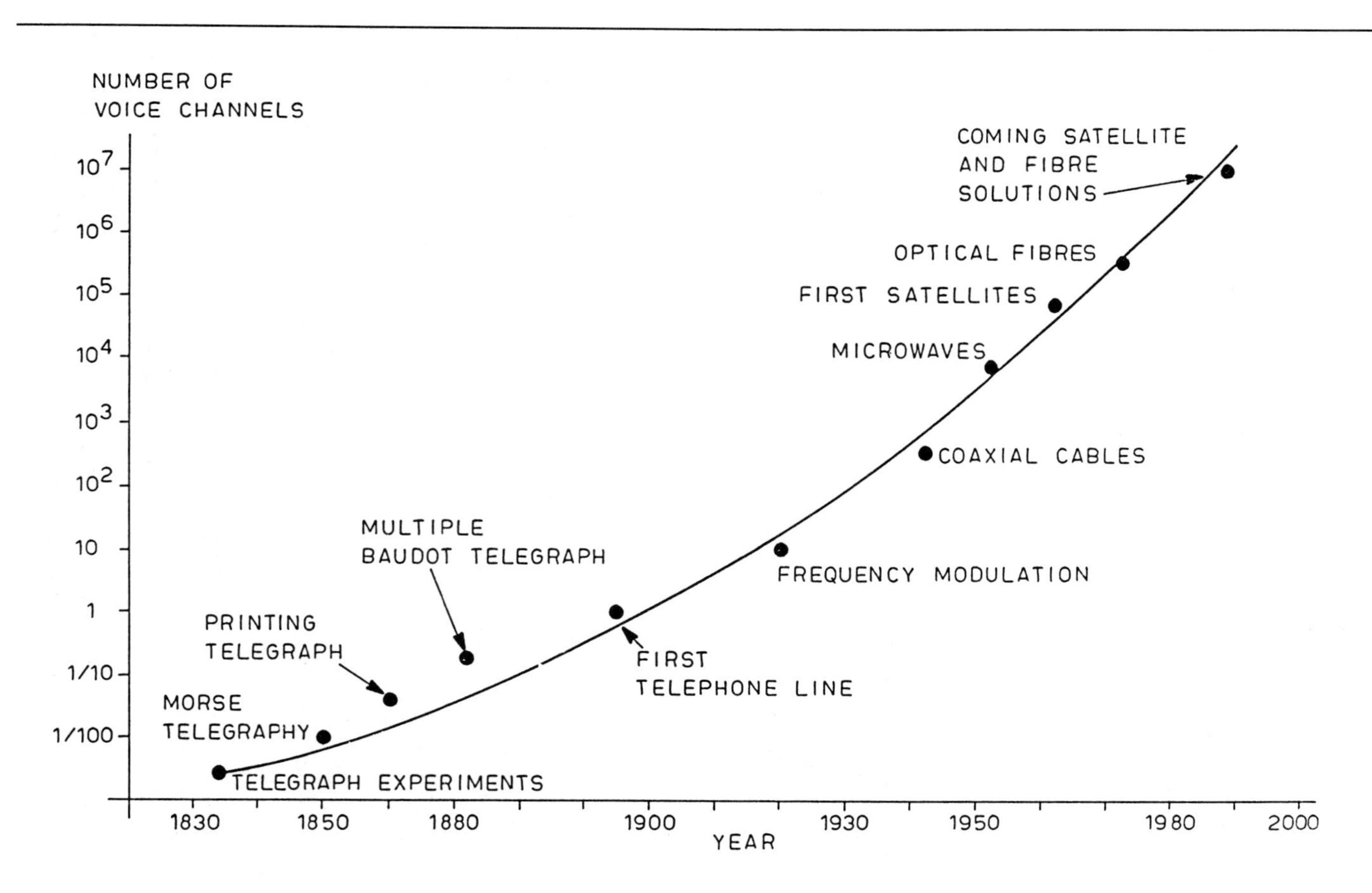

NUMBER OF
VOICE CHANNELS
10^7
10^6
10^5
10^4
10^3
10^2
10
1
1/10
1/100
COMING SATELLITE
AND FIBRE
SOLUTIONS
OPTICAL FIBRES
FIRST SATELLITES
MICROWAVES
COAXIAL CABLES
MULTIPLE
BAUDOT TELEGRAPH
FREQUENCY MODULATION
PRINTING
TELEGRAPH
FIRST
TELEPHONE LINE
MORSE
TELEGRAPHY
TELEGRAPH EXPERIMENTS
1830
1850
1880
1900
1930
1950
1980
2000
YEAR

Figure 2.3

intercontinental telephone calls rose from 3 million in 1965 to 200 million in 1980, progressing towards one billion, at a cheaper cost to the user.

Johnson & Johnson has carried out one of the most advanced space manufacturing projects to date: electrophoresis in zero gravity. The result has been the production of several hundred times the normal quantity of higher purity, expensive substances for medicinal use. Constructed by McDonnell Douglas, the apparatus was constantly improved over a series of Space Shuttle flights. It is being developed along the lines of a free-flying satellite factory which would only be brought back to earth after some months.

There is more to come in the future. Space stations in which the entire world supply of many pharmaceutical and photonics goods can be manufactured, in one or two weightless laboratories, imply a relative increase in productivity reminiscent of the first Industrial Revolution. The roster of products to be made in space includes:

- Medical applications;
- Pharmaceutical products;
- Communications;
- Lasers;
- Special lenses;
- Other optical and light-emitting devices;
- Semiconductors;
- High-precision mechanics;
- Energy sources;
- Flight control;
- Aircraft guidance;
- Satellite photographs;
- Launch rockets.

Constant availability of ultra-high vacuum, zero gravity, and the low density of materials in space, as well as cheap energy, permit both very fine products to be made and large structures to be assembled at a reasonable cost. Better quality is one of the aftermaths.

The General Motors Research Laboratories have been conducting basic research under microgravity conditions, including the behaviour of fire, combustion engines and lubricants, as well as welding and crystal growth, in zero gravity. Honeywell has experimented under microgravity on Mercury-cadmium-tellurium crystals for the optoelectronics industry. These are only some examples.

Why a Sustained Effort in Big Science?

Results already obtained from space exploration, and those which are forthcoming, help in answering the question: Why should the human society sustain a continuous effort in big-science, or in an area of research not directly dictated by the "needs of the people"? The answer, quite evidently, is that the results of science do answer peoples' needs through an impressive array of new solutions (Figure 2.4).

Large scale scientific projects are not undertaken simply by the wish to follow-up on the ambitions of astronomers and particle physicists. The purely scientific value of sending spaceships to get firsthand knowledge of the solar system has been proved many times. The same is valid of other big-science projects, as the preceding section demonstrated.

Many projects are competing for funds, most of them both urgent and important. Only the sketchiest knowledge exists, for example, about the interior of thc carth. An exploration of the earth's core would be very expensive, but would also yield scientific surprises and, quite possibly, practical applications, such as new sources of energy or minerals.

Just as significant are the rapid developments going on in solid-state physics, the science of solid materials. These, too, are most likely to have important practical applications in new materials, as we will see in Chapter 3.

The advance of science is unpredictable. Sometimes advances come more or less at random. Hence, attempts to direct research on the basis of possible commercial value are liable to distort the natural development of scientific effort, thus becoming counterproductive. Unpredictability applies to a broad range of pure research, but also well into the phase of development.

Furthermore, as we saw throughout this discussion, great discoveries in physics are interspersed by years of steady, but slow and painstaking, progress. In these periods of consolidation, the new discoveries are carefully fashioned into shape with practical applications results.

Sometimes a key discovery will come from a totally unexpected source. Radioactivity was discovered accidentally, when Becquerel noticed that a photographic plate kept in a drawer had mysteriously become fogged.

If experienced scientists find it difficult to decide a priori on the most promising projects, the inadequacies of our education system see to it that it would be indeed

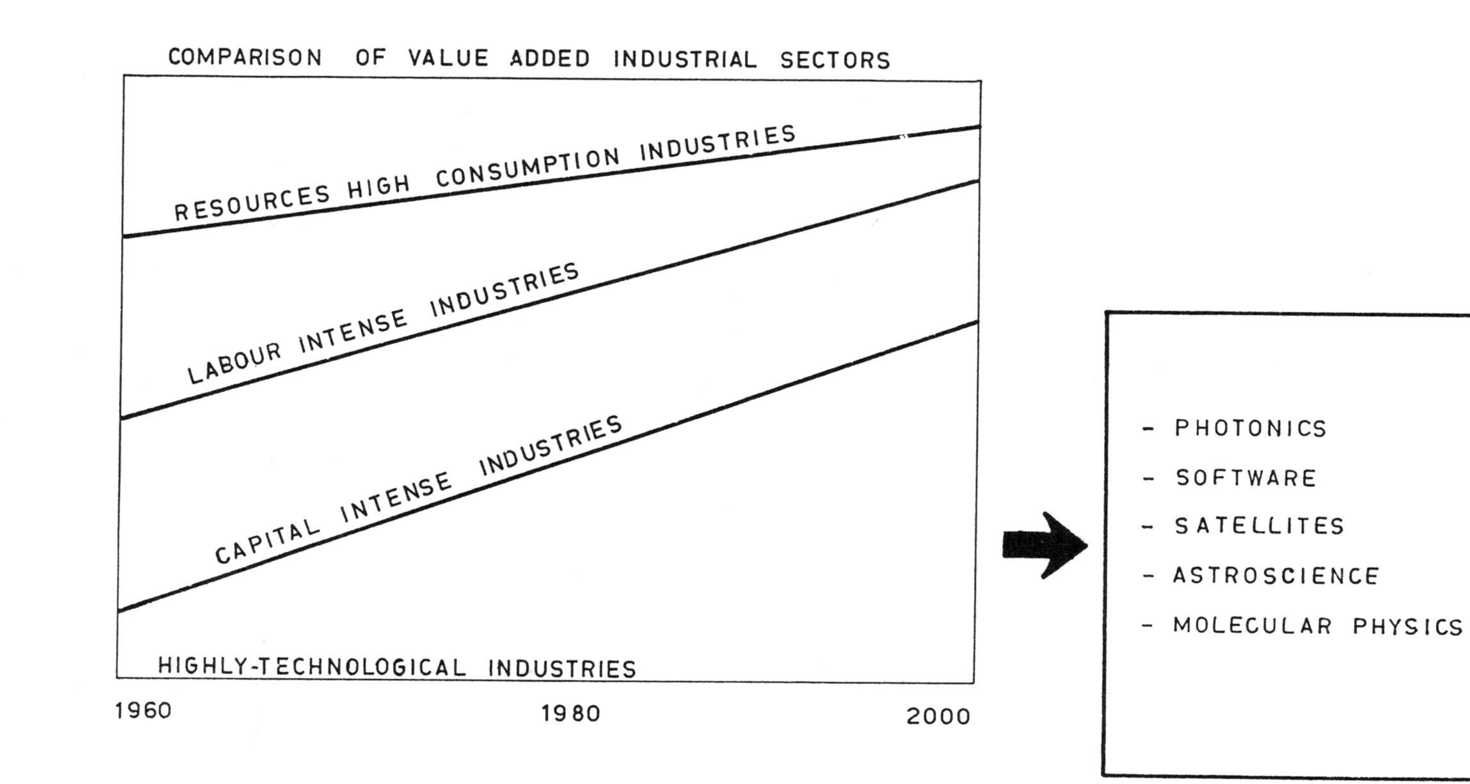

Figure 2.4 Progress in advancement of Industrial Structure

surprising if the general public could inform itself of the not immediately perceivable benefits. Only by looking at the record we can read the results. After World War II, when it came into being, big-science spawned useful innovations. For example, particle accelerators have led to the use of small accelerators for cancer therapy. Satellite communications owe an important debt to radio astronomy. Let us, however, always keep in mind the time lag. The time scale of major benefits from astronomy and particle physics is very long, perhaps 40 years or more - yet the practical results will eventually come.

Quite importantly, there is more to science than profit and loss from practical applications. The acquisition of knowledge has been an end in itself since the ancient Greeks debated the existence of atoms, the concept of infinity and the creation of the cosmos. If investigations into such fundamental problems today cost a lot more in investments, society is also more affluent and, therefore, it can afford it.

A scientist, a university, a society, an epoch, even a civilisation that stops inquiring about the universe would stop inquiring about other things as well. A lot of other issues might then die besides particle physics and astrophysics.

In spite of that, today pure science is rather underfinanced and, to a large extent, underequipped. Considering that pure research is the leading edge of advanced technology, it is extraordinary how labour-intensive such research actually is. In the United States, only 10 percent of the grant money paid by the National Science Foundation is spent on capital investment - though about a quarter of American spending on particle physics is devoted to equipment.

The fast pace of technology, and the sheer size of investments required to upkeep laboratories, see to it that the age structure of university departments is becoming rapidly older. It has become difficult for an outstanding young professor to build a research team around him and to purchase the advanced equipment that he needs. It is also hard to finance new fields of science.

One sensible approach would be to index budgets for basic research to GNP. Researchers would then be able to know that the budget available would grow over the medium to long term, and that it would grow only modestly. Facing modest growth, if they want to plan a lot of exciting new ventures, as they should, they would have to cut out dead wood to make possible better focused financing.

Plenty of dead wood exists with projects which slid from one department to another in the laboratories. At the same time, by staying around, non-productive scholars block the careers (and the jobs) of young, enthusiastic scientists. Most countries can now boast more second-rate research than has ever been done before, and some of their mentally-aged scientists are both backwards looking and dull.

True enough, there are some fields of science where productivity is high. Photonics, astrophysics and particle physics are examples. Here the average standard of both research and researcher is probably much above average, as these subjects attract some of the best brains, and also because competition is intense. But neither is immune from waste. To make feasible a sustained effort, scientists should improve their planning skills.

3

Materials for Man-Made Products

Introduction

In the distant past, for economic reasons, man started transforming materials. They were of natural origin: argil, stones, vegetable fibres, wood and the skin of other animals. The fusion of metals and glass manufacturing extended this interaction, but man's search for materials is more profound than the use to which he puts them. Human civilisation is closely linked to this search.

The earth has a wealth of materials, but they are not inexhaustible. The danger of some materials fading away is always present, as consumption has reached unprecedented levels. Leaving aside products of a petroleum origin, *each man on earth yearly consumes 10 tons of irreplaceable raw materials.* At the same time, the science of materials is making great strides.

About 10,000 years ago, man discovered that clay made of argilic substance got hard when treated by fire. This first purposeful transformation of a raw, natural, inorganic substance started a new beginning. A new material was created with properties due to human handling, and it began a process which during the centuries became very sophisticated. Let us not forget that terracotta has been used in Babylonia as a storage medium for scripts - as well as for invoices.

Today, the science of new materials is the cradle of a colossal field instrumental in determining tomorrow's economy. Countries which lag behind are out of luck with the future. Countries which depend for their living on old raw materials are already feeling the pinch.

At the baseline of this transformation is our scientific knowhow which permits us to fabricate custom-made materials for each of the goals we have in mind. This is a facility human civilisation never had before. From the general-purpose clay to the special-purpose alloys, there is a world of difference; and science is the reason.

But let us return to the beginning, with clay and *ceramics* as an example. Throughout thousands of years, argil, the first ceramics* used by man, served as a raw material - from pottery to bricks. The artist and the builder intensively used ceramics in antiquity. The scientist and the engineer profit today.

For the modern materials specialists the term ceramics signifies a solid which is neither metal nor polymer, but it can contain metallic or polymeric elements in the form of components or additives. High technology ceramics are elaborated from very different raw materials, some of which do not exist in a natural state. We also have basic knowledge of molecular structure, which was not the case in the remote past.

Some 3,000 years ago, Ancient Greek philosophers were the first to formulate the hypothesis that matter is constituted of minuscule invisible particles, all of the same nature. They could not prove what they said. However, this tentative statement helped man during the centuries first to understand, and then to structure, the science of materials.

Better materials permit us to build efficient engines. These engines are lighter and work at higher temperature, which means greater efficiency. Transportation offers an everyday example of the benefits: the weight of the cars which we drive has been greatly reduced over the years.

Since 1974, the weight of a typical passenger car in America has been cut by roughly 15 percent. Less steel and cast iron is used. Because of better engines and lighter cars, the average gas consumption has dropped from 18 litres per 100 kilometres in 1974 to 9.0 litres in 1987.

As the science of materials progresses, another reduction of 2O to 25 percent in car weight is foreseen for 1993. The main factor will be the intensive use of new alloys, and ceramics will be playing a key role. Thus we see that greater effectiveness in the materials we use in one domain greatly impacts upon other economic fields and their use of materials.

Man's ever increasing need for energy production is, in itself, a good example. In 1850, we derived about 90 percent of our (then quite limited) energy requirements from wood and water. Coal representaed the other 10 percent. Thirty-five years later, around 1885, coal and wood/water were at par as sources of energy production, each representing somewhat less than 50 percent. The difference (3 percent to 4 percent) was made up by gas and oil.

By 1925, gas and oil held a 25-percent share. Wood and water had dropped to 10 percent. The 65 percent was coal; it had reached its peak at about 78 percent of power production needs at the end of the 19th Century. Right after World War II the curves of coal and gas/oil crossed. The consumption of each representing 45 percent of energy needs.

---

* noun, singular

Gas and oil had their peak around 1973: standing at nearly 80 percent of power production needs. Then, in a matter of thirteen years, they lost 16 percentage points in the consumption scale. Eight of these have been picked up by nuclear energy; 6 by a resurgence in coal consumption; and 2 by wood/water.

Today, nuclear power energy contributes 8 to 9 percent of the world's energy production, wood/water 4 to 5 percent, coal 25 percent and the balance (some 62 percent) comes from oil and gas. Let us not forget that in the Western World only half of produced or imported oil is used as a power source - from power production to motor vehicles and other means of transport. The other half is used by the chemical industry as raw stuff to produce other materials, further underlining our civilisation's need for materials sources.

Materials are vital in energy production, and as components of man-made systems. In both areas scientists currently look for substitutes. During the late 1960s and early 1970s four major stimuli converged to propel the science of materials to new heights:

1. A solid technological infrastructure permitted a new leap forward in the materials sciences.

2. Space research and high-speed aircraft required sophisticated heat-resistant materials, not yet at hand.

3. Intensive competition in the world's unified market made evident the need for a leap forward in cost savings, not only in the products themselves, and their materials, but also in the processes which produce them.

4. With the realisation that the energy crisis was here to stay, new solutions had to be found - and this meant intensive materials for energy research.

While they had to work diligently to face the energy crisis and its aftermaths, scientists had also to address the thermic barrier and its challenge.

When in flight, the biplane of 1930 faced a temperature of 50°C. For the fighters of the early World War II years this stood at 100°C. It reached 420°C with the fighters-interceptors of 1980. The space shuttle endures temperatures of 1,000°C in its reentry to the atmosphere. It will be 1,500°C or more with the proposed stratospheric Orient Express to link San Francisco to Tokyo in a couple of hours.

Since the exploits of Chuck Yeager, man crossed the sound barrier. Now the new barrier to be crossed is the thermic. Materials will make or break this great effort. Hence, the intensive research which is needed.

An Infrastructure for New Materials

New engines, particularly those designed for space or interplanetary travel, would be constructed of materials which are very light and, at the same time, rigid, solid and resistant. Without doubt, these will be composite materials, most of which have not yet been discovered, but some seem to be around the logical conclusion of ongoing research.

The infrastructure of the materials sciences took many centuries to develop. We have spoken of clay made of argilic substances, and then baked, 10,000 years ago. We also said that it is only 3,000 years since man started making philosophical guesses on the possible structure of materials. What we did not say is that it has been for only a few centuries that man has really focused on this subject.

The materials effort might have been continuous also in the ambiguity, but the scientific element was not necessarily there. Increasingly, however, its need became felt. For all of human civilisation, the selection, modification and transformation of materials has been a fundamental cultural element. The statement is valid even though the discipline itself developed very slowly.

At the beginnings of materials technology, artisans used ceramics, then bronze and iron. Empirical methods for materials transformation were the first to see the light of day. Some findings were transferred through the practice of the apprentice. Many have been lost. Only after the invention of typography could texts be used to transmit more detailed, precise knowledge.

The invention of typography provided a valuable tool, though it did not by itself start a flood of publications. As a result, the scientific revolutions of the 17th and 18th centuries, plus the industrial revolution of the 19th, could benefit from no basic materials research on which to capitalise. The instruments at the hands of the specialists were still too rudimentary.

According to documented evidence, the era of the materials science started by the 19th century. That is the time when physicists, chemists and engineers got together to establish a new discipline: *Research on Materials.*

It was towards the end of the 19th century that chemists began exploring the structure of metals. Their work was enhanced through the microscope, which was invented in 1866. Until then, structural problems could ill be studied, and there was a mental wall between artisans and researchers.

By the end of the 19th century chemists and, to a certain degree, physicists were able to provide the engineers with theories explaining, or even furthering, their empirical observations. From then on, different disciplines would contribute to the materials science. The artisans had to wait some more years to reap the fruits from such scientific contributions.

Then things accelerated. In the 20th century, and more particularly during the last 40 years, physicists and chemists structured fundamental tools for research and development. In turn, these permitted progress at a high pace in the fabrication of new materials as well as improvements in the process of their fabrication.

An *internal architecture* was established. It specified how matter is structured at different levels. Theoretical and practical knowledge were integrated, and this permitted scientists to:

- Improve the treatment of natural materials, *and*

- Develop new ones which do not exist as such in nature.

Chemists, physicists and engineers could thereafter make great strides, because they finally understood the relations inherent to materials. They also had goals to meet - as we stated in the preceding examples. Precise goals helped them focus on the relations which exist between properties of materials and their performance. This opened new vistas in product development.

Figure 3.1 presents one of them. Plotted against the cost of a product and its weight are trend lines within a pessimistic and optimistic envelope. Quite importantly, all estimates point towards lower cost and lower weight. A new materials product is cost/effective if it finds itself below the optimistic trend line.

Equally significant is the fact that the technologies for the production of new

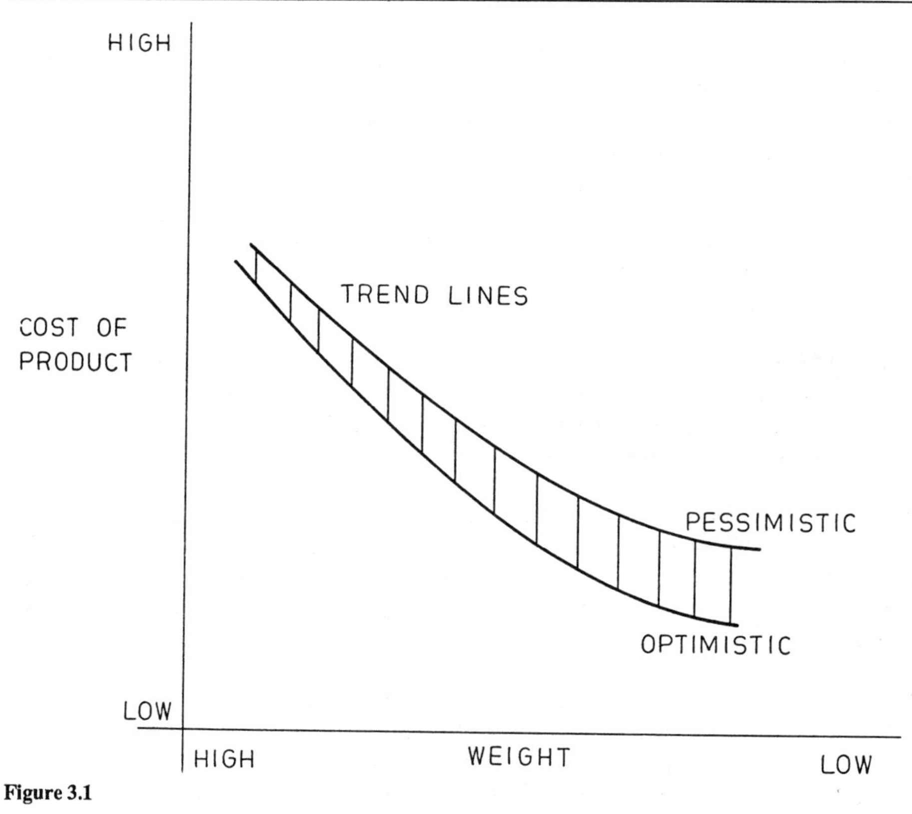

Figure 3.1

materials are no more capital intensive than the old. If anything, they call for less capital. But, they impose a mastery of tools and methods. Table 1 compares the percentage weight in terms of final cost using five key criteria. It has as a frame of reference the comparison of current forging solutions and the evolving powder metallurgy.

Science is, of course, far from having mastered all problems - environmental pollution being one of the most challenging. Generally, the production of materials poses severe problems of pollution and of security. This is as valid of the new technologies as of the old.

The discharge from an advanced manufacturing process for ceramics, of a factory for organic resins, or of the manufacture of semiconductors, are examples. Such effluent is not less poisonous to the environment than the effluent of a classical chemical plant, of a foundry, or of a factory for the treatment of leather products. Man has not yet succeeded in effectively protecting his environment, as current experience can tell.

Furthermore, it is quite proper to underline that, in reality, new materials are not final products per se. They are intermediate stages critical to the fabrication of more sophisticated and more useful new products as is the history of any goods, systems or processes made by man. The more sophisticated the transformations taking place, the

Table 1: Cost Comparison of Relative Weight of Key Factors Between Current Forging Solutions and Powder Metallurgy

Criteria	*Current Forging*	*Powder Technology*
Raw Material	38.0%	24.0%
Machines and Tools	22.0%	48.0%
Manpower	20.0%	9.5%
Capital	24.5%	11.5%
Energy	5.5%	7.0%
Total	*100.0%*	*100.0%*

more knowhow our technologists must possess, and the greater the computer power that must be placed at their disposal.

The story of new materials teaches us a lesson. While there is scarcely any place or subject on this planet which is not at risk of invasion by man and his technology, an irony of our society is that there are not enough technologists around to understand complex interactions on a multiple front. Yet, multiple fronts will characterise future progress, as the range of supporting technologies grows.

Figure 3.2 dramatises this fact. From theoretical physics to space travel, computers and communications are the supporting infrastructure, but there is a supply-demand

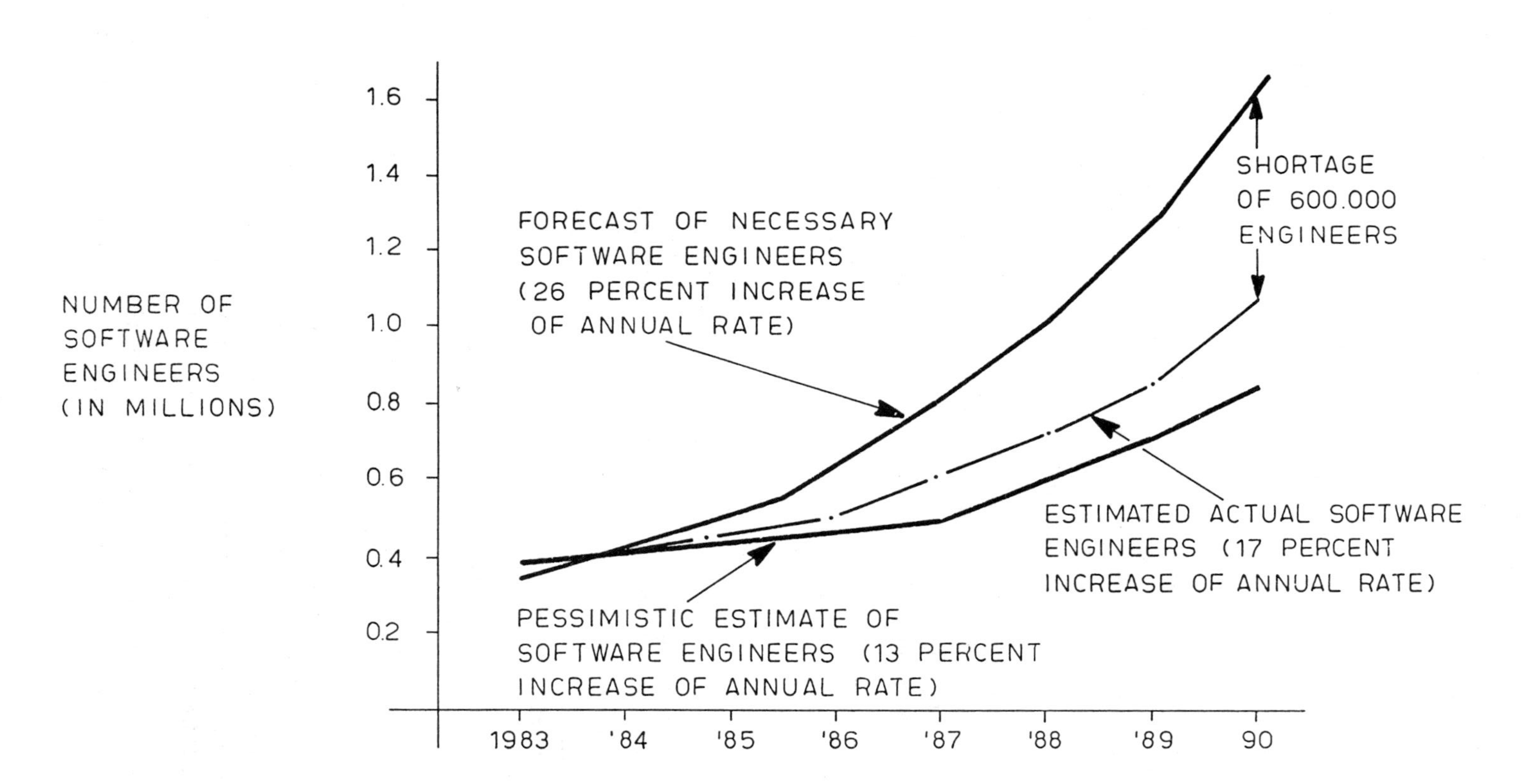

Figure 3.2 Forecast of supply-demand gap for Engineers in Japan

gap. Japan is one of the countries doing its best to close it, yet the official forecast predicts a shortage of 600,000 trained software engineers by 1990. In other countries the gap will be even more pronounced.

Similar statistics exist in America and in Western Europe. Shortages in trained computer technologists might not inhibit small advances, but could well put the brakes on major breakthroughs. And, while thousands of small advances may well go almost unremarked, it is the big, dramatic developments which catch the headlines.

There is a still more important development of which we must take account. Genesis says that God made man of clay, then blew life and soul into him. In the course of history, *the mastery of inanimate materials blended with philosophy to provide a transition from the arts, crafts and sciences to artificial intelligence* (AI). We will talk of AI in Chapter 4.

Should We be Betting on Biotechnology?

We have seen the bad news about biotechnology in the last section of Chapter 1. But there is also good news: the breakthroughs which are in sight. Breakthroughs have been traditionally taken as positive. If the news are bad, this is not necessarily in an abstract sense or due to an overall negative spirit, but because the human race *has not established the ethical values* necessary to gauge and calibrate its own accomplishments in the scientific field.

The good news starts with the fact that the biosciences are presumably on the verge of a series of new and exciting discoveries. While the significance of biotechnology can still only be dimly perceived, and the risks are those we have just described, in the minds of many its possibilities have acquired an almost mythical overtone in recent years.

Optimists point out that, with the help of biotechnology, the world food shortage can be eradicated; the energy problem solved; and precious metals effectively purified from ores. Such evidence is still scarce, particularly regarding the food shortage. All we know so far is that medical care might be improved dramatically, though we also have some positive evidence in agriculture.

Let us start with a definition. Expressed in the simplest way, *biotechnology* is the utilisation of knowledge of the living cell and its capacity to manufacture chemical substances. Since all living matter consists of chemical substances with an added dose of electricity, the biosciences set out to understand, and then harness, the natural life process itself. In a sense, biotechnology is as old as human effort to influence the substance of materials which can be found or created, then put at man's disposal. Science has long ago made use of microorganisms in preparing and preserving food and drink. The fermentation of fish and beer are examples. Nevertheless, modern biotechnology began with a series of scientific discoveries from the 1950s onwards, discoveries which have accelerated during the last decade.

Research results achieved in the past five or six years in the field of molecular biology, and especially in molecular genetics, have made the headlines. Simultaneously, hopes were raised and dangers conjured. Therefore, it is wise to briefly describe and assess the evolution of molecular biology and the effect it may produce.

Modern biology developed as an initially purely descriptive science. It remained so until the 19th Century, while two branches of biology evolved in parallel:

1. The biology of higher organisms, *and*

2. The science of microbiology, thanks to the invention of the microscope (by A. van Leeuwenkoeck, 1632-1723).

The evolution of biology from a descriptive to an experimental science has been dominated by three principles: Cell theory, Mendel's law of inheritance, and Darwin's theory of evolution. These theories were developed in quick succession respectively by Schleiden and Schwann (1839); G.J. Mendel (1870) - though Mendel's law was fast forgotten, only to be revived 35 years later by de Vries, Correns and Tschermak; and Charles Robert Darwin (1809-1882). They are all 19th Century developments.

Louis Pasteur (1822-1895) is the real founder of molecular biology and biotechnology. He is also the man credited with definitely refuting the theory of spontaneous generation of living things from inanimate matter. Robert Koch (1843-1910) in turn laid the cornerstone of our comprehension of the causes of infectious diseases. The work of both Pasteur and Koch made it possible to understand, and eventually combat, infectious diseases caused by microorganisms.

If the 19th century scientists laid the roots, their theories saw to it that well into the 20th century:

- Biology of higher organisms, *and*

- Microbiology

followed separate courses. But, at the same time, the expansion of knowledge about chemical processes in living cells permitted the creation of a new discipline: *Biochemistry.*

As acknowledgement of the existence of universally valid principles gained ground, it culminated in the discovery of genetic processes in a fungus by Beadle and Tatum (1941). Two years later, Delbrück and Luria for the first time proved the occurrence of genetic processes in bacteria, laying the foundation of *bacterial genetics.*

Quite important to the march of science has been the discovery of transformation in the case of bacteria. That is, the transferring of genetic information from one bacterial strain to another with the aid of a cell extract, but without direct cell-to-cell contact. The chemical nature of genetic information was found to be *dioxyribonucleic acid* (DNA).

Finally, in the early to mid 1940s, the isolation of microbiology came to an end. The understanding of cell biology, of genetics and biochemistry, led to a fusion of the biological disciplines. The new unified science was *molecular biology*, with scientists seeking to explain the processes of life at the molecular level.

Another most significant discovery has been the one which became the starting point of *genetic engineering*. Nobel Prize winners James Watson and Francis Crick demonstrated in 1953 that the genetic code, the central theme of life, could be deciphered. The cells could thereby be controlled, and induced to perform completely new tasks.

As tools on which focused the convergence of disciplines new and old, microorganisms became the basis for an entirely new technology, invisible to the naked eye. *Bioscience* is an experimental field crossing over traditional lines represented by biochemistry, medicine, agriculture, and microbiology.

The starting point with genetic engineering is the fact that living organisms can be used as chemical factories for mass-producing important substances in nature's own way. This is a process thrifty in the use of energy, not harmful to the environment, and efficient.

After thousands of years of engineering the earth's minerals, its cold remains into utilities, man set out to engineer the internal biology of living organisms. He did so in the hope of laying the base for a new world epoch: propelled by science, we are moving from the age of *pyrotechnology* to that of *biotechnology*. Here is where risk and opportunity meet.

In Japan there have been claims of developments which allow parents to choose the sex of their offspring. If this technique proves itself, it could be used for reducing hereditary diseases affecting only one sex. Or it could be used to manipulate man's future well before his birth.

Researchers in the biological sciences are making great strides in breeding bigger and faster-growing animals. They work, not by conventional breeding techniques but by genetic engineering. After having manipulated lower animal life, biotechnology will raise its goals. The risk is the eventual making of human clones as well as of supermen.

It is, of course, self-understandable that the sort of supermen sought out is *superbrains*. At the University of California, in Los Angeles, there exist well advanced experiments on mice to create superbrain features. In a way *the race towards more powerful brain capabilities is that of genetic engineering against artificial intelligence*, because, as we said, ethical standards have not been preestablished to guide man's hand in that race. The former case is more scary than the latter in terms of aftermaths.

For the time being, concrete results have been obtained in body-building. Australian scientists, for instance, have introduced the gene for growth hormone into a sheep's fertilised egg. This egg is placed into the womb of a surrogate mother sheep where a *transgenic* sheep develops. Several weeks after it is born the scientists activate the new gene by giving the sheep a dose of zinc.

Zinc acts as a catalyst. It switches the gene on to producing growth hormone, and the sheep grows faster and bigger than any of its brothers or sisters. The aim is that sheep fashioned in this way could become one and a half times as big as normal sheep.

There are also other plans to insert genes into sheep's eggs which code for better wool production, or resistance to disease. Sheep are not the only animals to be "improved" or *giantised* in this way. Scientists have already done it with mice. When is man's own turn coming up?

A Natural Information System, its Cloning and Associated Risks

What Watson and Crick have found is that genetic information is stored in the form of a long, filiform, double-stranded helical molecule of deoxyribonucleic acid. With the exception of a few viruses this is true of all living organisms. Hence some of the ongoing findings are portable.

The molecules consist of sugar: deoxyrbose. These are linked chainlike together by phosphodiester bonds. The sugars carry what essentially are four different nitrogen bases as free remnants:

● Adenine, A,

● Thymine, T,

● Cytosine, C, *and*

● Guanine, G.

Under the effect of electrostatic forces these bases establish specific linkages with each other. Guanine and cytosine, as well as adenine and thymine, have complementarity. Therefore, inside the double-stranded DNA, which twists like a helix, base A is always opposite T; G opposite C. A sugar/phosphate backbone delineates the outside of the molecule.

Because of the complementarity of the individual strands, DNA replicates identically. It also exhibits inheritance. It is transmitted to daughter cells. These are properties of great importance to further developments.

As the carrier of genetic information, DNA is a molecule universally present in cells of living organisms. It is confined inside the cell nucleus and divided there among the *chromosomes*. If we were to isolate the DNA thread from the bacterial cell (which has a volume of about one-thousandth of a cubic millimetre), it would be approximately one millimetre long. In the human cell, the entire DNA would form a filament about one metre long. In both cases, it will be very thin.

The unlocking of the secret of DNA provides a key to the mystery of how information is stored in genetic substance, that is the *genetic code.* This is based on a linear arrangement of the four bases of the DNA which we have been discussing: A,T,C,G. Taken together, three bases designated a triplet; they constitute a coding unit or *codon* corresponding to a specific amino acid.

These biological macromolecules perform a multitude of functions within the cell:

- They are the cell's structural elements, *and*

- As enzymes, they catalyse the chemical processes: the metabolism.

As a carrier of hereditary information, DNA performs the task of conserving data stored within it, transmitting usually error-free to the daughter cells during the process of cell division. This is known as *replication,* but this data is also called up and functionalised, and in this manner genetic information manifests itself.

As if in a man-made information system, functionalisa-tion takes place in two stages:

- Transcription, *and*

- Translation.

With the aid of the DNA, a changeover into a closely related single-strand nucleic acid molecule (RNA) occurs during transcription. RNA can exercise certain functions within the cell, or else act as a matrix for translation. The latter is often called messenger RNA (m-RNA).

Typically, the genetic code features a wide margin of redundancy. It is read via the transcription and translation of certain segments of the DNA. A segment that leads to a functioning RNA, or a corresponding protein, is called a *gene.* The human cell carries about 50,000 genes on its DNA thread. On a DNA thread length of one metre, information of about 1 million genes would be possible. Besides information for a specific gene product, genes store start and stop signals which control their own expression.

Another interesting reference is that, according to a hypothesis, the information of m-RNA occurs in higher organisms indirectly through protracted preliminary stages. This contrasts to what happens to lower organisms. In all, the universal mechanisms of storage, replication and functionalisation of genetic information guarantee the preservation of genetic integrity. This is both true of an individual and of populations held together through similarities in a taxonomical sense:

- Species,
- Genera,
- Families.

Procreation provides the basis for a continuous mixture of genetic substance within species. But this is not the only means through which changes and mutations materialise. Through the fusion of asexual or somatic cells it has been found possible to unite genetic information from cells of unrelated organisms in a single hybrid cell.

Interspecific hybrids have been achieved with selected plants. With animal cells, biotechnologists have not yet succeeded in triggering the development of complete living entities from such hybrid cells, though experiments conducted by cell biologists with egg cells - for instance of the clawed toad (Xenopus) - seem to show ''promise'', whatever that means.

Such a technique is extremely dangerous for the future of man, as it can lead to cloning of genetically identical individuals. Their present significance lies in the possibility they offer of studying the processes of differentiation taking place as the embryo develops. Their future impact is unknown, but it could as well be dreadful. Scientific sensation has also been a method known as in *vitro recombination* of DNA. It is based on the discovery by W. Arber, H. Smith and D. Nathan of the restriction enzymes and their function (early 1980s). Enzymes found in bacteria probably serve to protect the cell's own genetic integrity inasmuch as, thanks to an identification mechanism, alien DNA penetrating into the cell is broken down.

- Regardless of their origin the restriction enzymes identify certain sites on a DNA filament which is characterised solely by the base sequence.
- At these sites the enzyme cuts the DNA cleanly, or in such a manner as to create overlapping ends.

Back in 1973, P. Berg, S. Cohen and H. Boyer showed that, by this technique, isolated DNA could be cut in vitro into predetermined fragments. But, as the later research demonstrated, DNA fragments from different sources can be joined together, leading to recombination.

Experimentally, genes of the most widely differing origins have been put together in this manner, opening up new possibilities for genetic engineering, contributing to the understanding of the processes of evolution, but also creating the possibility of a 21st Century Pandora's Box.

As basic research has been succeeded by a number of experimental applications, biotechnology became an implementation-oriented science. It:

- Investigates the use of biological processes

- Interconnects with established technical procedures, *and*

- Aims to subsequently lead into industrial production.

In doing so, it exploits the chemical potential of living cells. The cell as a whole and individual enzymes, catalysts of the living cell, are being used as materials for specific chemical reactions, thus becoming parts of a transformation process.

Cell systems suitable for technical handling today are found primarily among the microorganisms - and for good reason. They are relatively easy to grow. They multiply at very rapid rates. They are easy to manipulate genetically. But what about tomorrow?

We have to admit that humanity is very short of ethical standards which might protect future generations from Nazi-type manipulation. Should this happen, the results will be orders of magnitude more disastrous than the Holocaust and atomic bombs. As no effort is presently made in promoting and then establishing new ethical standards, the most we can say for the time being is: "We shall see".

However, efforts are being made in further exploiting scientific aspects of the new-found knowledge. A comparison between biotechnology and synthetic chemistry, for instance, shows a number of characteristics which are common to both, as well as fundamental differences.

1. *The transformation of matter.*

While chemical synthesis is based largely on fossile, non-renewable, hydrocarbons, biotechnology draws mainly on components of the *biomass*. This is its raw materials.

2. *Reliance on the Biomass.*

Biomass is the organic matter which is continually generated in the biosphere, whether directly, as with vegetation, or indirectly, as with animals. In either case, biomass is a product of photosynthesis. As from a practical viewpoint, the energy of light is taken to be infinitely available, the renewal of biomass might be unlimited in time and capable of the regeneration in the natural cycles of the elements.

This is, of course, a gross overstatement. Man-made pollutants break the vital biomass cycle, as the process is not monolithic, and to materialise it has to transit through a number of vital steps: the *ecosystem*. Furthermore, we do not know what will be the result on the ecosystem of man-induced natural catastrophes - such as the ozone holes, due to the depletion of the protective ozone layer. The current hypothesis is that the delicate balance of the ecosystem will be the first victim.

3. *Overpopulation, pollution and destructiveness can have tremendous aftermaths.*

It is a false belief that the biomass will save man in the longer run from the long list of his failures, including the *modern plague: Man's own uncontrolled fertility.* What is true is that *if* the ecosystem holds together *then*, in contrast to petrochemical technology, which:

- Is exhaustible, *and*

- Places a heavy burden on the environment,

biological technology has longer-lasting resources and no detrimental effect in an environmental sense -except, perhaps, on man himself, which we really do not know.

In other terms, while plenty of valid reservations can be made as to the possible misuse of biotechnology in the future (see also Chapter 1), there are also some possible "pluses". Because of the biological origin of the raw materials and biotechnological products, environmental problems might be more easily solved than those caused by chemical synthesis. But we simply do not know whether or not biotechnology, too, may lead to an uncoupling of the natural cycles.

Also, blinded by his uncontrollable reproductive activities and great waves of waste, modern man -including the scientists - seems unwilling or unable to take notice of what happens around him.

- From the savage horde of pollutants: heavy metals, nutrients, chlorinated hydrocarbons, petroleum hydrocarbons, plastics, particulate materials, radioactive materials

- To the raw sewage, acid rain and greenhouse effect.

New York City alone throws into the Atlantic, and the waterways surrounding it, 8 million tons of sludge per year, roughly amounting to one ton per year per inhabitant. A small fleet plies the ocean 24 hours per day "releasing" this sludge.

These are statistics on human waste that biotechnology will hardly alter, but birth control might - if homo sapiens is determined to survive for another century or so on earth. It serves no survival purpose to exaggerate the possible (and largely unknown) benefits while minimising the probable negative effects. If anything, precisely the opposite should be done.

As far as science is concerned, what we do know is that, by contrast to chemical synthesis, biosynthesis has low energy requirements. Enzymes catalyse the chemical reactions of the living cell. Most of these processes take place within a temperature range of 20° to 50°C. A living cell is capable of synthesising:

- Highly complex macromolecules such as proteins, *or*

- Very simple compounds like ethyl alcohol.

It does so at high speeds. At a temperature of 37°C, one cell of the intestinal bacterium escherichia coli can manufacture about ten thousand protein molecules in ten seconds. Organic chemistry has nothing equivalent to this, although chemistry is superior in producing a range of low-molecular hetero-cyclic compounds.

At the end of Chapter 1 we spoke of the *megarisks* posed by bioengineering. They are class No. 4, the most potent of dangers which lie ahead. But let us not forget the others:

1. On-the-job risks caused by recombinant microorganisms.

2. Epidemics triggered by those manipulated microorganisms which escape control.

3. Disturbances of the ecological system, and of the natural evolution of the biosphere, which are feasible but unknown.

4. Deliberate use of genetically modified strains for purposes ranging from cloning people as an attempt to violate freedom in peacetime, to superbrains for production or war.

Theoretical studies have shown that risks No. 1 and No. 2 may be exaggerated. Medical microbiology and epidemiology might provide safety precautions, though concrete evidence is still necessary in this domain and might take centuries to obtain. Three Mile Island and Chernobyl are reminders of Murphy's Law: "What can go wrong, will".

Regarding risk No. 3, it is undeniable that, by becoming the dominant factor of the biosphere, man strongly influences the ecological system. This influence has recently been quite negative, and has assumed a giant global dimension.

It is not the simple existence of a process which creates the great risk. It is its massive impact. As General George Marshall said to the atomic scientists of the Manhattan Project: "The value of the military is in its ability to deliver. And to continue delivering". This is just as true of our industrial system and of its value - while being further compounded by man's lust and greed.

Still, the absolutely unwanted risk is a compound effect of genetic engineering and the lack of new ethical standards:

1. Moses has not updated his Ten Commandments to meet, or even foresee, the great technological leaps of mankind, and to direct the ethics of this epoch.

2. Necessary as they may be for those who need them, the self-important officialdom of today's religious authorities (any denomination) is more concerned to preserve the status quo and recapture the past, than to create a brighter future.

If it had been for scientists to decide, the atomic bomb would never have been dropped on Hiroshima or Nagasaki. But, science and scientists create a weapon, which politicians decide when and how to use, according to the whims and ethics of the moment.

Science has no morals. It is not immoral, it is *amoral.* Science is concerned with discovery, with creation. Morals are the guidelines of a human society, and as far as high technology is concerned, these do not exist.

Man would no doubt be faced with very tough decisions if he were to try to duplicate, or even to greatly improve upon, other men. Scientists will have a tough future manipulating germs, some of which are extremely dangerous indeed. In nature, life forms have evolved over millions of years. In man-made life evolution will take place in hours or minutes - and *the errors which can be done would be irreversible.*

Having stated the bad news, where the megarisks lie, let us now see the good news. Specifically, the positive side of genetic engineering.

What Can We Expect from Genetic Engineering?

One of the jobs done with genetic engineering is to transfer genes from one living organism to another. An advantage of this approach is that it enables scientists to mass-produce useful proteins. So far, the proteins they manipulate are only the ones that nature has already provided, though in small quantities, and sometimes not in the most helpful places. Hence the interest in this practice.

For these two reasons of *quantity* and *placement* genetic engineers aim to make wholly new mutant proteins from scratch. Mutant proteins can be made by a process of *site-directed mutagenesis.* This has three stages:

1. Pinpoint the amino acids, the building-blocks of which proteins are made, that need to be changed in order to produce a certain effect.

2. Write a recipe for these amino-acid changes in DNA, which will typically be a sequence of chemical bases.

3. Put the new DNA into bacteria, so that they can turn out the new protein.

Computers have a role to play in this process, as computer-assisted R&D makes the writing of such recipes easier and automates their manipulation. Today genetic engineers can reliably produce enough sequences to code a lower two-digit number of amino acids in a day. At the same time, bacterial strains that do not make corrections have been isolated through such processing, with yields rising more than twenty-fold.

Still, though by means of computer support productivity has increased, little is known about the factors that determine the structure of proteins. It is, in fact, difficult to predict the amino-acid changes that are needed to handle a protein in the right way.

However, new developments, particularly in computer graphics and in artificial intelligence, may provide a big boost in the protein engineers' work. The latter need good three-dimensional pictures of their tools. These are obtained through graphics, for instance, map crystals of protein subjected to X-rays which bounce off the atoms to produce a pattern that reveals the shape of the molecule (though not all proteins can be turned into crystals).

Another research method employs magnets and radio waves to produce images of proteins. But here, again, some regions are too big and complex for this technique. Out of tens of thousands of proteins in the human body, there are satisfactory pictures of just 350 or so. No wonder that these 350 proteins have been the subject of intense study.

In other terms, genetic engineering is a field of converging technologies. Computer-assisted research on proteins helps scientists understand their structure and the way they work. The results of this research enable scientists to draw up guidelines for the protein design, which leads to more research, and which in turn call for more computer-based experimentation.

Such studies have led to the conclusion that it now seems likely that enzymes and other proteins have evolved by aggregating different blocks of amino acids. Each block makes a predictable shape, and while each protein may have a unique amino-acid structure, parts of the protein may fold into one of these better known shapes.

In fact, genetic engineers hope that by looking for certain amino-acid sequences that might signal the appearance of familiar protein shapes (icons), they will be able to make three-dimensional pictures of proteins that cannot be examined by other means available today for protein research. At the same time, more ambitious experiments fuse large parts of one protein to another. Genentech has developed a potential anti-cancer agent by fusing one gene coding with another. Newer and safer polio vaccines have also been obtained by such fusion.

Other genetic engineering work, too, is exciting mainly because of its potential for clarifying some of the most mysterious events in the intricate process of living development:

- from fertilized egg,

- to human being.

In that process, scientists believe, lies the answers to many tragic diseases and other disorders. And scientific curiosity sees to it that the transplantation of genes in animals, and even the construction of artificial genes, have become relatively common.

An ''historic example'' was captured in a photograph, now famous in the scientific literature, of two mice, one almost twice as large as the other. They were from the same litter, but one grew to unusual size because it received a modified version of the human gene for making growth hormone. The risk, however, is that the next historic example will be mapped into a photograph of two Homo Sapiens, one twice as large as the other, and so on until variations in size exceed the order of magnitude -particularly in *brain cells*.

There are, of course, challenges still to be overcome before we reach that state. But the question is not *if;* it is *when.* So far scientists have not been so successful in directing gene transplants to a particular location on a specific chromosome. The transplants tend to go into the recipient's hereditary apparatus at unpredictable places. With time and effort, this too will come under control.

In Chapter 1 we said that genetic engineers are now targeting genes. This is true, the essential problem is that the inability to send a gene to a specific target sometimes interferes with the transplanted gene's ability to function. Serious damage could result if the transplant inactivated an important native gene or activated an oncogene to start the cancer process. But rejoice: Genetic engineering provides reasons to believe that it may eventually be possible to transplant, replace or modify almost *any gene.*

Many research teams have been pursuing related strategies to make targeting and correction practical. One strategy has been to devise ways of finding and collecting cells in which the spontaneous targeting occurred. Given time, there will be results. The key issue is not the sterility of the research. This research is very productive. The salient problem is one of:

- constraints,
- limits, *and*
- controls.

It is most urgent to provide valid, lasting answers, as the wider applications - and implications - of genetic engineering are just beginning to emerge. Random mutations could be introduced into specific regions of the genetic blueprint of a protein; the resulting mutant proteins could then be watched to see what they get up to, and the best ones selected for mass-production. But, again, who is to say what is *best*?

What does our society want in the first place? Research on mutations, and the targeting of genes, is not inherently bad. Selected mutations helped to revamp more penicillin than did Alexander Fleming's original work. What is *bad* is that we are behind in terms of socially acceptable research goals and overriding ethics.

Proteins and Biochemical Engineering

The positive developments of biotechnology depend on the integration of three *engineering* fields: (1) genetic, (2) protein, and (3) biochemical. A pivotal point is the production and use of enzymes which play a central role in most applications (Figure 3.3). Raw materials, such as sugars and starches, are readily available for the production of enzymes.

As stated in section 2, *enzymes* are proteins, produced by living cells. They act as biological catalysts and are composed of some, or all, of the twenty different naturally occurring aminoacids. They differ from, and are superior to, chemical catalysts in several aspects:

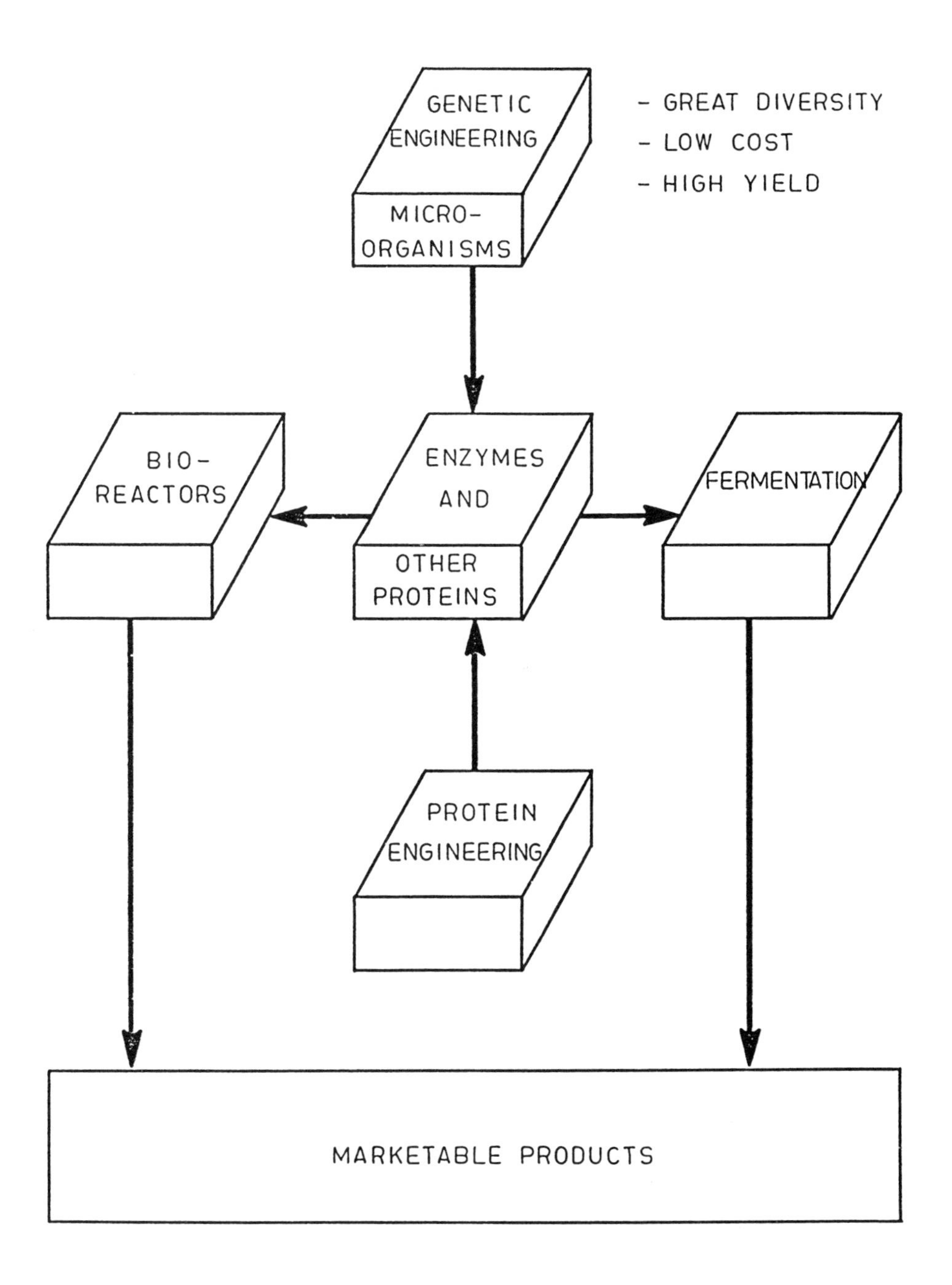

Figure 3.3 Marketable Products

- The temperatures at which they function best are relatively low,
- They are usually very specific in the chemical reactions that they catalyse; *and*
- They can be capable of catalysing the chemical conversion of tens of thousands of reactive molecules per enzyme molecule per second.

The point has been made that genetic engineering is the directed manipulation of deoxyribonucleic acid (DNA) molecules. They are the carriers of inheritable information in all living organisms. Genes, we said, are the segments of DNA molecules that, through the genetic code, dictate the structure of the thousands of individual enzymes and other proteins found in living cells.

The application of natural mechanisms to practical uses can affect the environment in which man lives. This is true of all of man's civilising functions. Both a productive implementation, and its policing, can be done fruitfully only at an interdisciplinary level, involving biologists, physicists, ecologists, agronomists, engineers, anthropologists, computer scientists and sociologists.

If the living cell is to become a part of man's general knowledge and materials in use, a very thorough, interdisciplinary effort would be the prerequisite condition. This should stress both *opportunities* and *risks* as well as examine thousands of possible mutations and their aftermaths.

In natural systems, DNA acts as a sort of language: the language in which all of nature's genetic information is written. Rapid methods of determining the substructure of DNA have been developed, known as *DNA sequencing.* Their object is reading DNA, making it possible to determine the complete structure of a gene.

Evolving methodologies for chemical synthesis of DNA molecules make it feasible to write in the language of DNA. Genetic engineering techniques themselves, known as recombinant DNA methodology, see to it that it is possible to edit the language of DNA. Through this editing, the naturally occurring text can be restructured.

Stated in a nutshell, the task of the genetic engineer is to modify the way the cell produces the enzymes responsible for the biosynthesis of compounds. The genetically engineered cell becomes a microfactory making the compound of choice more efficiently, at a faster rate and a higher yield than would otherwise have been possible.

In the way discussed in Section 5, *protein engineering* is a complementary process, but it has different goals. The use of many enzymes in industrial processes is currently limited by three factors:

1. Prior to the advent of genetic engineering, it had not been possible to obtain most enzymes in adequate quantity at low cost.

2. Scientists were not able to tailor existing enzymes to the requirements of specific industrial processes, imposed largely by economic considerations.

As a result, enzymes were not being used commercially as much as their broad spectrum of catalytic capabilities called for. This challenge is not yet over.

3. Protein engineering ultimately may allow many different enzymes in nature to be optimised for use as industrial catalysts in a wide variety of applications. However, knowhow is still quite thin.

This technology might enhance the development of enzyme-or antibody-based sensors for measuring chemicals and biological materials in industry and the environment. Prior to having clear ideas, we need to know a lot more about the product and the process.

Biochemical engineering has a different focus. Using *recombinant DNA* (rDNA) and other means of genetic manipulation, molecular biologists construct what is called a production strain of a microorganism, to make a desired product. Biochemical engineers then work to develop a process to employ the production strain effectively for actual manufacture.

As with other industrial processes, such manufacturing incorporates fermentation and product-recovery steps. In some cases the product is made directly during fermentation. In other cases, the fermentation step serves only to generate a biocatalyst: enzymes or whole cells.

The biocatalyst is used in a bioreactor. *Bioreactors* are vessels containing biocatalysts to which raw materials are fed and converted to the desired product on a continuous basis.

After the product is made, it must be isolated through a recovery designed to yield relatively high amounts of product at low cost. Such recovery must also ensure that the product meets or exceeds quality specifications.

Monoclonal antibodies is another major technique avoiding some of the problems of recombinant DNA, by simply combining the whole set of chromosomes into another, hybrid cell.

- If two cells are physically merged, often the genes in each of the original two cells will continue functioning as they did originally.

- If one of the original, or parent, cells happened to be producing antibodies (a portion of the immune system), then it will often continue producing the same type of antibody in its new hybrid cell.

Mono refers to the highly specific (mono-specific) antibodies that the new hybrid cell produces. *Clonal* denotes that the original hybrid cell and all its offspring (the clone) have exactly the same genes. Each cell will produce an identical antibody. *Antibody* identifies the protein product of this cellular production factory.

The term cloning is used in the so-defined context, but to confuse matters, it has a different meaning when it is employed with reference to gene or DNA cloning. With monoclonal antibodies a clone is a group of cells that came from the same parent. Each cell in the clone, by definition, contains exactly the same genetic information passed on from parent generation to first, second, third, and so on, generations.

This means that all the cells that came from the original antibody-producing parent cell will usually continue to produce exactly the same type of antibodies.

To manufacture monoclonal antibodies, antibody-producing B-cells (white blood cells) are primed with a target molecule, generally referred to as an antigenhat target, which may be anything - interferon, bacteria, even gold or other metals.

Monoclonal antibodies are already being used in clinical diagnostic products, mostly as a substitute for less-specific antibody mixtures. The first important new use of monoclonal antibodies is in the clinical microbiology laboratory, where current culture (growing) diagnostic techniques are time and labour intensive, and have inherent limits in terms of utility to the doctor.

Recombinant DNA and monoclonal antibodies are dramatic technological tools in the growing bio-technological arsenal. Biologists learn how to manipulate, recombine and reorganise living matter into different forms and shapes. For the time being, diagnostic and pharmaceutical products are the key areas of interest (Figure 3.4). But, the speed of the discoveries is phenomenal. New areas will surely develop.

Biological knowledge is currently doubling every five years, and in the field of *genetics* the quantity of information is *doubling every year.* We are literally hurling into the age of biotechnology, obliging ourselves to conceptualise the world in a fashion fundamentally different from anything we can readily identify with regarding our experience from the past.

Let us keep in perspective the fact that, as an industry, biotechnology is still in its early stages. In total, it currently offers a low two-digit number of products, very few of which have a market. The question may then be posed: If biotechnology is an industry that has no earnings and no revenues, why are we interested in it? The answer is linear: because of its potential, which can be realised in three major fields:

1. Health care

2. Agribusiness

3. Bioelectronics (Bionics)

A health care example is tissue plasminogen activators, enzymes that can dissolve blood clots. They will permit doctors to safely and effectively treat embolisms and circulatory occlusions, which now cause half the deaths in Europe and in America. Within the same timeframe, recombinant DNA will see to it that vaccines are developed for major parasitic diseases that afflict a large part of the world.

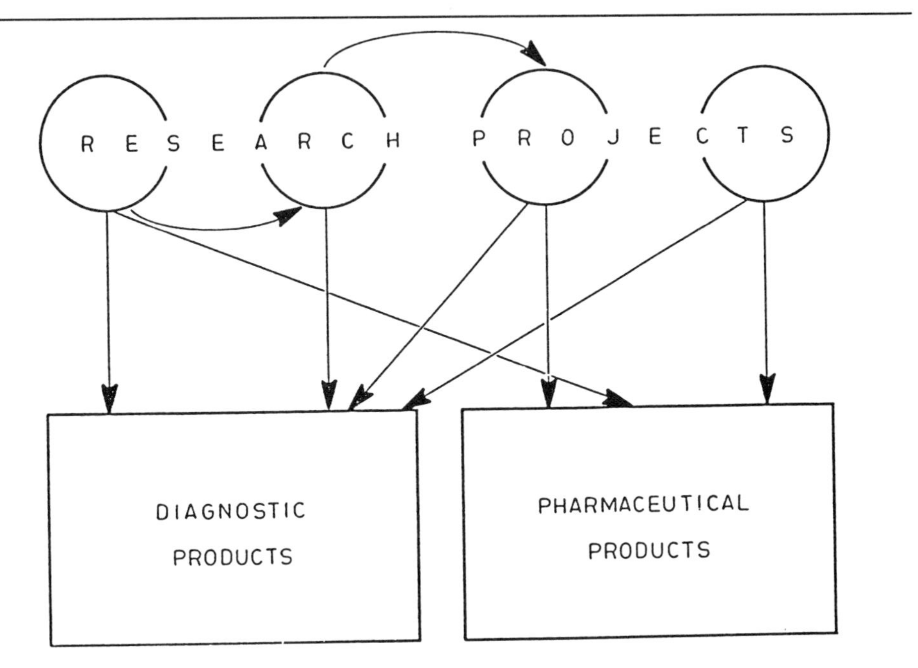

Figure 3.4

Health care and agribusiness are already well launched in terms of products - though what we currently have available is less than the tip of the iceberg compared to what is expected to come. Bionics is still at laboratory level, as we will see in section 6.

However, as I will never tire of repeating, the bioengineering experience will prove to be dramatic unless we approach it under *proper control*. The trouble is that we have not elaborated a concept for the *transition*.

From the new physics to biotechnology, what we know is that we need more power to see through the processes into which we have engaged ourselves and our future. The challenges that we have just described are taxing the human intellect, and it is cracking the foundations on which the current order of society rests.

Biotechnology and the Biochip

Science and technology have started showing interest in the interweaving of biotechnology and electronics. Biotechnology offers a means of exploiting *bioelectronics*. To understand what might possibly be done with bioelectronics, let us take a quick look at the electronics revolution - although in Chapter 5 we will treat this subject in detail.

Early computers were designed and implemented with vacuum tubes. In 1948 the transistor was invented. Eleven years later the first integrated semiconductor circuit was built. These are milestones in electronics. Today we have better than 1 megabyte chips, and developments will no doubt continue in the future with much higher integration.

This evolution not only has a decisive influence on the electronics industry and its products, but also impacts on society. The process of industrialisation as we have known it during the past 100 years had reached its limits conditioned through volumes, costs and marketing channels - but also pollutions. It is, therefore, only reasonable to look for new solutions.

Communications is a good example. Though faced with a fast-expanding market during the last 60 years, somewhat mid-range in this period, communications engineers moved up to a virtual wall made of transmission losses and less-than-wanted reliability. Only through the use of semiconductors was it possible to overtake this barrier by increasing the complexity of instruments and systems by several orders of magnitude.

Let us also recall that the first integrated circuits contained only a few elemental functions. Even so, their small size, low energy dissipation, compactness and reliability favoured their usage with military equipment, and in space programmes. Integration also had cost advantages. The more functions that are put on a chip, the lower the cost per function.

How might this development continue? Where are the limits? When will the limit of photolithography with visible light be reached? How far will monochromatic X-rays, or electronbeams, be used to expose highly compact structures? The answers are not precise, but cognisant people feel that the barriers are not too high. To many scientists, biomolecules and biosystems seem to be the way out.

Nucleic acids and proteins form the two most important classes of the biomolecules. We said that DNA stores the information on, and determines the composition of, all biomolecules: proteins perform the living functions. Nucleic acids and proteins are linear chains. The former consists of four, and the latter of twenty, building blocks. Their layout determines both the information in the nucleic acids and the structure of the proteins.

Consider, as an example, proteins consisting of 200 elements. If we form all possible combinations, building only one copy of each, we will fill the whole universe with molecules. A typical protein measures (in every dimension) several nanometers and is an extremely refined miniature engine, working on the border of quantum and classical physics.

This nanosystem has many similarities with solid bodies. Yet, because they are living matter, even simple biological systems can organise, repair and reproduce themselves. No solid body or chip possesses this property. As a result, biomolecules might be used in electronics at two levels:

1. Simple polymers serving as elements. Such employment will most likely happen before too long.

2. Whole systems, organised through nucleic acids and bacteria.

Number 2 is much more difficult than number 1, and the field will only then open up when the new physics and chemistry of the molecules will be better understood.

In terms of current research, for instance, there are presently experimental projects on fixing a monolayer of biomolecules on the semiconductor and metal surfaces. A photonelectron transduction has been efficiently performed in the visible light range with monolayer chlorophyll molecules on the model of photosynthesis.

Insight into the complexity of living systems encourages researchers (particularly in Japan) to explore the new frontier of bioelectronics. To their thinking, the molecular linguistic information system should be a reality. Instead of electrons a messenger molecule moves like a hormone and neurotransmitter in the living system. A target receptor recognises, by decoding, what the messenger means. This process may store supercondensed information and be accessible to the living matter (including the brain and nerves).

An initial stage of this research strategy is to develop a device for electronically stimulated release of messenger molecules, and for molecular recognition. For instance, simulating a synapse where the axon of one neuron comes into contact with another neuron.

Quite importantly, in the Japanese experiments the electrically stimulated release of a messenger molecule has been demonstrated with conducting polymers such as polyacetylene. The polymer confines a neurotransmitter within the matrix when a specific potential is applied, the neurotransmitter being released by an electric pulse.

Biosensors might be the first practical example of bioelectronic modules. They are being designed to recognise (decode) a specific molecule, with the resulting generation of an electric signal.

- Biomodules, capable of recognising a specific molecule, can bind the corresponding molecule to form a complex at a specific site of a structure.

- Biomolecules (proteins) are mostly immobilised on the solid support surface which is sensitive and quantitative enough to convert the molecular change into electrical signal.

Until today some 2,000 different enzymes from diverse biological sources have been isolated and described. Their molecular weight is about 100,000.

Let us remember that the technology of biomolecules is at its very beginning. While we are able to formulate many of the fundamental problems, the solutions which we are seeking are still far away.

We know, for instance, that the arrangement of the building blocks in the primary, linear structure determines the structure and function of the protein. But we are incapable of predicting, from the arrangement, the structure and function. This makes it possible to manufacture proteins tailor-made to specific goals.

To scientists this may look as frustrating as the successful manipulation of enzymes, antibodies and binding proteins. This has been demonstrated in varied manners over the past ten years. Gene technology also offers great assistance in the refinement of biosensors.

As usual, science is advancing step by step. Even if we know structure and function, we do not understand the connections one by one, and even if we suspect that non-linear processes play a substantial role, we cannot measure properly the processes themselves.

On the other hand, imperfect knowledge may eventually prove to be a blessing. Our current thrust to overcome uncertainties has led to the development of different biosensors:

● the Enzymelectrode

● the Enzymethermistors, *and*

● the Enzymetransistors

Their common characteristic is the physical proximity between the enzyme and the signal converter, whereby the latter is the product of the enzyme activities. Their advantages are a direct response and great sensitivity of the sensor based on the very small distance to the signal converter.

The tendency to bring enzymes and signal converter close together is an emulation of the MOS development. With biochips, proteins are used as electronic building elements. We are transferring, in terms of design, experiences with present-day trends in electronic circuitry.

Oxygen, ion-selective electrodes, enzyme-pH-electrodes, and thermic enzyme sensors (like enzymethermistors) are the most favoured signal converters. As most enzyme reactions are exothermic, heat represents the general means of measurement.

Just as important is to underline a rapidly-growing interest in instruments which connect semiconductor technology with enzymes or other biological material. The semiconductor elements often used in such studies are FET (field effect transistors).

Such building elements have been made sensitive to certain ions and chemical substances, and are known as ISFET or CHEMFET. In connection with an enzyme, the name ENFET is also used. In analogy to the Enzymelectrode, this building element can be called Enzymetransistor.

It is also correct to make reference to the use of piezoelectric crystals, as well as to ellipsometric approaches. In the latter, no immediate proximity between enzyme and signal converter is necessary.

The principle in ellipsometric solutions is the measuring of polarisation angle changes which occur if polarised light is reflected from a surface. The respective change of characteristics (layer thickness) depends on the surface. Hence, a change of the polarisation will occur if the surface is covered with a layer of biomolecules.

The addition of biomolecules, which undertake a complex linkage with the layer, leads to a further change. Its extent is measured with the help of a photomultiplier, determining the thickness of the layer.

The idea of using a piezolectric crystal as microbalance rests on the fact that the vibration frequency of the crystal is reduced when connection takes place on its surface. Research suggests that the same principle might be used in the case of enzymes, which function as an activity layer. This is one of the possibilities currently under study.

The New Microprocessors

Based on the considerations that we have been going through in the preceding section, we can say that proteins and DNA may be the next great frontier in electronics. Circuits of biological densities are projected for the end of this century.

With current integrated semiconductor circuits, an increase of complexity of nearly two orders of magnitude, and in functional speed of about 20, may be possible prior to reaching the physical limits. The way to the supercomputer on a chip for a unit price of $100 seems feasible based on technological projections, although the first chips, such as Intel's 486, will cost about $1,800 when introduced this year (1989) in the market.

The sharp cost reduction foreseen over the next couple of years is impressive, but it is not enough. Even if present-day technology presents speeds and storage capabilities unimaginable ten or fifteen years ago, the thirst for more power is difficult to satisfy. For some professions like:

- Nuclear engineering;
- Oil engineering;
- 24-hour banking;
- Risk management; *and*
- Space exploration.

even the fastest present-day machines are too slow. At the same time, the procedures for manufacturing and control of structures have not kept pace with product technology. This is even more pronounced in quality control, where the checking of big switching complexes is done through algorithms, with the control effort increasing exponentially with greater complexity.

Furthermore, contrary to biological systems, integrated circuits are missing the capability to self-structure and self-repair. In principle, integrated circuits are two-dimensional structures.

These are among the background reasons underlining the still tiny, but growing, interest in *biological microprocessors*, or *biochips*. The prevailing concept about their structure is that of functionally coupled neurons, and tissue for mechanical and metabolical stablisation.

Granted, the advantage of biological microprocessors, as contrasted to the semiconductor devices, is difficult to estimate because of the rapid development of the latter. But we do know that in the frame of a biological system the different biochips of a central nervous system can be optimised towards specific jobs.

The key issues are design and integration. Though the elementary self-standing biochip will be the first to come into existence, taken out of the context of a nervous system the biochip might only fulfill trivial tasks. We must think in terms of the network.

Each one on its own, the neural elements operate asynchronously, in an analog code. They can show great deviations in their operation, and have to be kept in a physiologically adjusted milieu. Biological microprocessors of the complexity of a cerebellum are currently not imaginable (though not necessarily impossible), and a coupling with technical sensors, effectors and other microprocessors will remain a commercially unsolvable problem for many years.

We are not around the corner of developing complex biochip systems. A great deal of work still needs to be done. But references to current state-of-the-art and associated difficulties do not exclude future technical usage of biochips. We would be short-sighted if we forget how often technology outpaces our best estimates - and we should be aware that a sort of biological microprocessor is being used today in medicine, while molecular clocks are in an up-swing.

In basic medical research the usage of biological microprocessor concepts such as neurobiological prothesis is followed with significant interest. As the nervous system is redundant and plastic, often the functions of damaged structures can be taken over by other functional regions. In this connection, a replacement of vital substructures through implantation of a biological microprocessor is exciting news.

Technically, such an approach is facilitated through the absence of normal blood/brain reactions. The central unsolved problem is the functional connection of the implanted biological microprocessor neurons with the nervous system.

Cut-off nerve channels of central neurons do not regenerate in an adult. At the high connectivity of the central nervous system it seems necessary to create an estimated 1,000 input/output connections per neuron. Still, initial successes in the implantation of embryonical biological microprocessors suggest the ability to create some functional outputs.

The interfacing of biochips also leads to basic questions of neurogenesis and plasticity, an important area of neurobiology which is developing quite rapidly. Another area of considerable interest gravitates around the knowledge of a molecular clock.

The fact that DNA can serve as a molecular clock marks a major evolutionary turning point. Using this method, scientists are rewriting the history of many species, including man, proposing a radical new basis for the system of scientific classification devised over 200 years ago.

Molecular studies show, for example, that DNA from human beings and chimps is about 98 percent identical. Such a degree of similarity usually occurs only among animals of the same genus. Yet men and chimps are divided not only into separate genei, but even into separate families: Hominidae and Pongidae.

The idea of a molecular clock is relatively new, but that of charting biochemical kinships among species was originated at the turn of the century by George Nuttall, a British doctor. Nuttall injected rabbits with human blood, causing their immune system to make antibodies, just as if the foreign blood were an invading bacterium or virus.

Proteins, modern science suggests, may function as a molecular clock. Such a clock, according to the prevailing hypothesis, is set in motion when two animals diverge. New biochemical techniques prompt researchers to apply this principle to the study, not only of many animals, including humans, chimps, gorillas, apes and monkeys, but also of birds, mammals, amphibians, fish, and even bacteria.

Studies show that, although proteins often behave in a clocklike fashion, different proteins evolve at different rates. One protein represents only one gene, and one gene only about a very small fraction (two millionth) of all the information in the DNA.

For this reason, DNA hybridisation aims to study DNA in its full span. This approach is based on the fact that the double stranded DNA helix can be broken into its complementary single strands by heat.

The procedure rests in combining samples of DNA from two species, boiling the mixture to break the double strands apart, and letting it cool so that the strands will recombine into pairs.

- Each strand tries to find its exact complementary partner.
- Some strands pair with partners from the other species that are not quite a perfect match.
- These form hybrid double helixes that are not as tightly bound together.

Researchers contend that the DNA clock must be calibrated with well-established palaeontological or geological reference points. Other researchers, supporting the molecular method, argue that this is much better than the study of fossils for inferences on the family tree of the species.

The thesis of the latter school is simple: ''We know our molecules had ancestors''. The counter-argument suggests that the molecular clock gives only part of the picture and that we have to combine the research on the ancestors (DNA clock) with that of the descendants (palaeontology).

While the interest in molecular clocks as research agents is very real, we should not forget that the clock action has been, since the beginning, a basic building block of computing machinery - particularly for synchronous equipment. Thus, to the databasing, data communications and data processing capabilities we can also add the indispensable clock action - provided that scientists know better about both the details and the aftermaths than they do today.

4

From Philosophy to Artificial Intelligence

Introduction

In Chapter 3 we spoke of biotechnology and of the risks involved in manipulating the genes in order to create superbrains. There is another avenue of development which is aimed at creating an intelligence greater than that which is possessed by man. Its origin lies in new types of software and non-von Neumann computers. Coined in the late 1950s, the term *artificial intelligence* (AI) serves as the common denominator of different, but complementary, disciplines.

The issues entering artificial intelligence reflect both a theoretical and a practical background. They are characterised by polyvalent approaches. As shown in Figure 4.1, AI is concerned with:

Pattern recognition; Symbolic reasoning; Natural language understanding; Automatic language translation; Intelligent tutoring systems; Voice recognition and synthesis; Artificial vision; Intelligent weapons systems; Knowledge-based computer aided design; Intelligent robotics; Brain emulation; and Expert systems.

Expert systems have been the first practical implementation of AI. Concerned with cognition processes, they are software packages that experts in specific fields enrich with their knowledge. They do so by distilling their expertise into a set of rules, entering such knowhow into the system. This is achieved by means of:

- Knowledge acquisition, *and*

- Knowledge representation

which are *knowledge engineering* disciplines. What we are after is practical solutions which are able to handle fairly complex problems, tackling issues requiring high levels of human judgment and expertise.

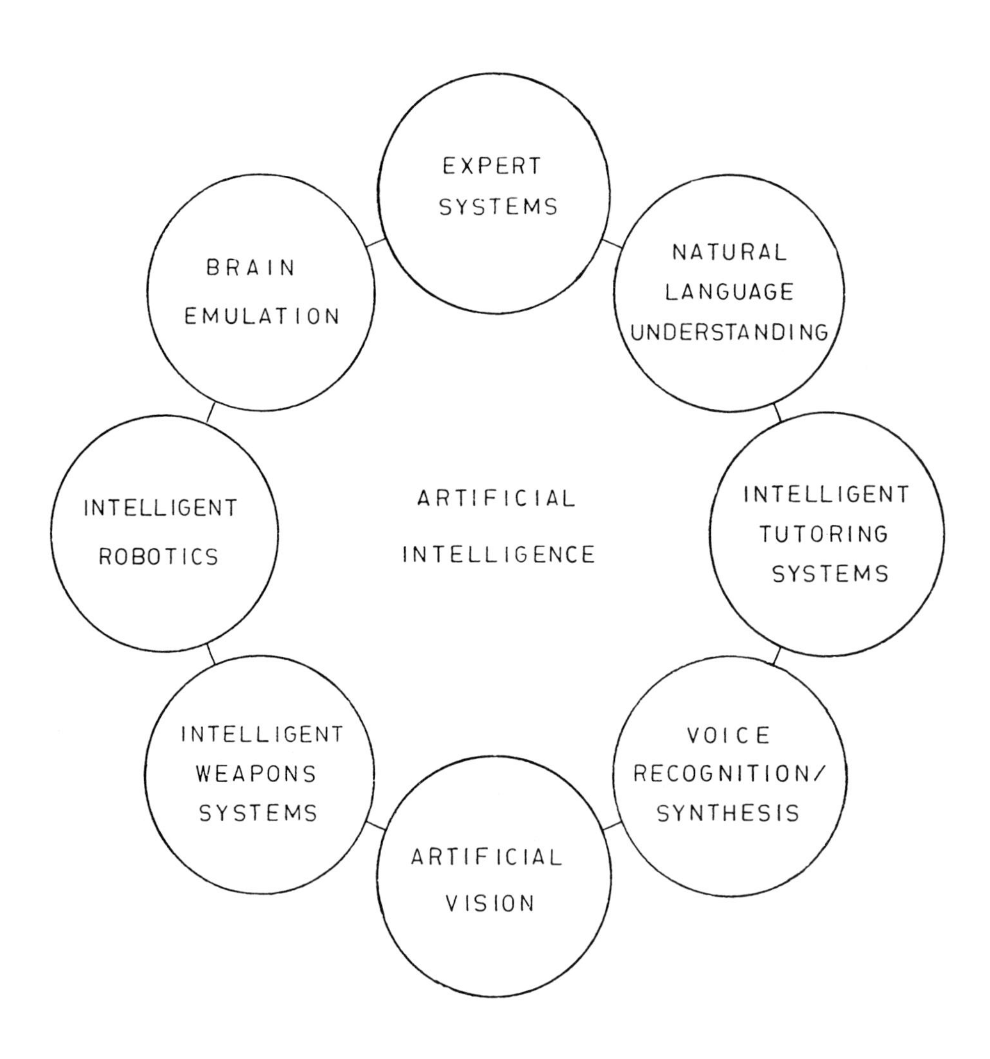

Figure 4.1 Artificial Intelligence

In a way, artificial intelligence is the continuation of physics. It searches for a *physical truth of the intellect.* But it also contrasts to physics as AI methodology involves concepts of metaphysics:

- Physics as we know it today has an admirable exactness.
- Artificial intelligence uses both vagueness and uncertainty in its reasoning, which are more representative of how the human mind works.
- Physics is governed by a criterion of certitude which is based on two pillars: rational deduction, and confirmation through experiment.
- AI employs more induction than deduction. Experiments can be undertaken in the limited domain current technology allows, not in a broader all-inclusive sense.

Experimentation and, therefore, certitude and clarity is the form of courtesy that the physicist owns. By working on the frontiers of knowledge, many endeavors in AI are akin to philosophical activity, which we endeavour to submit to basic analysis.

But physics can gain much from AI, as it finds itself today in a crisis of principles. This crisis is due to a deep and intense change. And there is no better sign of maturity in a science than such a crisis of principles, submitting what so far was taken as "basic truth" to *bold revision.* Out of such revision comes new principles which establish different departures than those we have followed so far.

Theory is nourished by doubt, and doubt is a basic ingredient in the AI constructs we presently develop. But the impact on physics from AI may be much greater than the foregoing paragraphs tend to suggest, and this for fundamental reasons of logic.

Any process of thought can be seen in a layered fashion, where each higher-up level is *meta* to those beneath it. It acts as steering and control on lower levels, placing boundaries and constraints upon them, but also directing their activity. It is *metaphysics* which regiments the physical science - not vice versa. And metaphysics has a great deal to do with the laws of thought into which AI integrates.

Metaphysics is philosophy, but the concepts which it promotes and develops in the higher-up layers (to which it addresses itself) influence other intellectual human endeavours such as intuitive perception and the world of art. There is an *intuitive perception* in the world in art, said Marc Chagall. It:

- Reflects the principles according to which the real world is shaped in our mind, and
- Impacts on the way that this real world is reconsidered with the help of fantasy, imagination, and fleeting sensations.

This intuitive perception helps the writer, the painter, but also the physicist and the engineer, to leave the traditional pathways, open new avenues and to do creative things.

There are many metalayers arranged in a hierarchical structure. The higher one of them has the power of imagination. As Chagall once suggested: "In order to do something substantial in art one has to begin everything from the very beginning. An artist is one who forever remains within the six days of the world's creation, and from them he draws his paints. Only those six days! After that, the world has been created and to create something new in art is impossible; after that one can only imitate".

Artificial intelligence today possesses no creative capacities on a massive scale. Man does. But there are also some beginnings in the AI field. The University of California, San Diego, designed an expert system to do paintings. It started with primitive art and tremendously improved by doing; learning as it went along. To say that it is prolific is to state the obvious. Less obvious is the answer to the query about what the future might bring in AI breakthroughs. This expert painting system can give some hints.

Philosophy and the Sciences

Philosophy and the sciences do not progress in a horizontal dimension. Their evolution is towards the depths, and their advance consists in questioning that which previously had not been questionable. This is precisely the reason why the need for philosophy and the sciences is inseparable from human life.

"I do not know whether or not what I call *my life* is important", suggests José Ortega y Gasset, "but important or not it does seem that it was here before all the rest - including before God Himself". It could not be otherwise. *Science and religion are two of the innumerable things that man creates in his own lifetime*, of which sometimes he benefits, and in other times he regrets.

"Life is what we are and what we do", Ortega suggests. "It is, then, of all things the closest to reach one of us. *Life is what we do and what happens to us* - from thinking, dreaming or worrying, to playing the market or winning battles". Yes, this is precisely the juice of life.

But life is not just the present. It contains all of our past, as well as *all of our future*. The present is short, very short. It is:

- Composed of a limited number of instants,
- Of things which happen now.

Yet, these are instants we must live. Nothing of what we do would be our life, if we did not take account of today, but also if we fail to understand that, though conditioned by the past, the present is mostly influenced by the future.

Unlike other things and beings in the Universe, man's life consists not of what he already is but of *what he is going to be*. Hence, of what he is not yet but may become.

We find ourselves by being our future. That strange privilege of occupying ourselves with the future is in reality what makes man, and what distinguishes him from other animals. Whether we like it or not, living our life is always a matter of having to do something: an occupation, a task, an obligation, or simply a decision.

All living philosophy teaches is *living with.* This means finding ourselves in the midst of an environment on which we act - but which also leaves its imprint upon us. We can renounce life, some people have done just that. But if we live we cannot forget about our environment. We can influence it over time - or we can choose a different environment in which to live.

To have a clear, factual and documented consciousness about anything in the world in which we live, we must give it our full attention. We must perceive it, conceive it, comprehend it.

- In order to see physical things we must turn our eyes towards them.

- But, to see logical things, we need more than consciousness, and this is understanding.

Precisely this consciousness leads to the incessant discovery that we make of ourselves, and that is *philosophy.* Further discovery primarily concerns the world around us, and this is *physics.*

A creative life implies a regime of strict mental health, of high conduct, of an imaginative personality, but also of constant stimulus and of *creative thinking* which keeps active the *consciousness of man's dignity.* In Chapter 1 we have spoken of thought definition. Among the things existing, or pretending to exist, in the Universe, there is no better example than being able to radically differentiate from all others and from what there was in the past. This is *thought.*

Life starts with awareness; then proceeds with the making of choices; and follows with taking action. Life is also perplexity, uncertainty, vagueness and the will to face the problems which steadily come up. "We don't really have problems in life. We have challenges", one of my professors taught me.

When, in the introduction to this chapter, we spoke of artificial intelligence we said that it approximates the human thought process better than other systems so far known, precisely because it can deal with uncertainty and vagueness. This emulates human thinking, not only in making decisions, but also in *using existing knowledge in search for more knowledge.*

Chapter 1 underlined that the search for knowledge is typically done in teams. "The isolated man", Alan Turing once commented, "does not develop any intellectual power. It is necessary for him to be immersed in an environment of other men, whose technique he absorbs during the first twenty years of his life". This is the role of the environment as cultural transmission agent, though more recently we have tried to accelerate the building-up of knowledge through AI tools.

Socrates considered search for knowledge to be inseparable from communal living and interaction with other people. When he was not teaching, he was in the marketplace (agora) engaging notables and common people in a person-to-person dialogue. When asked why he was so eager in conversing, he would answer: ''I learn as long as I live''. From antiquity to the present time, man has developed through interaction. The difference, however, is that what up to the 1940s was a man-dominated environment is today increasingly shared between men and intelligent machines.

- In the beginning, an overdose of machine intelligence may cause severe disruptions, even if great wisdom lies behind the system design.
- Trying times can indeed result. But most likely they will be followed by a symbiosis.

Though connectionist models do exist today, and in fact brain emulation was the first field to which AI addressed itself. There is not yet a real life AI brain model in existence, but most likely it will come in the period 1995-2000. By then we will most likely see simple neural systems on the market: silicon or photonic constructs used in toys, as a start, then in computers and robots. They will see, hear, feel, and remember in ways far superior to any artifact that exists today.

The intelligent artifacts will be artificial intelligence implementations beyond the expert system level currently available. But this raises another query: ''Should we be concerned with AI?''.

For academics and pure scientists the intellectual challenge of AI suffices. For financial people and industrialists, the preoccupation is in the form of dividends from the AI investments that they make. For people who think *in terms of consequences*, the issues involved are much broader. And for good reason.

''Run, run, comrade, the old world is behind you'', was the slogan on the walls of Sorbonne during the student revolution in May 1968. There are moments when man escapes his past to better address himself to the future. There are other times when man lives in his past and lets the future shape itself without man's influence. The latter are by far the most numerous - and the greatest danger is there. We cannot turn the wheels of history backward, as the 19th Century Luddite experience helps document. But we can channel future developments to our advantage - if and when we have control over them.

Artificial intelligence is only one of the fields where *excitement* and *risk* tend to merge. We said something similar of biotechnology: man's manipulation of the genetic code - and this on a massive scale, following the failure to reach a moratorium during the Asilomar conference in 1975.

As Chapters 1 and 3 documented, so far the risk has been well taken. But other issues are coming. One of the outstanding developments projected for the early 1990s, is artificial blood, synthesised from perfluorocarbons. It will be available for transfusion substitution and oxygen enhancement of tissue. It will be developed by biotechnology, and may find its way into the artificial intelligence realm - together with the biochips.

Breakthroughs projected for around Year 2000 involve both biotechnology and AI. Polymers will be merged with living cells to create artificial arteries, skin and certain organs, such as the pancreas and liver. At the same time, as we will see, we will be approaching AI constructs which surpass the intellectual abilities of man. Should we rejoice about the future? About the synergy?

The greatest risk in life is in complacency. Yet, taken by his day-to-day preoccupations, through the rituals of his ordinary life: dressing, undressing, eating, sleeping and playing, man rarely has the time or interest to occupy himself with his own future. At least not in an organised, thoughtful, well-planned way. So, life reserves surprises.

Philosophy started 25 centuries ago to reach its current stage. This has given time enough for many original thoughts to be polished, and others to pass the test of time. But *philosophic preoccupation with the emerging new sciences has not yet started.* Yet, we know that the logical and physical events behind them evolve very fast, and that the chances which we take are enormous. Does anybody care?

Imposition of Form on Matter

A synergy between artificial intelligence and biotechnology is not only likely, but also most probable. When it comes to synergy involving fields so far distinct from one another, history can teach us a lesson if we care enough to read and comprehend its messages.

The successive *knowledge revolutions* of mankind both influenced, and have been influenced by, the science of materials. Over the centuries, the development and use of new materials had *social* implications. At the same time, it helped to embed *technology's* breakthroughs within a given *cultural* concept.

The interaction of materials' inventions with the crafts and arts of their time became a *defining technology* and, therefore, a milestone in human culture.

- If clay and stone were the first materials man put at his disposition,

- The first technologists were those who knew how to shape clay into useful forms.

There was very little machinery to meditate between the human craftsman and the materials on which he worked. The craftsman's hand was the tool and the control of the production process. Animal muscle assured the motor power.

When man learned how to purify metals and cast them into moulds, the skills of the technologists were no longer the same as with those of the clay period. Something similar happened with arts and artists. Copper differed from clay in more than durability; the hardness of iron made possible the creation of more permanent tools. All the same, the change society experienced was not limited to the materials side and in the crafts.

At every epoch, philosophy and sociology bear the imprint of the predominent stuff, crafts, and tools of their time, be it pottery, weaving, or metal working. The mind's horizon somehow changes. *Human thought is fascinated by the imposition of form on matter.*

As new materials became available, and crafts developed to master them, man saw himself able to build to his image a tiny part of the world. His hands became creators. Language reflected this concept and eventually

helped man create logical constructs which emulated the physical property of form.

In Plato's writing the *Forms* are first mentioned in t*he objects of the soul's echo* when withdrawn from the senses. Contrary to man the craftsman, whose primary channels were the senses, man the philosopher* *perceived the Forms by his thought.*

In his writings, the ancient Greek philosopher makes a distinction between *Becoming* and *Being*. Socrates has used the same words in *Parmenides*. Intercourse with *Becoming* is by means of the body through senses; with *Being* through language, the soul and reflection. Thus, the concept is one of complete distinction between:

- The unseen, but intelligible, Forms, and
- The visible objects of body senses.
- The mind is akin of unseen entities.
- The body of the visible.

When the human mind studies things through the senses, Plato suggests, it is dragged down by the body and confused. When it is by itself, reflecting on entities, but without the senses, then it has wisdom.

In the Republic the lower kind of cognition is concerned with Becoming. The higher one being Being. Knowledge is seen as an intercourse of the soul with reality through reflection. Being is, in this sense, the meta-layer of becoming.

This division in upper and lower layers of cognition has been the forerunner of a popular (and sound) systems engineering approach that we have today, the ISO/OSI* recommended model. Here, too, we distinguish between a physical level, which is

---

* In ancient Greece, philosophy had often a wider meaning than today, covering all literal studies as well as analytical and cultural approaches.

**International Standards Organization, Open System Interconnection.

lower, and logical levels presented as different layers of sophistication over it. In the introduction to this chapter, we called these successive logical layers meta levels.

Plato's *Battle of Gods and Giants* was that of idealists:

1. Logic, soul and reflection, hence the Gods.

2. Versus materialists, the Giants.

There was a long philosophical tradition behind the two camps. Its imprint is seen not only in writing, but also in the daily life of people nicely mapped into an east-west distinction:

- The *Ionians,* in the east, had been seeking the real nature of things in some ultimate kind of *matter* or *body.*

- The Greeks from the *Italian* peninsula (Megali Hellas) in the west, sought reality, not in tangible body and matter, but in the *mind's interpretation* of things.

In this sense, the Ionians have been materialists, not merely pluralists. The Italian Greeks were essentially idealists - which today we will call logicians, the word "idealists" having, in the meantime, changed meaning.

According to Plato's definition, in the side of the Gods are those who think that *unseen things are the true realities.* In the part of the Giants are those who believe that real is only body and matter which they can touch and handle. The repercussions are much deeper than they may seem at first sight.

The contributions of Socrates, Plato and Aristotle are so very important because, for the first time in literature, *logic* was contemplated as an autonomous science with the task of ascertaining the principle of *affirmative* and *negative* propositions. In a way, the infrastructure of AI - and the mathematical tools that it contains - is a continuation and practical enhancement of the developments which started in logic.

This contrast is between *mind* and *matter.* To the flesh and, therefore, the materials, belong the senses along with bodily pleasures and appetites. The *mind* or *spirit* finds its function in *thought and reflection.* Performing this role, *it lays hold upon unseen realities.*

To dramatise the separation of body and mind, Plato speaks about the spirit's proper function as best carried on when it withdraws from the flesh to think by itself, untroubled by the senses. It is precisely along this line of separation that the *Socratic method* rests. The latter consists of:

- Raising inquisitive questions,

- Seeking definitions as the first step in the quest for knowledge,

- Constructing arguments by which such definitions, and other answers, will be tested.

Socrates was the first thinker in the western world who actively sought to mark off the boundaries through definitions. Two philosophical innovations may be attributed to him:

1. *Inductive reasoning* (arguments, analogies, generalisations), *and*

2. The *use of universal definitions* as instruments in the search for knowledge.

Both form the foundations of scientific thought. Arguments are the instruments of testing the truth of dominant ideas of each epoch. Definitions clear the concepts that we are dealing with.

Both physical and logical qualities must be defined but, of the two, the latter need clarification much more. Perception of physical qualities like hot and cold is made through the senses. This statement is not valid for logical qualities like just and unjust. In principle, l*ogical qualities* have no status in Nature. They are creations of conventions, ethics, laws, or public decisions - and need to be defined, elaborated, explained. To serve this purpose, Socrates was asking questions all the time.

- He greeted people with questions;

- Taught, confirmed and refuted ideas with questions;

- Left people with questions.

He also demanded answers that were short and to the point. If the other party declined to answer, he imagined a discussion partner holding a question-and-answer session with him.

The object of the discourse is *thinking*. This helped in clearing Socrates' mind, but it also enables the discussion partner to see the consequences of his own beliefs. The major contribution of the dialectic method is not only the elaboration of the art of asking questions but also, if not primarily, *the discovery of the art of constructing definitions* which can be used in substantiating the discourse.

Whereas the physician produces a change by means of drugs, the philosopher does it precisely by *discourse*. However, discourse can take many aspects. There is a contrast between *philosophy* and *rhetoric:*

Socrates: You will find a whole profession to prove that true belief is not knowledge.

Theaetetus: How so? What profession?

Socrates: The profession of those paragons of intellect known as orators and lawyers. There you have men who use their skill to produce conviction, not by instruction, but by making people believe whatever they want them to believe.*

---

*Quotation based on Plato's works, from F.M. Conford: "Plato's Theory of Knowledge," Routledge and Kegan Paul, London, 1935

Philosophy and rhetoric are not the same, and often come in contrast, which does not exclude the idea that they also have analogies. "The object of philosophy is the logical classification of thoughts. Philosophy is not a theory but an activity", Ludwig Wittgenstein once suggested.

Socrates, Plato and Wittgenstein

Socrates distinguishes between *knowledge* and *opinion* (or *belief*). So the statement: "I don't know" does not exclude "I hold an opinion, have a belief, or make a hypothesis". The distinction is fundamental. It places emphasis on our ability to differentiate between what *we know* and *what we do not really know*, though we may think we do or, more precisely, have a fixed idea about it.

Knowledge is generated by instruction or discovery, always accompanied by a true account of the grounds on which it rests. Knowledge is unshakeable by persuasion; it is possessed, Socrates says, "by the Gods and only a few among men". By contrast, *beliefs* are rather common. They are produced by persuasion. Hence the argument about orators and lawyers.

Beliefs are not based on rational grounds, and can be changed through persuasion without the unshakeable evidence necessary for a change in knowledge. But, as stated, true knowledge is possessed only by a few -while beliefs of some sort or another are held by all mankind.

This may be one of the reasons why Plato thought that neither in the crafts nor in the Forms of the mind are matters decided by mobs and votes. The key is a reasoned examination by men of knowledge having, on the matter under study, a learned basis for their thoughts. But knowledge alone does not suffice. The other quality, Socrates said, is *temperance,* the knowing of oneself. "Temperance is doing everything orderly and quietly", the ancient philosopher was suggesting.

In man, temperance is a rare, but most valuable, characteristic, with both logical and physical constructs to support it. These are correspondingly:

- Man's thoughts, *and*
- Man's tools.

Archaeologists, sociologists and technologists think that man's tools are several thousands of centuries old. Available evidence tends to suggest (but cannot confirm) that the first tool may be 1,200,000 years old. It was a cutting stone which could open animals. What man bought with it, is *time.* And this is, in fact, the story of agriculture:

1. The mastery of *fire* made man conscious of the physical system which, to a large measure, escaped his control.

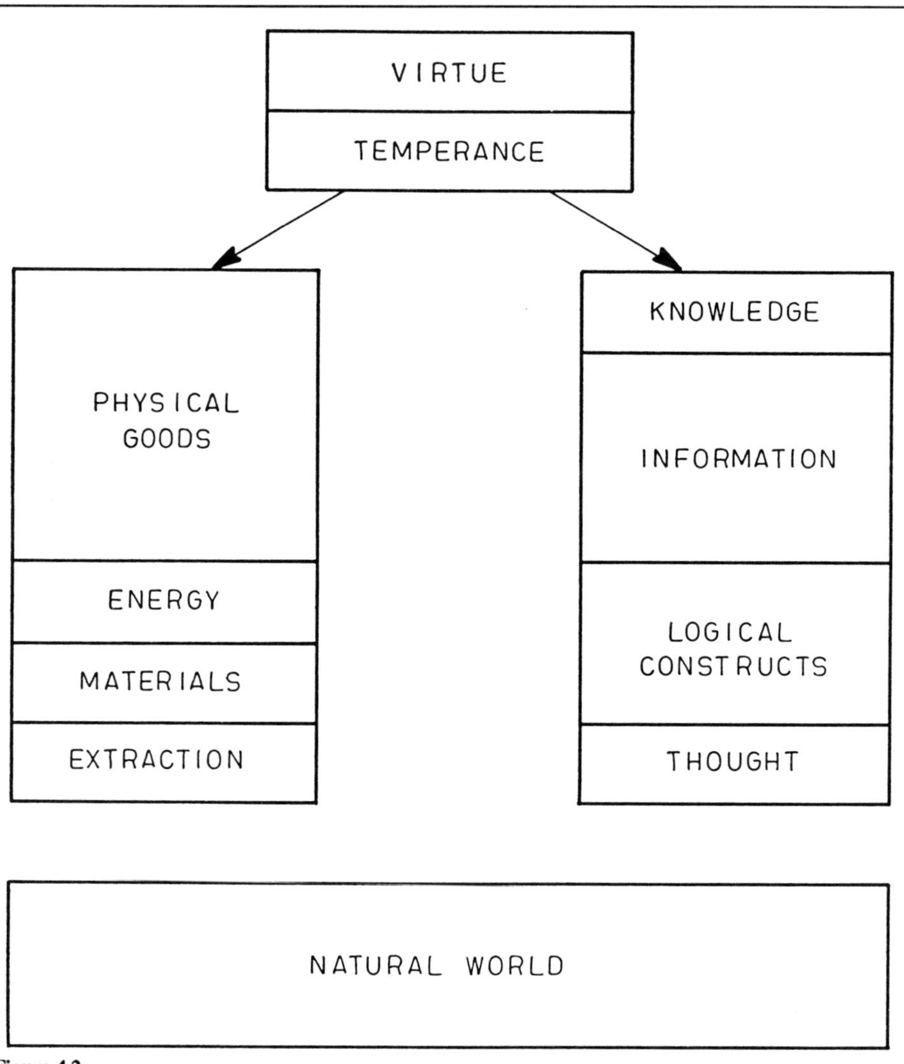

Figure 4.2

2. Through *agriculture*, which started about 9,000 BC at Jericho, man gained time.

3. He and his descendents were able to establish a *city* and organise it.

4. Command over the so-created artificial environment saw to it that man's own *identification* became a little more stable than it had been in the past.

This lasted through the years until the Middle Ages when fear, ignorance and theocracy combined to subdue the individual. But, the Renaissance again brought self-esteem. The fact that megasystems, and the rise of the masses, further destabilised the individual man or woman, and made him/her lose his sense of identity, will not concern us at this point.

It is not often appreciated that matter and body, the key ingredients in the process of *Becoming,* were the first to master the imagination of man, and his efforts to exercise control over them. Once the challenges they posed have been reasonably overcome - and this took more than 8,000 years of constant effort, leading from Jericho to Athens - it was possible to address the Being domain:

● Man's awareness of Being started through perception of *logical forms.*

● It evolved by means of a conception which defined the sense of *soul and reflection.*

● Then it crystalised through philosophical intercourse and, therefore, *metaphysics.*

Thus the Socratic advice for *temperance* applies in two domains, as Figure 4.2 suggests; one is *logical,* the other is *physical,* and both have to be brought under control. But the common man's actions are rarely characterised by temperance, as philosophers have suggested. What is more, the myriad of religious wars on this planet have long destroyed the myth that religion might act as *temporiser.* Only philosophy preaches *virtue* - and therefore its call is rarely heard.

This leads the thinker to *virtue*'s definition - *the knowledge which cannot be taught.* When asked, by Socrates, to define *virtue* most of his students made the same mistake in their responses. They offered a list of virtues rather than the definition of the *single Form* which is common to all of them. Socrates' answer can be easily structured in what today we consider, in artificial intelligence IF...THEN type statements through logic programming:

1. One who does not know virtue cannot know a part of virtue, for instance, justice, *and*

2. One who inquires what virtue is does not know virtue.

Ancient wisdom based virtue on a foundation of knowledge, but without properly outlining which one of the eternal sources of knowledge is the most dependable. If virtue, as Socrates said to Protagoras, is knowledge which cannot be taught, then there are neither teachers nor students of virtue. But there are teachers and students of knowledge.

There are also virtuous people. They build themselves up in their social environment in response to challenges and goals. Therefore, it is such a pity that while we have laboured to improve our geophysical, and most recently genetic, environment - we have done so little to better our ethical and logical structures which are the meta-layer over physical premises.

Stated in a logical, inner-directed sense, "I know" has a meaning other than "I am now perceiving." Part of the distinction lies in the nature of memory, and of the memory images which we have. They are instrumental in guiding our thinking.

The Protagorean notion that "man is the measure of all things" was not necessarily confined to immediate perception, but also included judgment. Perception contributes to knowledge, but is not synonymous to it. A great part of what is knowledge consists of truths involving terms which are not objects of perception. Memory, inference and judgment are examples. Therefore, perception cannot be the whole of knowledge.

The thinking mind uses terms like ''Exists'', ''Is different than'', ''If A Then B'', which are not necessarily objects of perception. They are reaching the mind through the channel of one or more senses, but are common to all objects of sense. These common terms are what Plato calls Ideas or Forms. They are the meaning of common objects, or the gateway to such meaning. If, as Socrates said to Theaetetus, ''Knowledge does not reside in impressions, but in our reflection upon them'', then we know something when:

- We had direct acquaintance with it, *and*
- An image of it remains stored in our memory.

We conceive an idea *in our own mind* but do not perceive it. While the object of perception should be purified through the process of conception, ''Exists'', ''If A Then B'', etc are the concepts identifiable to the process of conception. It is precisely in the field of objects, both known and perceived, that judgment twists about, turns around, and finally proves true or false.

But there is as well a different school of thought which maintains that in logic and, therefore, in the implementation of artificial intelligence, the whole process of *inference* is concerned with judgmental propositions which involve facts. In Wittgenstein's definition* for example, these facts are not atomic but molecular. A logical proposition is a picture (true or false) of a given fact and has in common with the fact a certain structure. But a logical proposition is not necessarily a judgment.

Logical propositions; predicates which we construct; rules and metarules that we elaborate; the facts, states and values (data) that we perceive or record; episodes in the larger domain, are building blocks through which is constructed a system of intelligence. This system can be natural, as we understand when we examine the human brain; or it can be man-made. That is the foundation on which rests artificial intelligence.

The introduction to this chapter defined the term and meaning of AI. But it did not elaborate on what it could or could not do. We will do so in the following two sections, including a response to the question: ''Can machines be more intelligent than men?''.

One more crucial query should attract our attention, whether we talk of natural intelligence or of man-made; that is, building up an expert system, generating a knowledgebank, creating an inference engine. What about decay and death in

---

* Ludwig Wittgenstein is Protagorean in his beliefs. His credo: "The world is the totality of facts, not of things; the facts in logical space are the world" resembles the propositions "perception is knowledge" and "man is the measure of all things."

systems, whether natural or otherwise? "Death", Wittgenstein once said, "is not an event in life. We do not live to experience death. If we take eternity to mean not infinite temporal duration but timeliness, then *eternal life belongs to those who live in the present*".*

Artificial Intelligence at Your Service

The mechanisation of labour for better productivity has been the base of the 19th Century Industrial Revolution. The concept was primarily economic, and secondarily technological. Machines substituted for muscular power while the control mechanism remained the human brain -from muscular reflex activities to higher-up intelligence.

Slowly, the build-up of industrial power made a change in control policy not only advisable but also mandatory. We could no longer live and work with the concepts which dominated the 19th Century (and the first half of this century) in terms of mechanizstion. The system became bigger than a human brain could control. Computers changed this perspective. Even so, the first 30 years in the utilisation of computing machinery were fundamentally an extension of the mechanisation concept - and, therefore, unexciting.

Manufacturing is not the only field of economic activity where 19th Century industrialisation had an impact. Mechanisation also affected agriculture. It did so the big way. Today less than 3 percent of the American population work in agriculture, and they produce a great deal more than what the nation needs. Something similar is true in Western Europe. Hence the huge American and European agricultural surplusses.

Advanced mechanisation also had an evident impact on mechanical, electrical and chemical industries with a corresponding decrease in the employment of the labour force. But, it affected very little the distribution and service industries, while other sectors such as education, finance, medicine and entertainment remained practically alien to its advances. Here is where artificial intelligence will have telling results.

Medicine was the first fertile area of AI implementation. Then it was engineering. Most recently, finance and management became the fields of focus. Wherever professional activities are employed, the domain is open to the implementation of expert systems, an AI technology for which now exists, providing a body of knowledge good enough to apply to daily work.

In many cases, artificial intelligence constructs sneak in unrecognised with a pallet of possible applications. They represent relatively small changes in what we were doing already through computers, seemingly aimed at better management. But, as the system is tuned up, the manager will find that the intelligent machine becomes the driver, or at least takes the co-pilot seat.

---

* Ludwig Wittgenstein: "Tractatus Logico-Philosophicus" Routledge and Kegan Paul, London, 1961

The classical case of accepting AI as a *peer* may well be in the entertainment domain. Its taking hold has come through *game playing* using computers, interactive videodiscs and three-dimensional (3-D) presentation. In fact, as the sophication of such systems increases, a journey taken in a computer may become more enticing than the real trip.

However, entertainment is not the only field with success stories. The automobile industry is among the first to capitalise on the emerging AI capabilities. It started by incorporating electronics in such areas as:

- *Navigational system* mapping on a video the auto's location, plotting the quickest route to the driver's destination.

- *Collision Avoidance.* Front and rear radar systems warning of an impending crash and alerting drivers to other approaching cars.

- *Drive-by-Wire.* Electronic controls replace the mechanical linkages connecting pedals to throttle and brakes. They are more accurate and more reliable.

- *Active Suspension.* Sensors spot bumps and potholes commanding wheels to be lifted over rough spots and planted back on the road.

- *Traction Control.* The tyres would not spin on slippery roads, among other developments.

- *Heads-Up Display,* including image enhancement, a feature which lets the driver scc through fog.

Fuelling this explosion in electronic supports is the affordable price of computer power, coupled with the automakers' new competitive drive replacing familiar mechanical and hydraulic components with more precise, chip-based systems. But electronics can only go so far without AI enrichment.

Costs do not seem to be a major problem. An electronic engine-control module is today a $200 device the size of a cigar box. It contains a microcomputer and a handful of other integrated circuits for regulating such functions as engine-spark timing, idle speed and fuel mixture. Capitalising on this infrastructure, tomorrow it will be equipped with expert systems support.

The culture necessary to use high technology modules on automobiles does not catch up as fast as the devices are introduced into use. Toyota estimates that more than 10 percent of the electronic mechanisms its mechanics replace are not defective. They are simply badly used. Also, the expert system-based diagnostics to serve cars these days are not effectively employed, because the training of mechanics in their use did not keep up with their production.

On the one hand the user population lacks the images and associated knowhow, and on the other the contributions technology makes are significant. The new antilock brake system from West Germany's Bosch contains only 4 percent as many components as the first prototype built 15 years ago. Delco's latest engine-spark controller is a single chip. The original module measured 12.5 cm by 17 cm.

Using advanced technology, Europe is mounting the most ambitious effort by far. A seven-year research programme is beginning, Prometheus, which will cost an estimated $900 million. The goal is to apply the technologies that made air travel safe to halve the toll of 55.-,000 lives now claimed by traffic accidents every year.

''It is cheaper to pay for advanced electronics than to spend millions building new roads'', says Trevor Aspinall, research division manager at Britain's Motor Industry Research Association.* But the new technology, when ready, would have to wait for a mass effect to leave its impact. On the consumer side, for example, communications among cars and outside computers will be effective only if many cars, perhaps 30 percent of those on the road, are equipped with the necessary gear.

Let us not forget, for that matter, that the imaginative competition for new technology implementation and AI assistance is only now starting. Computing, including parallel computing, is still in its infancy. We are three to four decades away from machines having the complexity of the human brain. But we do have man-made devices which are 10 million times (107) faster than the brain itself.

The neuron in the human brain works at a slightly faster speed than 1/1000 a second (10^{-3} to 10^{-4}). But because of delays at the synapse and other reasons, the effective rate is 1/100 of a second (10^{-2}). Today, available computers work at the nanosecond level (10^{-9}), and experimental machines reach a multiple of that speed. That is electronics. The oncoming photonics switching promise 1,000 faster computing, raising the difference between the human brain and the machine to 12 orders of magnitude (1012).

However, speed is not everything. Also very important is system behaviour, accounting for processing elements (PE), cycle time, bits per cycle per processing elements, bits per second per PE and bits per second per system. Also, compactness and energy consumption play a key role. All counted, today's parallel computers are 1/10,000 as powerful as the human brain (10^{-4}). But they are constantly improving.

If man-made systems continue to double in capacity every 2 to 3 years, it will be possible to build a computer system of power equivalent to the human brain in about 30 years. It is, however, less clear whether we can solve the software problems within this timeframe. Hence the three to four decade estimate which will be made in section 6.

---

* Business Week, June 13, 1988

The Android Brain

Several developments will be converging toward an Android Brain. Many of these will be surprises. Some will be good. Others might be dreadful. Breakthroughs can take many possible courses, and it has been quite unclear since the start what may come at a later date. Nuclear power production can serve as an example. The problem is not in the rare exception of implementing a technological breakthrough. It is in the massive usage to which, afterwards, man subjects all his discoveries. Figure 4.3 plots

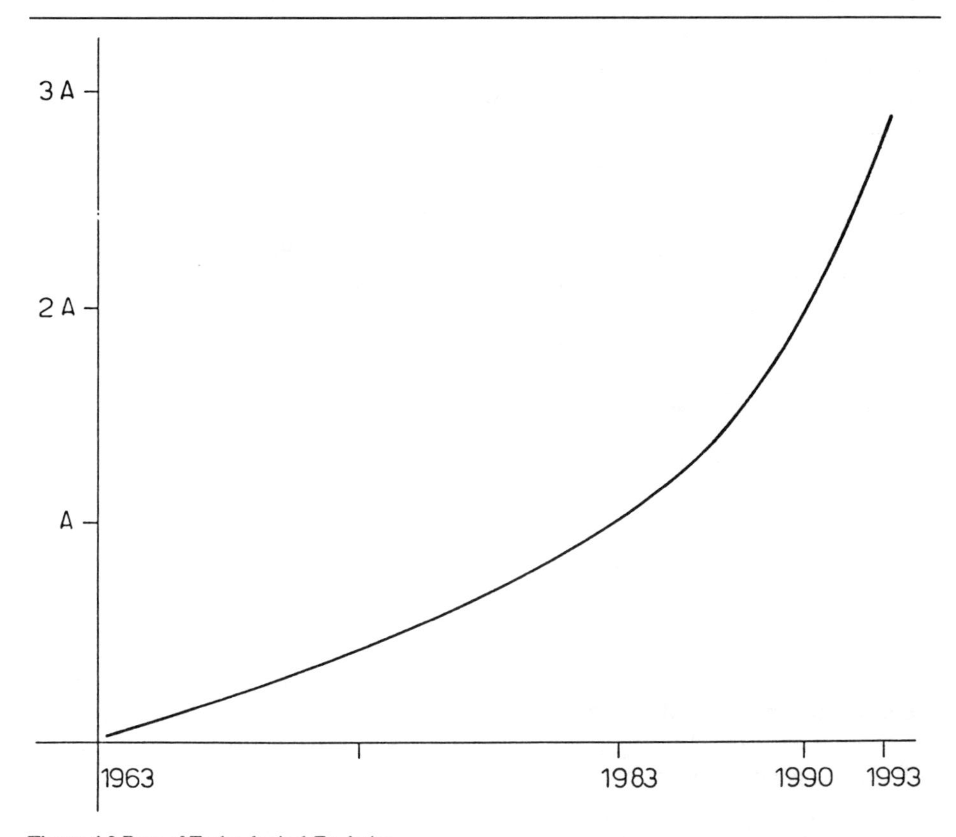

Figure 4.3 Rate of Technological Evolution

the rate of technological evolution in the computer field. As in all domains of science, it took some time to create momentum. The 1953 to 1963 period has seen a slow growth. Momentum started picking up by the mid 1960s. Thirty years of computer implementation reached, by 1983, a result which we will call "A". A study done at the time projected that during the next seven years (1984-1990) the mass of results will reach the "2A" level, and this projection is validated by the facts.

As the momentum of science and technology accelerates the forecast is now made that the growth curve will still bend upwards, reaching "3A" by 1993 and then continues growing before it teppers off. Probably, by the beginning of the 21st Century, we will have available:

- Circuitry equating to biological densities.
- Hardware and software adaptive to environmental stimuli, beyond the level of expert systems.
- Teachable computers leading toward the use of brain type metalanguage.
- The implementation of automated intuition.
- Brain augmentation through high level intelligence.

Within this evolutionary perspective, projects focusing on the development of Android Brains will be, by far, more important than landing on the moon (1964-1969), or the Manhattan Project (1940-1945). Artificial intelligence constructs will make feasible the industrialisation of knowledge. Combined with scientific breakthroughs in biotechnology, *knowledge cloning* risks becoming everyday business. And *that is a big risk.*

Regretfully, nobody has yet seriously considered what this will mean to the human race, though some ideas do exist. The android population will have peculiar merits - and also demerits. It will take advantage of: gigastream communications, memory-to-memory exchanges, large distance coverage, very fast switching speeds, biological densities, full cloning, fast reproduction and ever increasing AI capabilities.

It will no longer be necessary to wait for the 9 months that a physical process takes, and for another 20 years to educate a person through arcane approaches. Admittedly, the men and women of the year 2030 may regret the day they lost the old-fashioned ways of education. This being said, here is a probable scenario:

- By Year 2000 there will be available a machine of nearly human complexity.
- In the 2020 to 2030 timeframe the software challenge will be solved.
- Then, between 2040 and 2050, intelligent man-made systems will be more powerful than humans by one to two orders of magnitude.

We do not talk of energy sources. Machines, by order of magnitude more potent than muscular power, were developed during the decades which followed the industrial revolution. Now the reference is intellectual power in the century following the development of the computer.

Nobody can really tell how realistic the forecast I just gave may be in terms of timeframes. But this is the *direction* in which things are going. As other experience suggests, technological progress is irreversible. We should aim to channel it rather than to stop it, as stopping it cannot be done.

Events may move faster than we expect, or they may be slower, though the former probability is higher than the latter. The problems in meeting what seems to be a tough timetable are mainly intellectual. Once they are solved, developments will take enormous capital investment - but they are not impossible.

The road the next 4 to 5 decades will transit is necessarily vague. Key developments can be expected to come from the United States, Japan, Germany and Britain in that order. Every major breakthrough we will experience in science and technology will be followed by a dual uncertainty:

- In terms of its effective implementation, *and*

- In regard to its possible aftermaths further out in the horizon.

There will also be uncertainty in regard to social acceptances, but the rise of the masses partly waived that constraint, as the masses can be conditioned. Democracies proudly feature a benevolent conditioning mechanism, in contrast to the malign conditioning that prevails under dictatorship. Fortunately, all four of the above-mentioned front runners are under democratic regimes.

There *might* have been a different course in the individualistic societies which followed the ancient Greek Middle Age and the Middle Age of Western Europe. As José Ortega y Gasset aptly suggested, one of the mechanisms of uncertainty which plays a key role in an individualistic environment is the appreciation that, when we hear something, and an elemental objection occurs to us, this may also have occurred to other persons - but we are not sure about it. In a sense this means ambiguity, and it is based on doubt or scepticism (skepsis).*

Doubt opens the crack which proof will fill. The problem is that, for better or for worse, we do not have proofs available on what the megatrends we talk about may bring. However, the father of truth is the desire for learning. To learn the future, we must think and investigate, rather than remaining passive observers, or stumbling backwards into the things to come.

Looking back to the science-fiction of the 19th Century, Jules Verne for instance, we see that many technological events were properly forecast. By contrast, examining the science-fiction of, say, the 1930s, its authors seem to have:

- Underestimated the science and technology breakthroughs

- While they overestimated the social and political impact

To successfully forecast the future we should not only project on the coming decades, but also use the lessons from the past which could be helpful in the future. The lesson

---

* The Greek word skepsis means thinking, hence doubt. such as the development of a unified world government, or the resistance to developments which did not abide to a well-perceived social content.

from the Industrial Revolution is that, once muscle power was cheaper than human muscle, it was extensively used. The aftermath was an enormous explosion in goods production.

With artificial intelligence may come an enormous explosion in services. But, just as in the case of the Industrial Revolution, the Knowledge Revolution will see a large wave of instability in social structures. It will be of a magnitude that no political scientist ever endeavoured to treat.

The new Karl Marx will talk of Human Capital and of Android Brains. His thesis will be a nemesis on the highly uneven way in which countries, companies and people develop artificial intelligence and exploit it. The rewards and threats related to AI constructs will be so enormous - he will say - that they might become either:

- A unifying factor, *or*

- An explosive trigger.

This having been said, there are many questions whose answers escape us. Some are funny: "Will androids be bisex, unisex, or middlesex?". Others are serious: "How would humans react if androids become independent, with their own will and decisions?". Further still: "If they conflict with humans?"

Will artecrafts behave as humans behaved throughout history, in terms of destructiveness? What about, if becoming more intelligent than humans, machines built other artecrafts much more intelligent than themselves?

There is no precedence which permits us to make educated guesses and give them as an answer. What we know is that the 21st Century will be the era of machine intelligence. It may also be the end of man's journey into the Galaxy. Intelligent machines will probably make the exploration for us.

- The passive way to look at the subject of the future wave is to admit that human beings are nothing else than stepping stones to higher intelligence.

- The active philosophy maintains that the only way in which to find out about Android Brains is to design them, build them, use them and try to keep them under control.

The true motor behind the active approach is man's self-consciousness. Through AI we may be able to understand the human brain - but we may also end up by fusing it into a larger system. When this happens, the fault will be our own. And the same is true of the nemesis.

5

Giant Strides in Electronics and Photonics

Introduction

One of the developments in computers and communications that we will experience during the last decade of this century is very large scale integration, with the 4-megabit chip, and then the 16-megabit chip becoming the standard. Another is universal broadband communications through satellites and optical fibres.

The first reference is *electronics,* the huge, dynamic field of technology which has exploited the powers of the electron, capitalising on the theoretical work done in the early 19th Century by Maxwell and Faraday. The second is *photonics.* Photons can be employed as more compact carriers than electrons, storing and transmitting massive amounts of text, data, voice and images at low cost. Eventually they will also be used for switching.

The early great step in electronics was the vacuum tube. The next milestone was the transistor, which was more or less immediately followed by integrated circuits. In fact, the invention of the transistor in 1947/1948 has been one of the major contributions to man's industry. Transistors offered a faster, smaller and cheaper way of performing the various functions of the vacuum tube. Therefore, it can be said that, while the era of electronics timidly started with vacuum tubes, it grew and flourished with transistors.

Bardeen, Brattain and Shockley at Bell Laboratories invented the transistor, a solid state device with a couple of significant abilities. It could:

- Amplify an electric current, *or*
- Act as a simple switch.

The early transistors were made of germanium, a material of the same chemical group as silicon. However, germanium is fairly rare, and germanium transistors had a limited temperature range. Silicon had most of the advantages of germanium, and few drawbacks.

The first silicon transistor was made by Texas Instruments in 1954. Silicon is cheap, plentiful and easy to handle. It has an oxide which is useful in the fabrication of semiconductor devices, and it can also be used as a dielectric.

But, while silicon continues to master the centre stage of electronics technology, researchers try new methods to improve ways of compacting even more devices onto a single chip, by increasing the speed of the device while using less power. A good example of step-by-step improvements is silicon-on-sapphire, involving the laying of fine silicon tracks on a slice of sapphire, thus providing a better combination of speed and power than conventional silicon chips.

The age of electronics is still young. The electron was identified only last century and the microchip, on which today's information technology depends, has been around for only a couple of decades. The successes crammed into these few years have created the impression that electronics is a field capable of limitless improvements. This is not so.

Electronics will give way to a superior technology based on light. Physicists did not realise until early in this century that light came in separate packets, called *photons.* However, science made great progress in manipulating photons. A photonics revolution is already in the making.

The new optical technology is moving rapidly into place, providing sensitive nerve endings for devices from smoke detectors and blood analysers, to large scale computers. Scientists in America, Western Europe and Japan are pushing hard towards a still-in-the-future optical computing engine that uses photons, rather than electrons, for power and efficiency. A massively operating optical brain may be the means to achieve the immense computing capacity needed for the Strategic Defense Initiative, or Star Wars. The race is on, and no one wants to be left behind.

The first shot of the photonics era was the invention of the laser in 1960, again at Bell Laboratories. Until then, those trying to work with light had to cope with disorderly wavelengths. Lasers created a source of light with a uniform wavelength. Each wave has been moving in step with those around it, giving birth to a tool of potentially immense power.

Photonics in the 21st Century will be what electronics represented in this century. This is realised so much that corporations and national research laboratories around the world are now spending billions of dollars to harness the new optical technology.

The first fruits of such efforts are already apparent in conveniences like fibre optic telephone lines, laser printers, compact disks, credit cards bearing holograms, and laser pricetag scanners in supermarkets. The forecast is that every area touched by light technology will see advances over several generations - and these advances will show one or more orders of magnitude improvements over what we know today.

Research on Conductivity

Chapter 3 underlined the fact that there is a lead time, often significant, between the physical discovery and its technological usage. After three decades of familiarity with the basic principles which characterised semiconductor materials we are now well placed to appreciate the impact of the small, low-power amplifier that replaced the large, power-hungry, vacuum tube - in business, industry and our daily life.

Back in 1839, Becquerel discovered that a voltage could be generated by shining a light on a liquid that conducts electricity. Protons in the light cause electrons to be emitted in the liquid: this is the basis of the photoelectric effect. In 1905 Einstein explained the phenomenon, using Plank's new quantum theory, and by the 1930s this became a known means for describing the behaviour of electrons.

Physicists have been working on such premises for years. With regard to what concerns organised R&D on conductivity, the early 1940s saw the first significant event of which to take notice. Marvin Kelly, director of research for Bell Laboratories, witnessed at that time a demonstration using silicon, a then little-understood ''semiconductor''. This demonstration was given by Russell Ohl.

The researcher showed a small black rectangular block with two metal contacts: when light illuminated a narrow region near the middle of this piece of silicon, a photoelectromotive force (emf) of about O.5 V was developed. This was the great grandfather of the p-n junction transistor.

Six years later, scientific research on the capabilities presented by semiconductor materials was given a boost because of Bell Laboratories interest in the eventual use of semiconductors in circuit devices, and the possibility of a solid-state amplifier. Headed by William Shockley, a theoretical physicist, the research team included Stanley Morgan, a circuit expert, John Bardeen, Gibney, and Walter Brattain.

Their mission was work on semiconductor properties: materials with conductivity between that of an insulator and a conductor. There is a barrier on resistance to the flow of electrons, though the strength of the barrier can be varied by manipulating an electrical field. This ensures that an electrical field applied to a semiconductor might be used like a valve, to regulate the flow of a current.

The goal was challenging, and it was realised that, in spite of all the work done before and during World War II, physicists were still far from a real understanding of the semiconductor phenomenon. Copper oxide, and other semiconductors on which the early work had been done, are very complicated solids. Silicon and germanium are simpler; hence the decision to try to understand these first. Research was thus directed towards fundamental problems concerning the effects of impurities.

In the beginning progress was slow, but in November 1947 Brattain and Gibney showed how one could overcome the blocking action of surface electrons with a

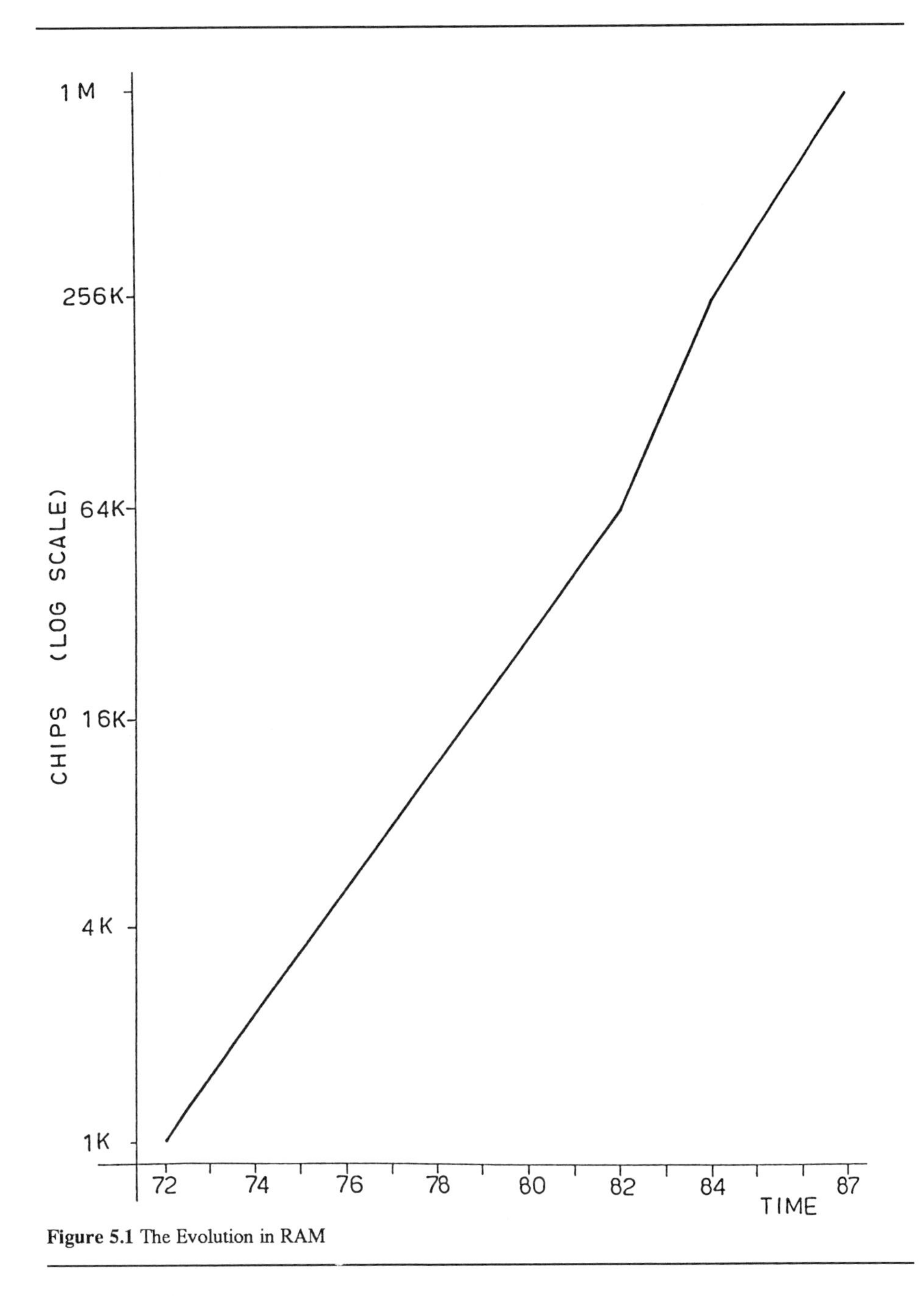

Figure 5.1 The Evolution in RAM

strong electrical field and an electrolyte in contact with the semiconductor. Within a month, a breadboard transistor was born. Bardeen and Brattain invented the point-contact transistor. This involved a plate of n-type germanium, and two contacts of gold. Shockley contributed the work that led to the junction transistor: no point contacts, just a tiny rod of germanium with a p-type layer sandwiched between two n-type layers.

The concept of the junction transistor evolved during 1947 and early 1948, though it required three more years of effort to make it work. Ironically, when the invention was made public, in June 1948, it attracted little interest: engineers brought up on the vacuum tubes were not eager to switch, and, at the time, the general public could not care less about the bolts and nuts of electronics.

Price was also an element. In 1953 a single transistor sold for $21. Thirty-six years later, with integrated circuits, the cost has become less than 0.1 cents, a very tiny fraction of the original price. Figure 5.1 dramatises the evolution in chip technology for random access memory (RAM) semiconductors between 1972 and 1987. Orders of magnitude of greater chip density have been achieved in these 15 years.

However, the market for transistors did not take off immediately. In 1954 their production just reached one million. Three years later, by 1957, 30 million transistors a year were being made - and the price was falling. By 1978, there were 10^{12} semiconductor "bits and gates" manufactured, and the number became 10^{18} by 1984. To survive, the semicondutcor industry needs to increase its yearly business by a significant margin and, given the falling prices, this means a yearly growth in physical units of 50 percent or more.

In terms of laboratory work, research has focused on two directions. To:

- Reach higher levels of circuit integration, and

- Substitute silicon with better materials which would enhance the end product.

Because silicon has its limitations, interest focused on gallium arsenide (GaAs), which can run twice as fast. At the present day, gallium arsenide is probably the most important of the newer semiconductor materials. Research into its properties and potential is being undertaken at a series of commercial laboratories throughout the world.

The Bell Laboratories have, for example, built an integrated circuit from gallium arsenide that incorporates 16 tiny lasers along its edge. That chip is used in converting electrical signals to light. The high power lasers employed for communications through fibre optics utilise complex compounds, such as GaAs, to achieve the correct wavelength with precision.

While the laboratories' work continues steadily to sharpen the cutting edge of technology, management also appreciates that scientific achievements is one thing; mastery of the world market is another. Based on original American inventions, USA companies were the first to control the chip market on a global basis. But, things have changed.

Led by Nippon Electric and Fujitsu, Japanese chip makers began pouring money into semiconductor research and development, and also into manufacturing engineering, aggressively attacking the overseas markets. By the late 1970s Japan had grabbed a 40-percent slice of the 16 kilobit (16K) random access memory (RAM) market - and it has edged up its share ever since.

In the 1980s the Japanese became the market leader in the generation of 64K chips, commanding a 70-percent market share. Then the race focused on the 256K and 1 megabit (1M) RAM. The former can store 256,000 bits of information, or enough to memorize 5,000 words of text. The latter handles four times as much. These are the chip markets of today.

Semiconductor designers also strive to pack the components of the silicon chips closer together, because the electronic circuits can then operate more quickly, since signals have less distance to travel. With memory circuits, the aim is to cram more storage space into a smaller area.

Integrated Circuits

The history of chips is one of increasing density: from the planar transistor, to the integrated circuit on a single chip (which originally included four transistors along with other components), followed the a bipolar logic array. Then came 1K, 4K, 16K, and the entire central processing unit. The microprocessor was born, and chip density continued to increase.

In terms of technology, the number of components per chip doubled annually in the 1960s. Around 1972 designers ran out of unused space on the chip for additional components. As a result, the rate fell somewhat. But, according to some projections, gigascale integration (GSI) may be achieved by the Year 2000, which means one-billion-components per chip. Figure 5.2 presents the trend and possible range of variation in superdensity. We are not there yet.

The milestone of 1M RAM, which has been crossed successfully, brought home further the issue of integration. Since the invention of transistors, the problem has been to connect them to various components in an electrical circuit. This could mean a lot of connections: a circuit with a hundred thousand components could easily require one million different soldered connections linking them to one another. Efficient solutions had to be found.

In the beginning, the only machine that could make the connections was the human hand. Then, in the late 1950s, Jack St. Clair Kilby and Robert Norton Noyce separately formulated and exploited the idea of miniaturising the components and placing them in a single medium, thus making multiple devices on a single piece of silicon.

This made the interconnections between devices a part of the manufacturing process, thus reducing size, weight and cost per active element. Eventually the idea yielded the

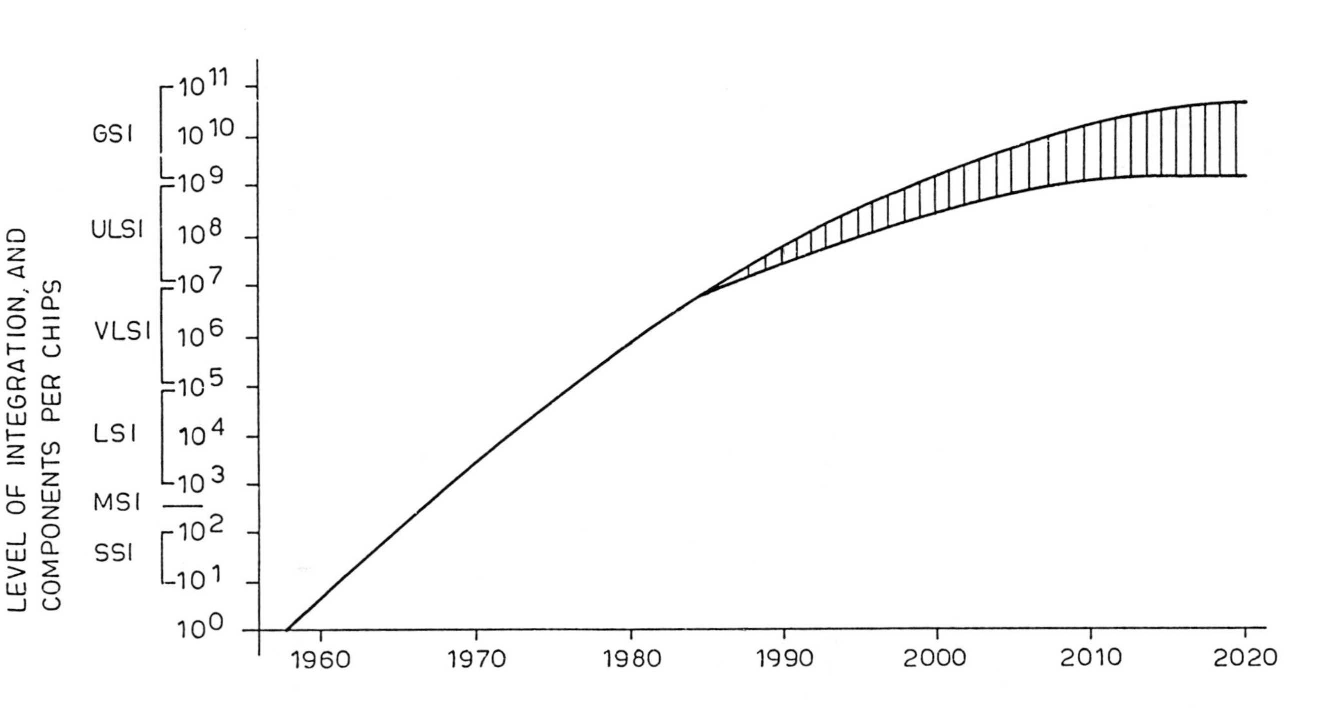
CURRENT AND PROJECTED CHIP DENSITY
LEVEL OF INTEGRATION, AND COMPONENTS PER CHIPS
GSI
ULSI
VLSI
LSI
MSI
SSI
10^{11}
10^{10}
10^{9}
10^{8}
10^{7}
10^{6}
10^{5}
10^{4}
10^{3}
10^{2}
10^{1}
10^{0}
1960
1970
1980
1990
2000
2010
2020

Figure 5.2

integrated circuit (IC), the semiconductor chip, or microchip, of which we have already spoken.

The development of integrated circuits called for expertise in two processes, etching and diffusion:

- Etching makes feasible a circuit pattern on a wafer of silicon.

- Diffusion fires impurities into the silicon, so as to achieve positively and negatively charged regions.

The goal is that of making a surface able to overcome the need for wires joining one circuit to another, by depositing a thin layer of material that would connect the circuits. Intel's fortune was built on the invention of Dr. Noyce, its co-founder.

As with every invention, key developments were required for the great step forward. By the mid 1950s, engineers had learned how to define the surface configuration of transistors by means of photolithography: etching was a developing discipline. The principle of solid-state diffusion for introducing the impurities that created "p" and "n" regions had also become known. It is, therefore, of no surprise that batch processing of many transistors on a thin wafer, sliced from a large crystal of germanium or silicon, began to displace the earlier technique of handling individual transistors.

In essence, the concept of the integrated circuit began to take shape a few years after the invention of the transistor, as a way to further exploit the characteristics of semiconductors. The body resistance of the semiconductor itself, and the capacitance of the junctions between the positive and negative regions, could be combined to realise a complete circuit of resistors, capacitors and amplifiers. From this simple concept developed the ability to pack an increasingly larger number of components per chip.

Noyce's applied research, and that of Kilby, fit properly in a technology which regarded the miniaturisation of components as an increasingly important goal. Communications, computers and weapons systems were, are, and will continue being built with thousands of parts of various types. A pressing need is to make things smaller in order to save space; to simplify the business of wiring all the components together; and to increase reliability.

At Texas Instruments, Kilby's invention of the IC cost about $25,000 up to the time when it was conceived, and another $75,000 to show that the concept would work. The first integrated circuits were demonstrated in September, 1958. A patent application was filed in February 1959, and the invention made public. Once again, the reaction was sceptical; few people realised at the time what a breakthrough this was.

The integrated circuit, as was conceived and developed in 1959, accomplishes the separation and interconnection of transistors and other circuit elements, electrically rather than physically. The required elements are interconnected by a conducting film of evaporated metal that is photoengraved to leave the appropriate pattern of connections. An insulating layer separates the underlying semiconductor from the metal film, except where the contact is desired.

The basic approaches were known for some time, but the new concept altered the engineering way of looking at things. As technology reduced costs, improved reliability and made so many more functions feasible, the pace of events accelerated. While, in the mid 1950s, semiconductors substituted the vacuum tubes as the main logical elements in electronic circuits, function per function, the 1960s opened broader avenues.

In the early 1960s the manufacturing technology made a big step forward. It became possible to realise 20 (elementary) logical functions in one circuit. With this, small scale integration (SSI) was born. SSI was a stepping stone to greater achievements; the process to higher and higher integration had just begun. This brought about a tight coupling between R&D and manufacturing engineering. It is not enough to develop the product; just as important is the development of the specialised production machinery.

Similarly, the applications perspectives have been broadened. This was helped by the fact that the integrated circuits not only answered the demands of the then current market, but also opened a much larger market that was latent at best. Computers, communications, and weapons systems require very large numbers of active circuits compared with products characterised by analog amplification (such as radios); hence, the cheaper and more reliable the component is, the further one can go system-wise.

But, technology has its believers and disbelievers. When some chip manufacturers first predicted 1-gigabit RAM chips, able to store a billion bits of information - or about 4,000 times as much as today's standard 256-kilobit chips - many said that this was beyond the physical means. Today, the 1-gigabit chip is not only considered feasible, but projections have it that it will be on the market in less than 10 years. As an example of what this means, a 1-gigabit chip could hold 4 1/2 hours worth of digital sound.

Not long ago, at the International Solid State Circuits Conference (February 1987), NTT showed a prototype of a 16-megabit RAM, able to store 64 times as much as the 256-kilobit chip. At this same Conference, USA manufacturers introduced prototypes of less powerful 4-megabit RAM chips. These, too, were considered impossible in the late 1970s.

The Wave Toward Megaintegration

Japanese technologies are ahead of their American counterparts in terms of *megaintegration,* because their government and their industry focused in time on the right infrastructure. The name is X-ray technology, with the No. 1 country being Japan and the No. 2 West Germany.

As a result of a headstart, Japan is rapidly pulling ahead of the United States in the development of the crucial X-ray technology that will be used to manufacture computer chips in the mid-1990s and beyond. At stake is American competitiveness in a number of vital areas, from military technology to consumer electronics.

To understand the impact of the new thrust, let us recall that chips are now made using light to etch circuits onto silicon wafers. The new developing technology, called X-ray lithography, can make denser chips that will ultimately be able to store three orders of magnitude more data.

Just because the challenge is so great, American computer experts are now urging the creation of a national research program to ensure that USA manufacturers are not shut out of the world semiconductor market. The background issue is that megaintegration technology is considered to be too expensive for individual companies to develop alone. The Japanese, for instance, have already set up a joint industry- government programme that will spend nearly $1 billion on X-ray lithography.

Contrasted to American investments in this field, $1 billion is a lot of money. The United States has spent an estimated $50 million to $100 million, and is planning to spend an additional $100 million. That makes it 20 percent of the corresponding Japanese funding, at best.

IBM, among others, seems to be so concerned about this oncoming challenge of the mid-1990s that its officials are rumoured to have approached American chip makers and offered to share costly equipment from IBM's own X-ray technology research. ''Investment in this vital technology is not a matter of choice, and it must happen soon'', said Jack D. Kuehler, IBM vice chairman, and the company's highest-ranking engineer. ''Other players in the industry, particularly in Japan, have already discovered this and are on their way''.*

In fact, as of this moment, IBM seems to be the only major USA manufacturer with a real commitment to developing the advanced chip-making technology, and it is said that the cost of developing X-ray lithography was already straining its resources. Experts estimate that the cost of developing a chip-making plant using X-ray technology may jump to $500 million in the middle of the 1990s, from the less than $100 million that it costs today.

The Bell Laboratories, of AT&T, should have been another key player in the X-ray lithography field, but in 1987 it scaled back its support for such research. The future challenge is further magnified by the fact that:

1. Semiconductor experts think that industry is close to exhausting current chip-making technology. Hence chip production will be forced to turn to X-rays, which have a shorter wavelength than light and which, therefore, can be used to draw finer electronic circuit lines.

---

* International Herald Tribune, December 14, 1988

2. The most promising source of X-rays for advanced chip making is a synchrotron storage ring, a costly circular particle accelerator that generates radiation at the correct wavelength and intensity.

The Japanese seem to have 10 synchrotron storage rings under development for use in making the megaintegration chips, and five more in planning. By contrast, in America, only IBM is so far building a ring as an X-ray source for making commercial chips.

At the Japanese Photon Factory, a research laboratory, a number of companies, including NTT, Hitachi, NEC, Fujitsu and Toshiba are working cooperatively on X-ray lithography. Another two storage rings are being developed at the Electronics Technology Laboratory, a national research effort. The conquest of the world markets near the end of this century is well launched!

As with megaintegration, the cost per bit drops sharply, the Japanese vision of the future of RAM will put megabits of memory into major appliances. To reach into the 21st Century markets the Japanese are focusing on four major technologies that they want to add to their high-end appliances:

● Optical storage,

● Voice recognition,

● Video image processing, *and*

● 16-megabit or more powerful chips.

While the 16-megabit RAM integrated into major appliances is still to be seen, let us not forget that chips made the $7 pocket calculators possible. They have already brought electronics into house appliances. They made feasible significant advances in a range of other technologies. They became basic building blocks in all man-made products, from watches to medical instruments, and the space shuttle.

Chips did not eliminate the central mainframe computer, but shrank it in size and in cost. Most importantly, they helped spread-out low-cost computer power onto the users' desks through personal computers (PCs). This provided both infrastructure and stimulus for advances in business, such as decision support and expert systems.

Using the same chips as basic components, computers and communications merged. It became apparent that they served each other very intimately. IBM got engaged in satellites (SBS), then merged SBS with MCI, a communications company. It also bought Rolm, one of the foremost telephone equipment manufacturers. After deregulation, AT&T got into computers from the micro-and minicomputer end, capitalising on its experience from telephone switching centres.

Some companies made a fortune with computers: IBM and Digital Equipment Corp. are the better-known examples. Others lost a fortune, e.g. RCA, GE and Xerox, to the tune of half a billion dollars apiece.

All companies, winners and losers in this tough, restless competitive market, kept on investing in chip technology. The USA Department of Defense has been a great sponsor. More recently, the frontiers of integrated circuit technology have been expanded by the Defense Department's very-high-speed integrated circuit (VHSIC) programme, which found its way into supercomputers.

This multiphased VHSIC effort accelerates the trend towards higher speed, denser circuitry for microchips, through innovative design, and fabrication techniques. The VHSIC chips permit complex functions, such as those in sophisticated signal processing and artificial intelligence, to be built into compact systems.

VHSIC developments are in line with a general trend in making the basic electronics circuitry faster, smaller and cheaper. These aims are always at the top of a semiconductor designer's list. As stated in the preceding paragraphs, since the early 1960s it has been the pattern that every two or three years the complexity of a silicon chip doubles while the price falls - though the most dramatic price drop with semiconductors has been experienced during the late 1960s to early 1980s (Figure 5.3).

As technological progress continues at a significant pace it is expected that, by the early 1990s, the fantastically thin lines on integrated circuits will shrink to invisibility: roughly 1/200 of the diameter of a human hair, or 0.5 micron. That is too small to describe with today's chip production equipment, so many companies are betting on X-rays to do the job.

Since their wavelength is much shorter than those of the light sources now used to print chips, X-rays can easily handle submicron lines. Japan and West Germany have launched major efforts to develop X-ray lithography, in hopes of leapfrogging Silicon Valley, and they have made a head start in their efforts.

Microprocessors

As scientific advance altered classical design concepts, there has been a major shift towards digital systems propelled by advances in microelectronics. An analogue equipment cannot handle large numbers of microcircuits, whereas a digital system requires them. A pocket calculator contains 100 times as many transistors as a radio or a television receiver - the emphasis is on *processing power*.

Such emphasis led a new thrust to applied research. As with the integrated circuits, Intel and Texas Instruments co-invented the microprocessor. The basic idea was to put some extra functions onto a single chip. Within a decade, this simple need created a torrent of opportunities.

In the summer of 1969 a Japanese firm asked Intel to design integrated circuits for a new family of calculators. Ted Hoff studied the Japanese design. It was a rather complicated arrangement, with a lot of logical functions being distributed around the calculator. As he examined his alternatives, Hoff was impressed by how much of this could be centralised on a memory chip.

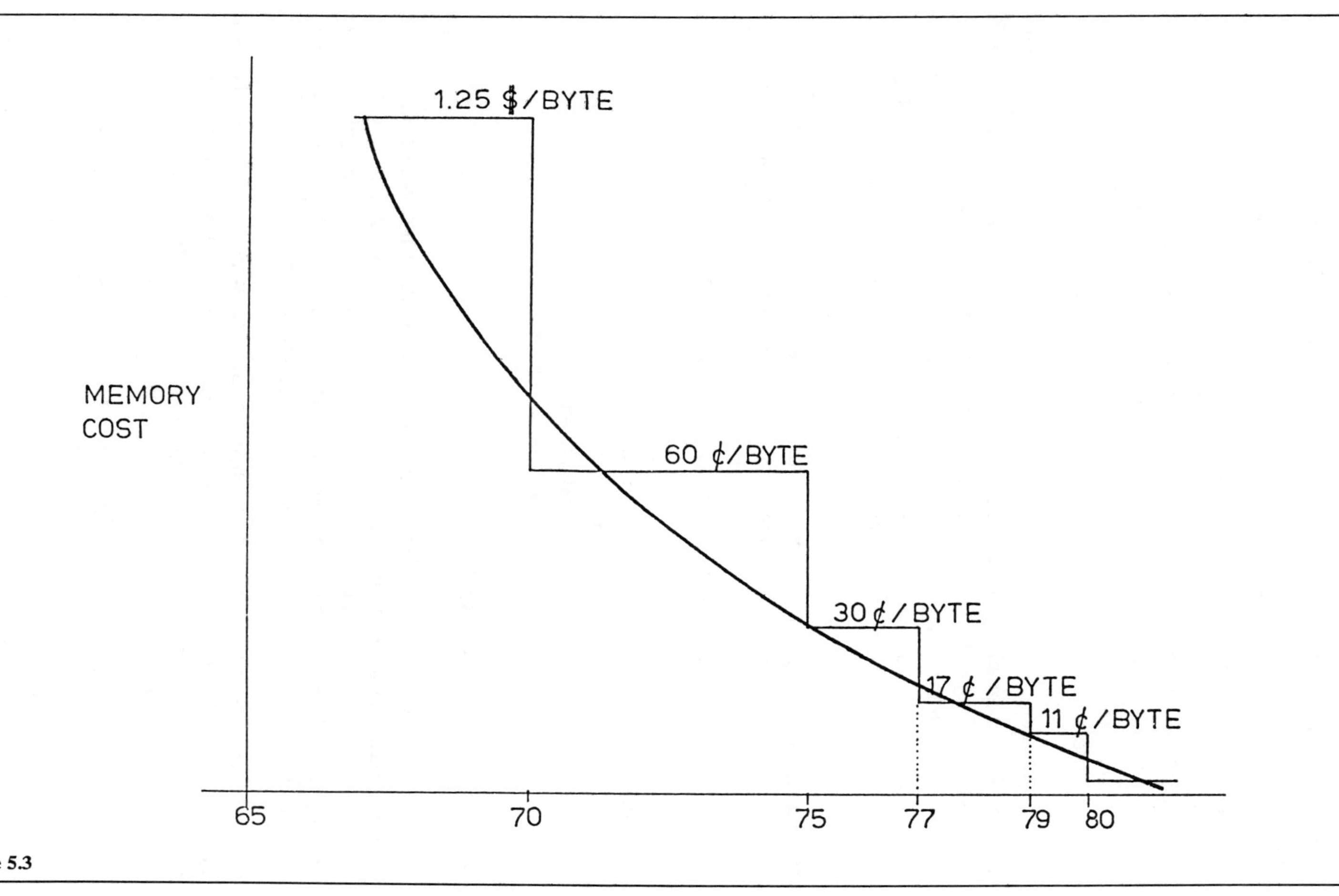
1.25 $/BYTE
MEMORY
COST
60 ¢/BYTE
30¢/BYTE
17 ¢/BYTE
11 ¢/BYTE
65
70
75
77
79
80

Figure 5.3

The Hoff solution was presented to the Japanese clients of Intel as a way of meeting their cost budget, but also as a way of making the system more elegant. Yet, in spite of the ingenious idea, not much happened until the next spring when Frederico Faggin arrived from Fairchild. While Hoff designed the computer architecture, Dr. Faggin developed the circuits.

Let us recall that, at that time, calculators typically needed 10 chips. The Hoff-Faggin solution took three. Meanwhile, in late 1969, a Texan company went to TI with practically a similar request. The Texas Instruments'engineers, too, came up with a single-chip processor and patented their chip.

The *microprocessor* was born to simplify the computer circuitry, and to cut costs, but in the process it opened a new horizon for progress. The men who engineered the great advances in the semiconductor line, like those behind the new physics, the quarks, gluons and leptons, have not been working in an ivory tower. They are practical scientists applying themselves to problems with a reasonable chance of being solved.

Hoff, Faggin and the TI microprocessor designers have been ingenious risk takers, willing to make mistakes, to recognise them and to learn from them: this is how the avenues of further progress are being paved. But, the risks were reasonable. Imagination and hard work is the best policy for getting results without waking-up the Frankenstein of biotechnology, and committing errors which may be irreversible.

The technological benefit has come in waves. In the Intel line of microprocessors, the first wave took 14 years, and led from the original 4040 to the 80386. In the process it became one of the two dominant engines in the market (Figure 5.4), only to be superceded by the 80486 announced in mid 1988.

The 486 inaugurates a new era in microcomputing: *personal mainframes.* It is packing more than 1 million transistors, delivering the number-crunching of a low-end IBM 3090. With the 486 hitting the market in 1989, the first microcomputers incorporating it should show up around 1990. They will be rather expensive, and the first deliveries the 486 will cost around $1,500. But prices will soon drop.

The greater impedence of a chip of the power of the 486 is not its price. It is the basic software which will run on it, particularly the operating system (OS). As of early 1989, OS/2 still exploits the power of the 286 chip - a 1984 announcement. The OS is practically two generations behind the current state of chip making, and catching up is not easy. Unix and Mach would most likely be the OS solution.

The mid to late 1980s have also been characterised by a time of uncertainty, where the Reduced Instruction Set Computer (RISC) seems well on its way to becoming the dominant design for some years. VHSI will most likely play a role during the next decade. After this will probably come the time of photonics, and of the biochip.

It is, however, proper to underline that, while design approaches change as technology evolves, corporate efforts during the last three decades of this century

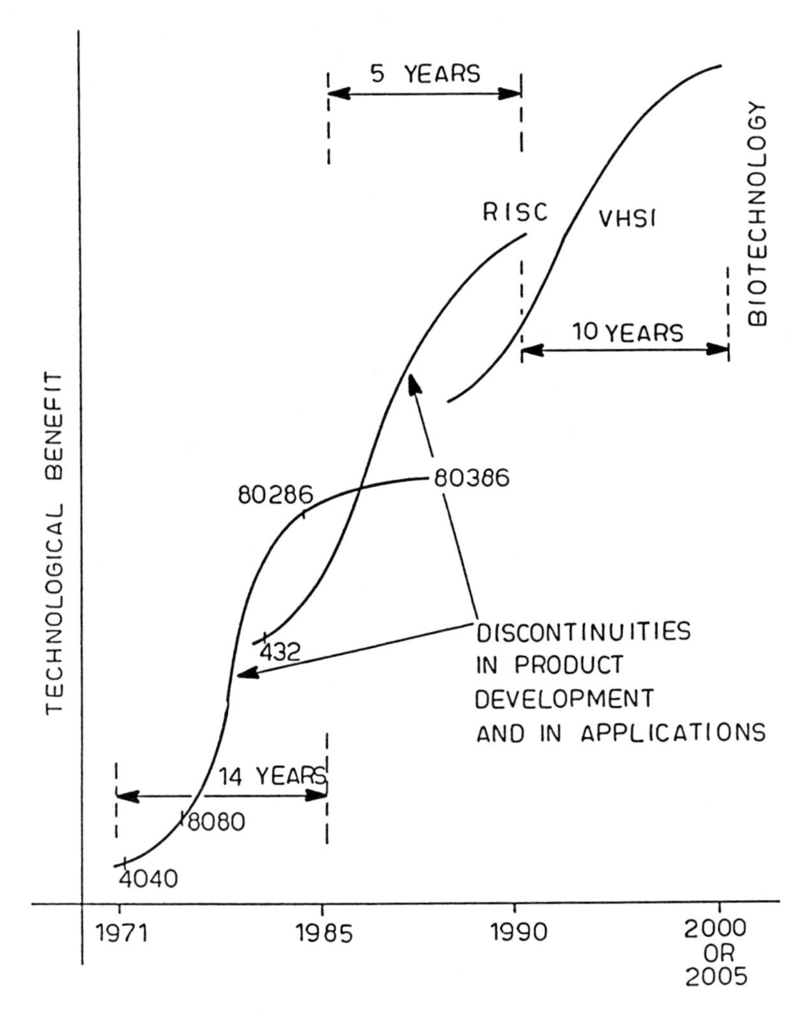

Figure 5.4

greatly capitalised on the power of microprocessors. Since the early 1970s, these computers-on-a-chip marked a new phase in electronics and computing - but also in a myriad of products which have become the motor power of our civilisation.

Microprocessors lowered the cost of computing to a level where it could be integrated into almost any application imaginable. And they set the grounds for the appearance of equipment as diverse as electronic games, new aircraft navigation systems, aids for the disabled, heart pacemakers and a totally new generation of industrial controls.

Nevertheless, the microprocessor's main contribution does not lie in the cost reduction that it makes possible, nor in the fact that it serves as an aid to miniaturisation. Its top strength lies in its potential for taking computing power and placing it in the hands of ordinary people, at an affordable price.

Who controls the microprocessor markets, and who stands the best chance for the future? Microprocessors are not the largest market segment. That distinction belongs to the memory chips. But micros are used in many different products, and are often sold with a bundle of associated chips. Besides, many chips that are separate today will tomorrow be fabricated with the microprocessor itself, in one piece. CPU-over-Memory at every node of a hypercube is an example.

We said that the Japanese have 70 percent of the market for computer memory chips, but Japan's chipmakers hardly dominate the world market for microprocessors. They account for 30 percent of microprocessor sales (which is a lot), compared with 63 percent for USA companies and 7 percent for the Western Europeans. Even these numbers overstate the Japanese strength, because 99 percent of all microprocessors sold were designed by American companies and are made under second-sourcing licence.

Intel's microprocessor designs account for 60 percent of industrial sales. Motorola's have about 30 percent of the business. The remaining 10 percent is split between chips designed by National Semiconductor, Zilog and others. The evolution of the Motorola product line from its original 68.000 chip to the 68.020 is shown in Figure 5.5.

Developing and bringing to market a new and different family of processors costs a lot. For its V series micros, NEC has already spent an estimated $500 million to get the six members of the series off the ground. But money alone would not buy market share. Becoming a key player in the market of the micros calls for:

- A great deal of software, *and*

- First class marketing skills.

Potential customers want to know which micro runs the most software. Software sells hardware. Not vice versa. Software is a growth market on its own merits. In the years to come, solid state software will provide a major advantage in selling microprocessors.

The Day-to-Day Struggle for Dominance

American semiconductor companies have kept Japan's microprocessor makers at bay with innovative designs, but the Japanese have countered with leadership in the complementary metal oxide semiconductor (CMOS) chipmaking process which, over the years, gained ground. NEC is also betting on the fast-growing market for 32-bit microprocessors.

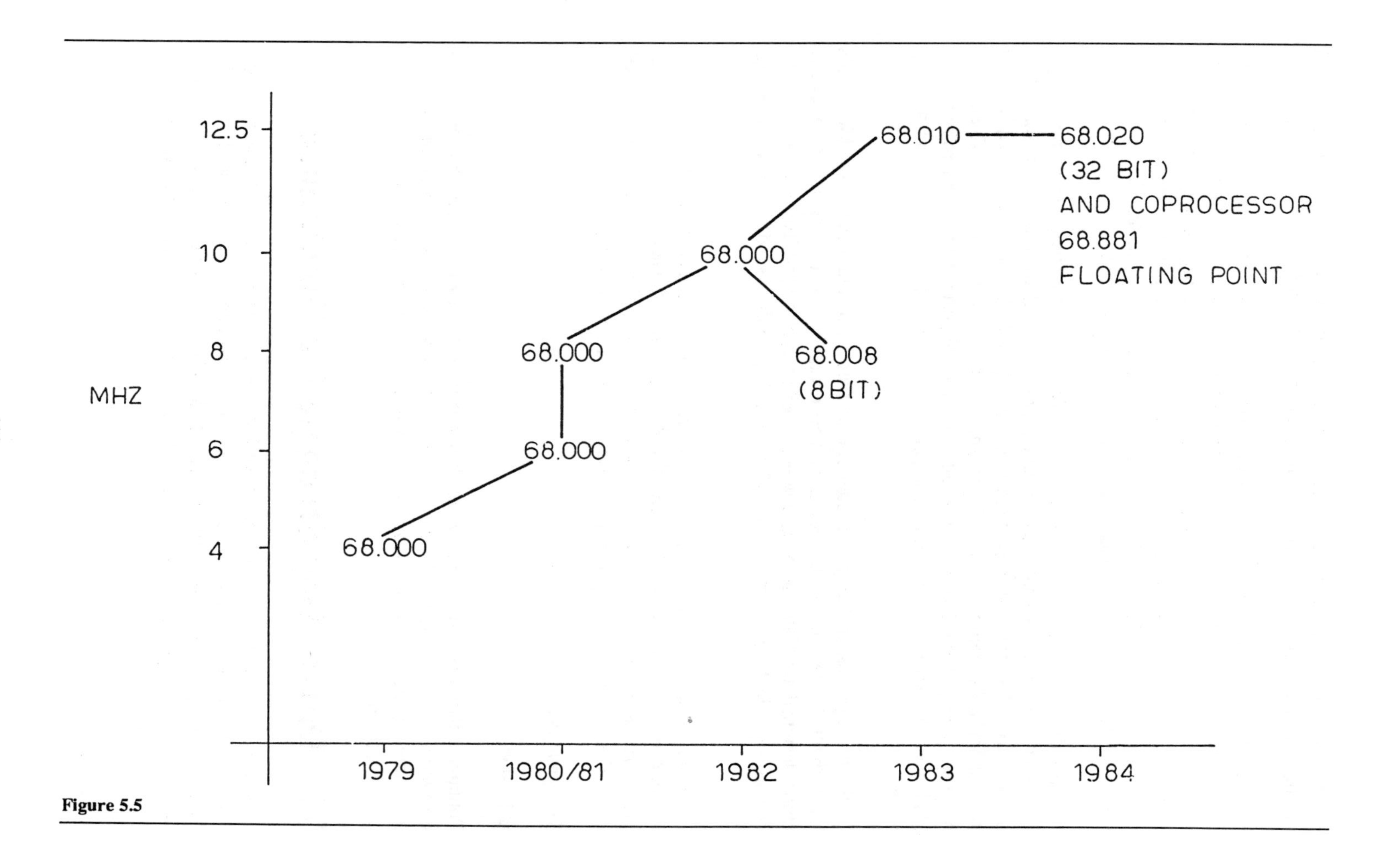
MHZ
12.5
10
8
6
4
68.000
68.000
68.000
68.000
68.010
68.020
(32 BIT)
AND COPROCESSOR
68.881
FLOATING POINT
68.008
(8BIT)
1979
1980/81
1982
1983
1984

Figure 5.5

Since Intel is making its 32-bit micro compatible with its older products, it has to compromise the new chip's design. NEC is freer to target maximum performance. Besides:

- 32-bit chips are somewhat easier to program than previous generations, and

- Their implementation keeps in mind the need to switch software from one type of 32-bit micro to another.

This raises hopes that designers will not be as locked into the particular chip they select as they are today. Something similar can be said of RISC technology.

USA chipmakers see another compelling motive for Japan's interest in micros. "The Japanese want to win in semiconductors, but they also want to run the computer industry", some sources tend to assert. There is, of course, a close connection between user needs, the drives in a competitive environment and high technology.

But nothing is static in the computers and microprocessors business. Every 3 to 4 years the major manufacturers bring out new lines of computers which are more powerful than the ones they replace. Typically, these need better microprocessors. If a company falls very much behind in producing the latest engines, it will have a hard time catching up. But, also, failure to endow its machine with plenty of sofwtare has serious consequences.

Among other reasons, software support is so important because, without doubt, the most important contribution of the microprocessor is that it broke down the psychological barriers that computers had inadvertedly created at the time of their introduction. It also changed the behaviour of highbrow computer professionals who, over the years, got into the habit of looking at users as underlings.

In the early 1950s computers inspired widespread awe. The press dubbed them "giant brains". They were massive, but very slow by today's standards. Some high priests knew how to handle them, and jealously kept the secret to themselves. This was no policy designed to gain the users' trust - or, for that matter, to guard the high priests from ossification.

For more than a decade computers were very expensive, and few companies could afford them. Typically, one data processing centre served the entire organisation. Its machine lay behind a glass window, people in white robes were attending it. Those who wanted to learn about the beast had to look through the plate glass, as in the zoo.

To get a solution to their problem, users had to reach the resource through intermediaries; organisation experts, systems analysts and programmers proliferated. And so did the costs. Users felt like supplicants, and were often slowly served through central staff. They still are.

So, users were very receptive to new solutions. In the late 1960s and early 1970s more flexible approaches were provided by minicomputers. In the late 1970s and early 1980s microcomputers became the best tools for user service. Small in size and in cost, the new engines became powerful, and they connected online to central databases - both private and public.

Through Fourth Generation Languages (4GL) microcomputers are programmable by the user himself. But, while the languages are the visible part of the system, the real motor in this race toward competitive strides has been the microprocessor and the basic software which runs on it - from the operating system to database management systems (DBMS).

Microprocessors, computer systems, basic software, languages and a valid applications library are interwoven. They are also dynamic subjects. Things change fast in this industry. In the computer business, your market is your fate. And the microprocessor's market is the whole world.

Today, the microprocessor is much more than an industrial strength engine. In a knowledge society, it is a key to dominance. It is also a big driver in the market place. No wonder all Western nations try to promote their own companies in the spotlight of microprocessor development and sales.

But few players succeed in this effort. The entry price is very high, and so are the stakes, as well as the risks. Sometimes, the winner takes all. In other cases, there is no clear winner. Design technology has been making enormous strides, and each new development brings forward impressive facilities. It also uses a great many of the tools high that technology makes available. Changing boundaries on the floor plan of a chip is like changing the borders of a country - it requires a lot of negotiations and diplomacy; but also knowhow, and a little bit of luck.

New methodologies are being developed to help in this mission, while others known, and then forgotten, are reinvented. An example is *reverse engineering*, propelled by the Russians. Through an electronic microscope, they take a chip apart, study its properties, and then emulate or improve upon it. As it happens, "improve upon" is the speciality of the Japanese.

While the basically developmental trend is expected to continue, we are also at the very beginning of a totally different era. The *why* is self-evident, through the steady process of evolution. The *how* calls for a reference to another basic invention which is seemingly unrelated to electronics and yet, in reality, it constitutes part and parcel of the same major trend.

A New Era of Photonics

Photonics is widely expected to form the backbone of much of the next generation of information-based technology. The particle of light can attain greater efficiency in storage, processing and transmission than electrons ever achieved.

Photons travel *faster* than electrons. They have no mass and:

- Unlike electrons, which interfere with each other,

- Photons can be made to pass through each other unperturbed.

Also, light behaves both as a particle and as an electromagnetic wave. This means that optical devices could be based on much the same operating principles as those already used in electronics.

Quite importantly, we are discovering the limits of electronics. One is the speed at which electrons travel through semiconductor material. So long as electrons remain the information carriers and switching elements of computers, this sets an absolute limit on the speed of computing, and hence on machine power. Electronics has not reached that limit yet, but it may be drawing close.

The alternative is *lasers*, the acronym for *light amplification by stimulated emission of radiatio*n. This is a means for amplifying beams of light by focusing light waves of high quality in a *narrow, intense beam.* In this context, light includes the invisible bands, infrared and ultraviolet, which border on the visible spectrum.

A laser beam differs from ordinary light in three particular respects: (1) All its waves have the same wavelength; (2) They are in phase with each other, thus one wave crest cannot neutralise the throughput of another wave, as their motion is synchronous; (3) Laser beams can be switched on and off in extremely short intervals.

The last quality opens significant possibilities for switching circuitry, as we will see in the next section. Three areas of photonics technology implementation exist under the current state of the art. The chronological order of their development is:

- Transmission

- Storage, *and*

- Switching

So far it is the transmission side that has dominated the use of the *photons*, employing them more as compact carriers than electrons, transmitting massive amounts of data as laser pulses through very thin glass fibres.

Optical communications networks are beginning to spread within and between cities around the world, replacing less-efficient copper wires. In a similar sense, optical disks first complemented, but will soon start replacing, magnetic disk technology, which is now 30 years old.

It is no coincidence that information transport through photonics has been the first to take off. When the laser was invented in 1960, the immediate idea was to utilise its powerful transmission capacity. However, it soon became evident that it is not practical to transmit laser beams through the atmosphere in the same way as radio waves. The air contains dust particles, moisture, often rain or snow crystals, which can be a considerable hindrance to laser transmission.

A further obstacle is that air frequently forms layers. Reflections occur at the intersections of temperature layers, distorting the laser beam. A laser beam can travel several hundred metres and, in good weather, several thousand. But, the optical attenuation becomes so great that the signals are no longer readable when they reach the receiver.

In 1966, research scientists advanced the idea of propagating laser beams through glass fibres. The latter were already known in their function as waveguides, mainly as endoscopes and for laboratory tests. The endoscope disperses light inside the organ being examined, transmitting a picture to the doctor's microscope.

The problem, however, was that the waveguides available at that time were suitable only for very short distances. The glass being used was not transparent enough to permit transmission over a longer range. Thus:

- One of the prime requisites for the development of optical fibre communications was high purity glass, *and*

- The same was true of an infrastructure based on optics.

Glass has been known for 5,000 years. It is generally made by fusing quartz sand at high temperatures. In the physical sense it is a liquid with properties which make it rigid at normal environmental temperatures. But, high-purity glass is another matter. It is not made by melting.

For high-purity glass technologists used silicon chloride gas. Combined with oxygen, and a certain quantity of germanium chloride vapour, silicon chloride is heated in a tube. The silicon combines with oxygen to form silicon oxide. The powder obtained is heated further until it liquifies into glass. This is one of the processes possible for making high quality glass.

There is also the need for advanced solutions in optics, as we will see when we talk of optical disks (Chapter 10). Fibre optics technology is based on the controlled reflection of light within the optical fibre. Light only travels at 300,000 km (186,000 miles) per second in a vacuum. In all transparent media, it is slower. The material it is required to traverse poses obstacles to the stream of photons, which slows it down.

The use of metrics has been just as vital. We use the term *refractive index* when defining a medium. This is the ratio of the velocity of light in a vacuum to the velocity of light in a specific transparent medium. Expressed mathematically it is:

velocity of light in a vacuum Refractive Index = velocity of light in the medium

Within the implementation domain of laser technology, the reflective index is metrics of quality, and a very important one in media choices.

Once the technical problems have been solved, the use of photonics as communications carriers took off. Today, in America, more than one third of the cables connecting telephone city centers to the different user groups are made of optical fibres. This share is growing fast, propelled by cost/efficiency, convenience and reliability. Older transmission materials, such as copper, are paying the price.

The first transatlantic cable made of optical fibres was laid in 1988. It can transmit 40,000 simultaneous conversations - or four times as much as the last copper cable installed in 1975. It is significant to note that, in the middle to late 1970s, fibre optic technology (still in its infancy) could transmit only 90 simultaneous conversations.

It is also interesting to record that AT&T projects for the fibre optic cable in the Atlantic a life cycle of 20 years, with two recoveries during that period. ''However'', said a senior executive, ''by the time we first recover it, in about 7 years, the currently embedded technology will be so obsolete that we expect to be faced with a replacement problem''.

The next big step is bringing optical switching and data transmission inside computers to speed up calculation. This will first be achieved with chips that combine electronic with photonic switching and transmission. Scientists in many laboratories are experimenting with chips of that sort.

Under today's technology, semiconductors are not only utilised as receivers, they also play a role as ''semiconductor lasers''. Current solutions are a hybrid of electronics and photonics. In such devices, a light ray is generated by a complex layered construction of different semiconductor materials, such as gallium arsenide and gallium aluminium arsenide.

In other words, to open up the domain of photonics implementation, laboratories had to conquer a major disadvantage of the photon - its inability to carry an electrical charge. The answer has been to channel the light through waveguides etched onto chips made of materials with unusual optical properties. These materials change their ability to conduct light when an electric field is applied to them. Using lithium niobate, engineers have been able to make a wide range of optoelectronic modulators, switches and other devices.

Chips that allow photons to be switched would eventually lead to ultrafast computers that work entirely through optics, although some scientists are still sceptical about this. However, as we will see in the following section, optical computers are still a decade or more away. When available, they could work 1,000 times faster than their electronic counterparts - and we all know what a major achievement it is to improve computing speed by three orders of magnitude.

GaAs is a different story. As early as 1962, scientists proved that its crystals could generate the coherent light emissions essential to lasers. However, the first prototypes could operate only in liquid nitrogen cooled to minus 328°F. For the invention to have any practical value, it had to work at room temperatures.

The challenge was enormous, but Hayashi - a Japanese scientist working in the USA - built, after four years of effort, the world's first reliable and practical semiconductor laser. This has been as important to optical science as the transistor is to electronics. Today, as technical director of the Optoelectronics Joint Research Laboratory, sponsored by MITI, Hayashi is progressing to the next level of laser research, but section 8 also tells of the strides made at Bell Laboratories.

The creation of hybrid optoelectronic integrated circuits that combine:

- The signal-transmitting functions of photons powered by lasers

- With the signal processing abilities of electronic circuitry

Such a solution presents considerable advantages not only with regard to the end goal but also in terms of the intermediate steps. Smaller, faster integrated chips will be incorporated in everything from consumer electronics to the next generation of very high speed computers, thus creating a new quantum step.

Research at Bell Laboratories

We said that optical switches, known as *transphasors*, could, in theory, be able to operate 1,000 times faster than electronic ones. But, for the present, transphasors are primitive. They still have to be actuated by too much light, and they are bulky. They have not yet been miniaturised on chips in the way that electronic switches have.

The preceding section underlined the fact that optical switching works through the hybrid approach of using gallium arsenide. The most advanced laser technology at Bell Laboratories is based on this medium, with two variants:

- Channel substrate (CSBH) by chemical etching, *and*

- Capped mesa (CMBH) with active and grading layers.

At Bell Laboratories, wafer growth followed distributed feedback laser (DFB) technology. The laser attained is smaller than a grain of salt, like a P/N junction, though the latest advances are, for the time being, mainly applicable to land-based systems. Intentionally, in the trans-atlantic fibre cable older technology has been used, for greater reliability.

Another frontier to be crossed is the massive development of production capabilities vs. laboratory solutions. At Bell Laboratories, current production facilities are fully automated, run by a robotics system. ''Human reaction is not fast enough for 1 Micron structures'', suggested the lab director.

In July 1988, AT&T was also proud of the headway its researchers were making in their work. The statement has been made that, back in 1985, Bell Laboratories were about 2 years behind the Japanese but, since 1988, they have achieved a significant advance. Now the Japanese come to Murray Hill to learn the latest in laser technology!

Furthermore, an important place in the list of demonstrations at Bell Laboratories was given to *optical switching* proper. Optical switches are studied from a multiple perspective. Most interesting is their possible contribution to an optical computer far more powerful than the electronic machines that we use today. The new technology now in advanced research:

- Features some of the shortest pulses of light ever made, *and*

- Supports 10-12 switching speeds vs. 10-9 for gallium arsenide.

The difference is the stated three orders of magnitude, with all the consequences that this could lead to.

The medium is *trans-Polyacetylene* (CH)x, and the process is known as *charged-solution dynamics*. This is a:

- Thin film, single crystal approach

- Acting as fast optoconductor

though scientists are still trying to understand why. *Solitons* may be responsible for the associated conductivity.

In terms of physics, trans-Polyacetylene is unusual because of its degenerate ground-state configurations. These result from different phases of bond alternation, between single and double bonds. Transition regions between the domains of these configurations are thought to be mobile and soliton-like in nature, having analogs to several other branches in physics.

At the Bell Laboratories, the research is led by Dr. L. Rothberg, who may one day become as famous as the inventors of the transistor. There are still many problems to be solved, one of them being that picosecond switchings are not air-stable. Oxygen attacks the structure.

Still, progress made so far in optical switching has given exciting results. Both direct photogeneration of charge-soliton pairs and neutral- to charged-soliton conversion have been experimented with.

- The former are created by intrachain absorbtion and decay on a subpicosecond time scale.

- The latter are formed following interchain photoexcitation.

These constitute the dominant population of charged solitons at times greater than 20 ps.

Confinement of these excitations to one dimension helps to make potentially useful nonlinear optical materials. The nature of the resulting *excitons* and *polarons* in the organic semiconductors is also of great physical interest.

Work on interchain excitation of electron-hole pairs in oriented materials is in progress. It promises to be important in interpreting the photoconductivity of Polyacetylene. The gates to the future 10-12 or faster switching capabilities are not yet wide open. But, ongoing research is cracking the tough nut of physics.

Global Sense of Competition

The list of contenders in photonics research and development reads like a *Who's Who* of high technology. AT&T is a giant in the field. The Bell Laboratories, with a research budget of some $2 billion annually, now conducts more research in optics than in its original core pursuit of electronics. In the USA alone, the list also includes IBM, 3M, Texas Instruments, NCR and GTE.

Quite understandably, the optical competition is global. The USA spent at least $1 billion in 1985 on optical research and development. It grew to an estimated $1.6 billion in 1989. Is this too little or too much? The answer can be found in the competition. In the same period. Japan spent roughly $3 billion, and is considered to be the international leader in most optic fields.

Early breakthroughs in the optical research battle have also been achieved in West Germany, Britain, France and Canada. Russia may be conducting the largest optical computing research program of any nation, spending on that domain five times as much as America - according to some estimates.

The most ambitious competition is the race to build the first optical computer which, as we said, was considered a practical impossibility only a few years ago. AT&T took a significant step towards that faraway goal in June 1986 by introducing the first optical equivalent of a transistor.*

If the AT&T, and other similar, projects are successful, we will have computers that run on light instead of electricity. The advantages of computing with light are many, the first being speed. The second great optical advantage comes from the fact that photons have no charge or mass, as has already been stated.

- They are often thought of as waves rather than particles.
- Unlike electrons, photons have little effect on other nearby photons, and can even pass right through each other.

This phenomenon excites computer scientists because multiple beams of light in an optical switch could remain separate, providing a significant opportunity to apply

* Time Magazine, October 6, 1986

innovative computer architectures for parallel processing, which is the currently dominant design concept.

Another advantage is that optical switches might be able to operate in more than the off/on (O, I) states of transistors. Additional functions could be created by having increasing, but discrete, levels of laser brightness in an optical switch, thus leading to richer, multivalue logical systems.

Thus there is an impressive list of reasons why the key to the future is to go from electrons to photons. Bell Laboratories has a reputation for innovation, having invented the transistor, the maser, the laser, and many other devices that have changed the art of manipulating light. Therefore, it can be trusted to present concrete results with the optical wave of technology.

As Section 7 demonstrated, considerable effort focuses on creating an optical analog of the transistor, which is still the muscle behind computation. In its first conception, it would switch light on and off in a way similar to that in which a transistor switches electricity. Advances are already being reported in the creation of flashing switches, the transphasors, that would lie at the heart of optical computers.

Progress takes place by steps. Today, the universal goal among scientists in the field is to shrink optical switches and pack them tightly together. We have seen what is being accomplished at AT&T. None of the other major players wants to be left out of the race, and surprises in terms of results which come faster and go further in accomplishments cannot be ruled out at any time.

But, for at least the first half of the 1990s, the toughest competition will be to conquer the market of the already-established fields of photonics implementation. Both optical fibres and optical disks represent a tremendous market potential. By 1991 optical disks are expected to capture 20 percent of the $40 billion market for computer storage devices. And this is only a beginning.

The optoelectronics market is still in its initial phase of development. For 1989, the industry is expected to bring in revenues of about $20 billion. By comparison, the electronics industry anticipates about $250 billion in revenues.

Projections regarding the coming decade bet on photonics. Optics is expected to grow at a pace of 30 percent to 50 percent annually for the next several years, reaching sales of more than $50 billion by the early 1990s. In contrast, the market for purely electronic products will most likely grow by just 5 percent to 10 percent annually. By the end of this century, *photonics will be eclipsing electronics.*

6

The Need for Implementing Big Systems

Introduction

The search for better solutions to the information problems facing a modern organisation is not only technology driven. It is primarily promoted by the desire to improve management's ability to respond to market challenges, serve the company's clients in an able manner, provide its own people with timely information for their decisions, and lower its production and distributed costs. These are actions necessary for survival.

The thrust to newer, more efficient, technology is market driven. This is true all the way from the stimulus to invent the microprocessor, to the demand which currently develops for supercomputers. As my recent research in America indicated:

- *Supercomputer hardware spending* is expected to increase to *$1.8 billion*, in the USA alone, by 1992.

This is a projected growth of 257 percent from the estimated $700 million worth of supercomputers sold in 1988.

- *Supercomputer software sales* are expected to reach *$2.2 billion* in the USA by 1992.

That is a 250 percent increase from the $880 million of software sold in 1988.

Taken together, software and hardware expenditures indicate that user organisations have been spending, in one year, $1.58 billion for *supercomputers*. Given the nascent mass effect, any organisation is advised to ask why the drive toward supercomputer power? Where are we standing? Which are *our* goals in technology? What are we going to do about capitalizing on the ongoing advances?''.

Big systems are not small systems which have outgrown their original size looking for a new shell. The applications requirements have changed, sometimes in a radical way, as we will see in Chapter 7 when we talk of *Realspace*. Big systems call for a different approach in terms of:

- Concepts
- Design
- Implementation, *and*
- Maintenance

than the one we knew with smaller systems during the last 3 or 4 decades. To properly design big systems, and then exploit them, we must re-educate ourselves. Otherwise our investment will have trivial returns.

Thirty-five years of experience with computers and communications gives plenty of evidence that able solutions are not only a matter of technological investment and of making the right choices about where to spend the next billion dollars. The two pillars on which rests a successful communications and computers system are educational and organisational, and we must prepare for their role.

In a 1986 study of 16 of the foremost "factories of the future", Richard Walton of the Harvard Business School found that, in companies with a serious commitment to information technologies (IT), corporate activities became more interdependent. At the same time, faults and contributions became more anonymous and, therefore, more difficult to control.

The introduction of advanced information technologies, and the reorganisation which should both precede and follow large investments, means that news moves farther within a firm. Top levels of management currently know more about problems at the lowest levels. This increases organisational visibility. It also demands better systems functionality, in a software, hardware and organisational sense.

There is a difference in the manner in which developers and users of computers and communications look at system functionality. For instance, since deregulation, strategic plans in the banking industry became attuned to the provision of high value-added financial services. This means sophisticated solutions, adding an extra stimulus toward supercomputers and artificial intelligence (AI).

Sophisticated solutions must be supported by the most modern equipment available and by a huge amount of knowhow. They also should be renewed as more advanced concepts, software and hardware, develop. This means still larger investments. A leading financial institution was to suggest that the $500 million backbone of its new network is being designed to connect to more than 230,000 terminals.

Increasingly, in applications environments, the user's viewpoint dominates. Technology adjusts to it. In the manufacturing industry corporate focus is shifting toward three issues:

- Costs must come down;
- Tolerances must shrink, *and*
- Quality must improve.

Pricing alone would not bail anyone out of this fiercely competitive market. High quality and cost control is the key to profitability.

Within user organisations the priorities have changed. While many managers typically try to justify computers by showing how they would eliminate jobs, few have faced the fact that the jobs cut might be their own. However, those organisations that have gone the furthest in using information technologies tend to have only about half the managerial layers of their rivals. Technology flattens hierarchies.

This awareness of the need for restructuring, using technology as a sharp knife to cut corporate fat, has been building up during the last four to five years. Another basic understanding by top management is that current solutions will not deliver the necessary breakthroughs. New departures are necessary.

New departures will be served through Fifth Generation Computers (5GC) and AI. Policy-makers in a user organisation do not care about how hardware parallelism will be constructed. What each executive cares about is a documented answer to the question: ''What can they do for *my* job?''. Even more clear-eyed is the question: ''What can they do for *our* company?''.

Technologists have another viewpoint, but their mission is different. To answer management's query in a factual manner, they must reflect the relative standing of each product. This is prerequisite to the ability to evaluate it in terms of interest of implementation.

Conceptually, the managerial and technological approaches are converging in the sense that 5th Generation Computers (5GC) and artificial intelligence (AI) have come in a most timely manner. The first available 5GC machines of:

- *Coarse Grain Parallelism*: such as Teradata's Database Computer (DBX 1012), *and*
- *Fine Grain Parallelism*, for example, the Connection Machine,

have already appeared in the market, and have been successfully installed, passing the implementation test. So, they can do a great deal in answering managerial questions in terms of cost/effectiveness - provided that their strengths in terms of *applications perspectives* are properly studied and understood.

In this chapter we will not be concerned with what makes a 5th Generation Computer. Neither will we be talking in any detail about photonics media. Both these topics are covered in subsequent chapters. By contrast, the immediate effort is to give perspective, for without the proper perspective we will fail in the implementation of even the most powerful tools.

The Need for Parallelism

The need to provide effective integration of all computers and communications resources available to the organisation, and the ability to share the database in a global sense, are examples of the pressures under which DP/MIS operations find themselves in order to produce results. The urge to do more, and do it better than what was already the case, has always been present. The new factor is the speed of new requests, and the magnitude of the pressure. A rapid look on developments in information processing is most revealing in regard to the change of the landscape over three decades of DP practice:

- From batch to online;
- From mainly calculation to the integration of data processing and word processing (DP/WP);
- From centralised to distributed systems;
- From DP/WP to intelligent constructs - that is, enriched through AI; * From simple online systems to large multifunctional computer networks;
- From emphasis on data communications to multimedia, integrating text, data, graphics, image and voice.

Organisations able to sort out their priorities are adopting as their objective integrated computers and communications networks with AI capabilities. It is under this dual aspect that the use of non-von Neumann computers should be considered. As we will see in the appropriate chapter, 5GC rests on three pillars:

1. Intelligence machines

2. Parallel-type operations

3. Very high cost/effectiveness

That is why the Japanese PSI is 5GC, while it is a sequential engine. For the same reason, the Teradata database computer is also 5GC. Although it does not incorporate AI, it is a fully parallel database-oriented system.

A wise approach to the design of the communications and computer systems of the years to come is to rethink and restructure the way in which we apply our knowhow.

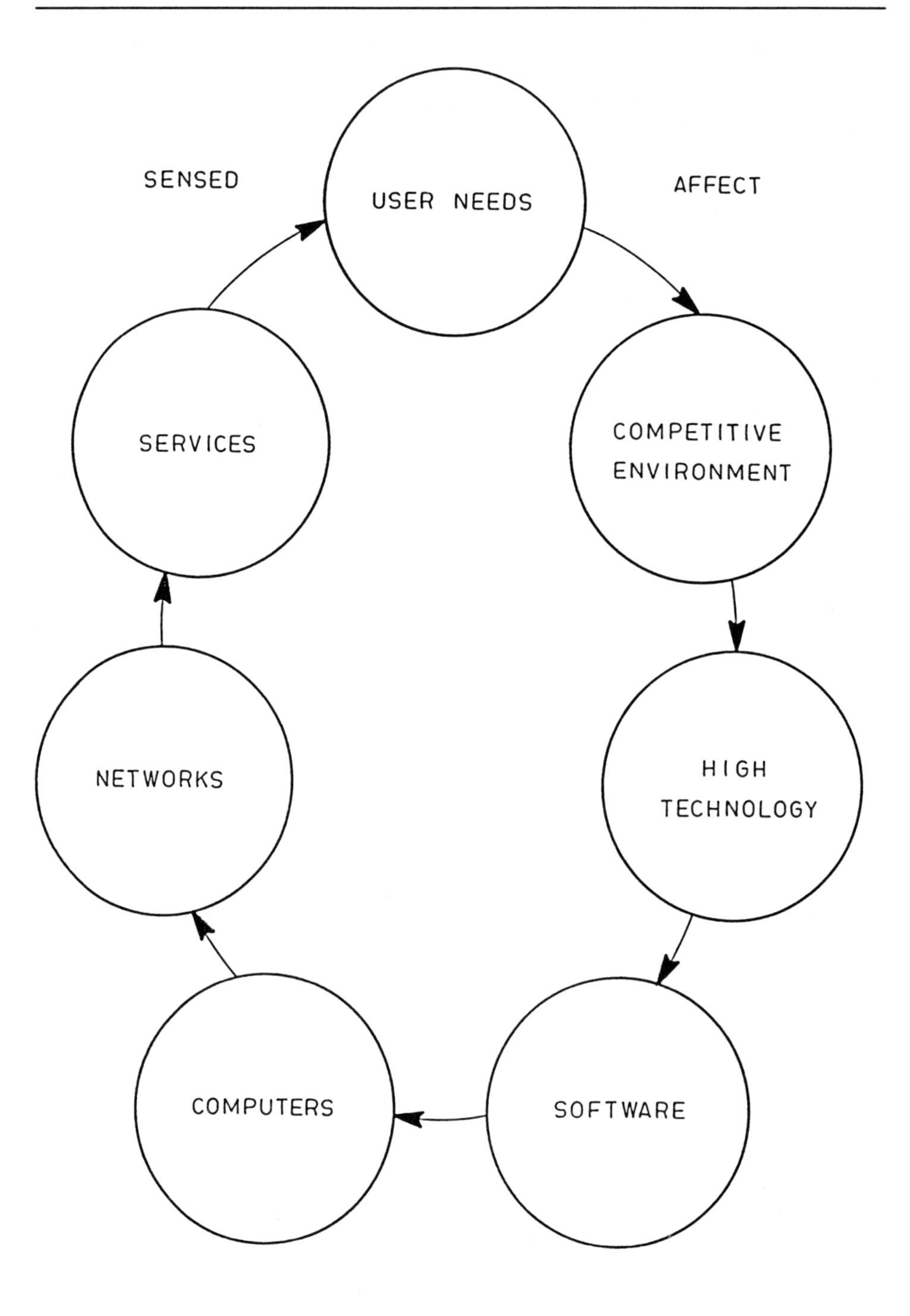
SENSED
USER NEEDS
AFFECT
COMPETITIVE ENVIRONMENT
HIGH TECHNOLOGY
SOFTWARE
COMPUTERS
NETWORKS
SERVICES

Figure 6.1

New approaches are necessary in order to successfully invest the financial means put at our disposal for the implementation of technology. Figure 6.1 suggests such an approach which has been successfully tested by leading-edge organisations. What Figure 6.1 suggests is that first and foremost we should be highly sensitive to user needs. User needs have never been the hallmark of the old - and by now deadly obsolete - Data Processing organisations. The large majority of failures we have experienced with computers during the 35 years of their business usage is due to this very simple fact.

Unfortunately, even today the all-important "Electronic Data Processing" (EDP)* officialdom is more concerned to preserve and recapture the past - not to create a brighter future. This leads to sheer conservatism and blindness in front of opportunities.** Since the 1970s, EDP decided to turn its back to technology and to cultural change. As a result, it has become ossified. Soon it will be on its deathbed.

Properly sensed and clearly appreciated, user needs must affect in an important way the solutions that we will provide to assure that our organisation survives in a competitive environment. This is the role that high technology is required to play. Software, hardware and networks are tools. The crucial issue is the nature of the services they can be made to provide.

Such services must not only be punctual. They should also be rendered at a reasonable cost. In the case of computer hardware, this reference regards the cost of MIPS.

- Fifteen years ago, in 1974, the cost of 1 MIPS mainframe equipment stood at the $2 million level.

- In 1989, its costs 10 percent of that money -yet it still costs $200,000 for one MIPS supported by traditional mainframes, but only $1,000 to $3,000 for one MIPS on 5GC.

With parallelism and artificial intelligence playing a key role, the change in costs is two orders of magnitude - and two orders of magnitude is quite impressive. In fact, so high is the cost/effectiveness that good management would jump at the opportunity. With its 65.536 computers which are operating in parallel, the Connection Machine sells for $3 million. It gives 2,500 MIPS steady computing power and 10,000 MIPS peak power.

These are 1989 figures. While they will improve over time, the discrepancy between

---

* "EDP" used to mean Electronic Data Processing. In the majority of cases, today it stands for Emotionally Disturbed Persons.

** The word "conservatism" should not be confused with the name of a political party. Under Margaret Thatcher, the British Conservative Party has become the agent of change, while Labour regressed to the role of the guardian of the past. Precisely like communist Russia and most of its satellites, or, for that matter, the EDPiers.

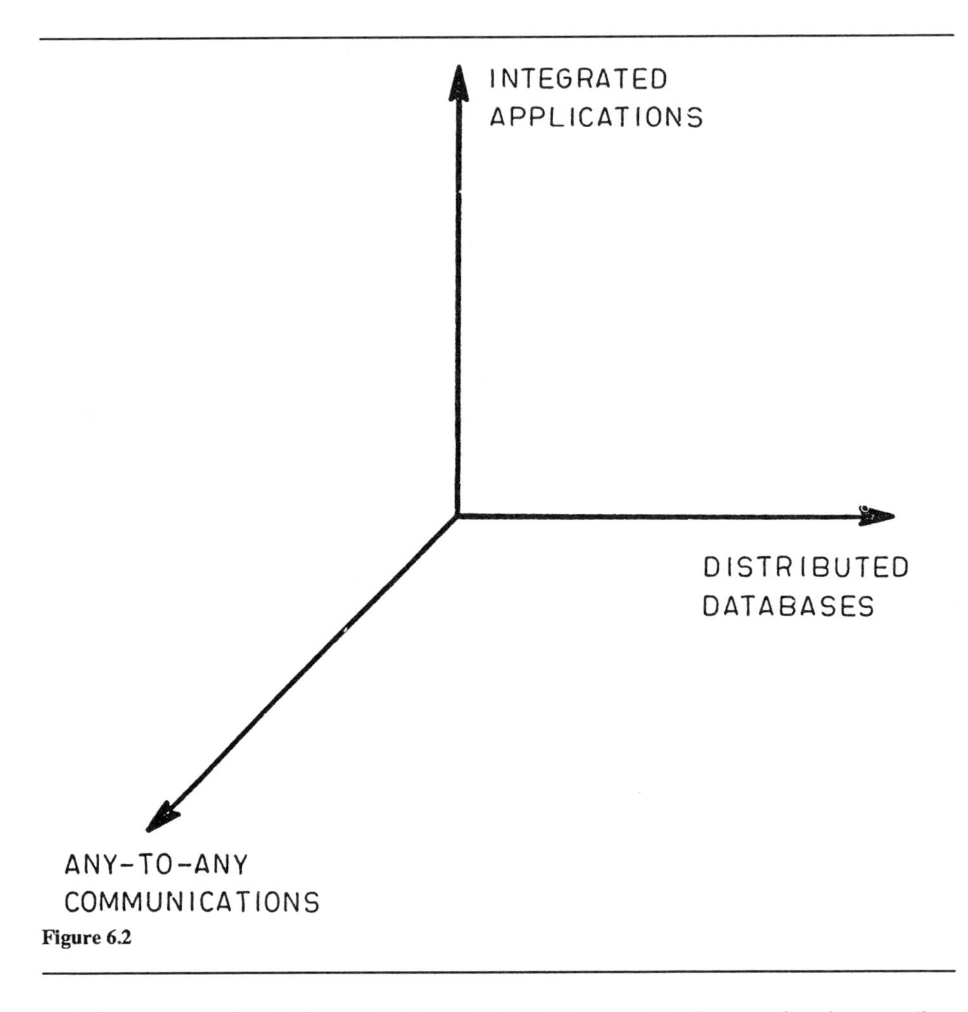

Figure 6.2

mainframes and 5GC will prevail. As such, it will generally characterise the next 5 to 7 years. By the mid-to-late 1990s cost figures will change again, and by a considerable amount. The reason is photonics, provided that current research bears results. But, for the time, being we are not talking of photonics computers. Our evaluation should regard electronics engines in two classifications:

- Sequential vs. Parallel, *and*

- AI constructs vs. dumb data processing solutions.

While a lot of transactional type operations, as well as many decision processes, seem to use a sequential scheme, quite often there is confusion between information collection, information processing and decision making. Serial processes have classically been paper based. For more than three decades computer usage closely emulated this serial approach - and in the process kept paper alive.

Modern large scale systems do not work that way. They work in parallel, and expand in the coordinate frame of reference suggested in Figure 6.2:

1. Integrated applications,

2. Distributed databases,

3. Any-to-any communications.

Among these axes run fully parallel operations, and the more the system's structure approximates to this necessary parallelism, the better it is for the end result. Furthermore, old-type software with its procedural limitations cannot answer the requirements for highly flexible and adaptable realtime processing posed by the system. Artificial intelligence is the answer. As with a person:

- The expert system may get all the information it needs, then apply it in parallel to reach a decision, or

- It can decide sequentially, as if using consecutive filters, with the last piece of information it received being the decision key - but the action to be taken is implemented in parallel.

Sequential techniques may need to ask fewer questions than parallel approaches, but they also need to be better tuned, and are much more cumbersome in software support. In this sense, there is a dual approach to consider as a new discipline evolves out of the merger of:

1. Very high power computing due to parallel architectures, *and*

2. The broadening implementation of artificial intelligence in what used to be data processing/word processing functions.

This dual approach is an *enabling technology for information understanding*. As such, it is much more important than the transition we went through as we moved from data processing to information systems. *Our mental frame of reference is now radically changing.*

The new definition in computers and communications sciences focuses on the development of constructs able to understand unstructured information. This is happening for the first time, opening up totally new perspectives for implementation - but also demanding a much higher calibre of skill.

Parallel machines do not just perform many steps at a time. They *digest information*, then use high resolution presentation schemes for communication to users. This is a facility which has never before been available. If, and only if, we capitalise on its potential we can push forward the frontiers of knowledge for years to come.

The information systems of the future will be aggregates of integrated components: pointing to intelligent problem solving; designed for specialised processing jobs such as image enhancers; available for shared information management; and able to provide interactive user access to the system by invoking intelligent application programs.

Supported by natural language processors and very powerful workstations, knowledge-intense applications will utilise a shared base of data and knowhow. A new generation of database systems will be required to assure integrated access to a variety of heterogeneous repositories; from security/protection to integrity, concurrency control, recovery and, above all, agile support to the enduser.

Large Scale System Requirements

The 1980s have seen a fast growth of the computers and communications resources. But we know there are limits to growth with von Neumann machines. IBM mentions four as being topmost: limits to single processor performance; differential between processor and I/O speeds; availability limitations of a single system; and complexity of system management - including necessary extension.

We can easily identify the growth items which are

putting current computer solutions under stress. Five factors are topmost:

Storage capacity

Most major organisations now feature over 100 gigabytes of disk storage; some have reached or exceeded 1 tera-byte (one trillion bytes). This grows by an impressive 25 percent to 50 percent per year, with the average being about 40 percent per year.

Processor power

Many organisations feature an installed base of 100 to 200 MIPS, with an impressive 50 percent to 60 percent growth rate per year. IBM projects, with this growth, a 1,000 MIPS system in many organisations by the early 1990s. Quite likely, it will be much more than that, given the demand for power by the more sophisticated AI constructs and the fact that supercomputer power is now offered at affordable cost.

Continuous, non-stop operation

This requirement has led to fault-tolerant systems, but also to the need for obtaining online maintenance and configuration changes in a non-disruptive manner. The same is true for the introduction of new equipment and of new software. One of the requirements for continuous operation is simplified system management in software and hardware. 5GC points toward simplified system operations, as well as to significantly greater uptime. Large scale system operations still pose another requirement which von Neumann computers can ill-serve:

Single system image

Because of the way in which they are currently practiced, database splits not only slow system growth, but also make the use of the available information elements uneven. The global database concept is missing in most installations. Coordination also leaves much to be desired, as we will see when we discuss the fifth key reference:

Increasing use of artificial intelligence

This impacts both on the available computing power and on the global database to be accessible online. AI seeks to expand the business of computers and communications in order to solve problems traditionally considered to require human intelligence. That is why, not only is a greater computer power required, but also a quantum leap in the sophistication with which available systems are employed.

This can be done through AI. As stated in Chapter 4, artificial intelligence includes such domains as cognitive sciences, natural language, language translation, artificial vision, robotics and expert systems. Expert systems manipulate knowledge. They also help us to explore the contents of large databases. Design-wise, they can be:

- Inductive, developing their own rules and methods as they find appropriate, or
- Deductive, with rules assigned in advance by the developer, but kept dynamic by the system.

But, under the current state of the art, they are specialised in a chosen domain. Contrary to currently available microprocessors, which can be seen as jacks-of-all-trades, artificial intelligence chips are narrowly focused to master particular tasks. Expert systems use this approach. That is one reason why they are so superior to classical data processing applications.

While simple, adolescent expert systems easily fit on currently-available personal computers, their growing sophistication to hundreds and eventually thousands of rules demands greatly increased machine power. The unit of measurement is no longer million of instructions per second (MIPS) but:

- thousands of floating operations per second (KLOPS)
- millions of logical instructions per second (MLIPS), *and*
- billions of instructions per second (BIPS).

Between MIPS and BIPS, the difference is *three orders of magnitude*. Novel architectures of electronics components - such as the hypercube (see Chapter 8) - provide BIPS power today. But, we also said in Chapter 5 that the first implementation of photonics switching will improve speeds by a factor of 1,000.

Then the unit of measurement (metric) will become *trillions of instructions per second.*

Even at the workstation level, the need for much more machine cycles than are now available begins to be felt and, with this, the ability to handle *predicates.* A good example on how to fulfill these requirements is the increasing use of supercomputers in America as well as the personal sequential inference (PSI) engine developed by ICOT. We will study it in greater detail in the Chapters on 5th Generation Computers.

The rules and methodology of an AI construct find themselves in its *knowledgebank.* The knowledgebank is not a database in the data processing sense of the word. It is a knowledge base which requires for its handling an *inference engine.* The latter acts as the compiler of the knowledgebank.

We also said that current expert systems present on-the-job expertise and intelligence only within a well-defined (hence, limited) *domain of knowledge.* The expert system depends on its domain-oriented knowledgebank for reaching conclusions and giving advice, as in this knowledgebank have crystallised the rules into which is mapped the way in which a domain expert thinks, acts and works.

However, modern information systems also require large, global databases which are both:

- Secure *and*
- Easily accessible.

The database is a term often used to denote an organised aggregation of information elements: files, records, fields, bytes and bits. Such information elements may contain data, text, images or digitised voice. They are properly defined through system analysis, but stored, retrieved and presented by the computer.

Databases have themselves become large scale systems. They are distributed and intercommunicating. We also said that they are expanding at a rate which currently stands between 25 percent and 50 percent per year - with 35 percent per year being a representative enough growth rate. Some very large financial institutions and industrial organisations have already reached the one trillion bytes level (one *terabyte*) in databasing, and they are fast progressing beyond the terabyte level.

As a metric of how big the terabyte is, let us recall that only 28 years ago, when the first disk storage was introduced (the Ramac 305 machine), it stood at 5 megabytes. At the time, this seemed a ''colossal'' memory size since, until then, drum memories typically stored only a few thousand words. Databases bigger than a terabyte stand at six orders of magnitude greater capacity than the biggest storage engine available in the first decade of computer usage.

To drive terabyte storage capacity we need lots of computer cycles. Hence, we add to the large scale system requirements. What is more, we require that these very large databases manage themselves, not only automatically, but also in an intelligent way. Hence the need for *idea database* implementation, which essentially means AI constructs to run the database engine.

Designed to manage the knowledgebanks, *knowledgebank management systems* (KBMS) will act as a metalayer to the now more-classical DBMS. But, the information in the database must be made available to the man with the problem. This means that we need to develop very efficient man-machine interfaces which are agile, user friendly and enriched with AI. A better term is *human windows*.

The system is built for the *enduser*: the manager and professional in the field of its applicability. The human window is necessary for the enduser to effectively communicate with the intelligence inside the machine -as well as for his acceptance of computer support in the broader sense of the term.

Efficient human windows are supported through AI. It is much easier to converse with an intelligent system than with a dumb one. But, intelligence consumes machine cycles. As in the case of large databases, we need more computer power to drive increasingly more intelligent systems.

Inference mechanisms, distributed databases and human windows at the endusers' workstation need to communicate. Networks are designed for this purpose, but networks, too, are becoming more intelligent. They are now equipped with AI constructs which not only help in transmission proper (including error detection and correction) but also give interpretative analysis and consultative advice in terms of diagnostics and maintenance.

In fact, expert systems approaches are used all the way from network design to implementation and network maintenance chores. Inference, data processing, databasing, human windows, transmission and networking at large imply large scale systems with lots of requirements for cycles and inference capabilities. This brings into perspective new prerequisites and definitions.

Automating the Nervous System of the Organisation

The reference to the ever-increasing application needs and the results to be obtained through 5th Generation Computers and artificial intelligence brings into perspective the limits currently existing to greater automation. At a meeting which took place in Tokyo, Oki mentioned t*hat 70 percent to 75 percent of all essential office functions have not been automated, and cannot be automated with von Neumann type computers*. Other Japanese organisations confirmed that finding. MIT did a similar study a few years back. The USA, too, finds a 75 percent level of non-fulfillment of necessary implementation premises.

In the banking industry, for instance, backoffice solutions have evolved into enduser processes. However, in many cases such applications are limited and ineffectual because data is currently spread over a variety of machines and is manipulated by different, incompatible, non-integrated software packages. This keep employment high at the backoffice while the financial institution also invests heavily in technology, from which it gets less-than-anticipated results.

At the same time, attempts by data processing managers to put the data into the hands of the endusers result in complex applications because they involve twisting old, inappropriate systems rather than developing new ones. Such "developments" typically involve not only cross systems communications but also very large and untidy software modules which are difficult to structure, manage and maintain.

For these non-automated, or little automated, work areas a different solution from those which have been tried in the past, and have failed, is necessary. As Figure 6.3 underlines, in this solution:

1. Communications is the core.

2. The next layer is multimedia: digital image, voice, text, data and graphics.

3. The link between this kernel and the outer layers is artificial intelligence.

Some of the more-enlightened system researchers understand that in AI we now have enough knowledge to permit us to abandon the surface (qualitative behaviour) and go to the layers below. This permits us to define the underlying behaviour of intelligent systems -solving our problems accordingly. Two more levels of reference must be brought to the reader's attention:

4. AI will support a wider range of job-specific personal services.

5. The outer layer in Figure 6.3 is the modules and procedures necessary to automate what has so far escaped automation efforts.

This is precisely the 70 percent to 75 percent of essential office functions. Not only can AI be instrumental in this work, but it should be used to assure the needed user friendly interfaces.

All of this points to the fact that we currently are at a delicate phase of transition. The old Data Processing is dying out, but its high priests fail to understand the new environment, and/or they are afraid of it. Therefore, they resist the change. Yet the wave of change will eventually sweep away obsolete Data Processing managers and Cobol programmers alike.

By the mid 1990s the ossified programmers will serve in nothing else but banal old program maintenance. AI-enriched computers will communicate with their users in natural language constructs, and they will program themselves:

- Today, computers manage information.

- Tomorrow, they will both use and manage knowledge.

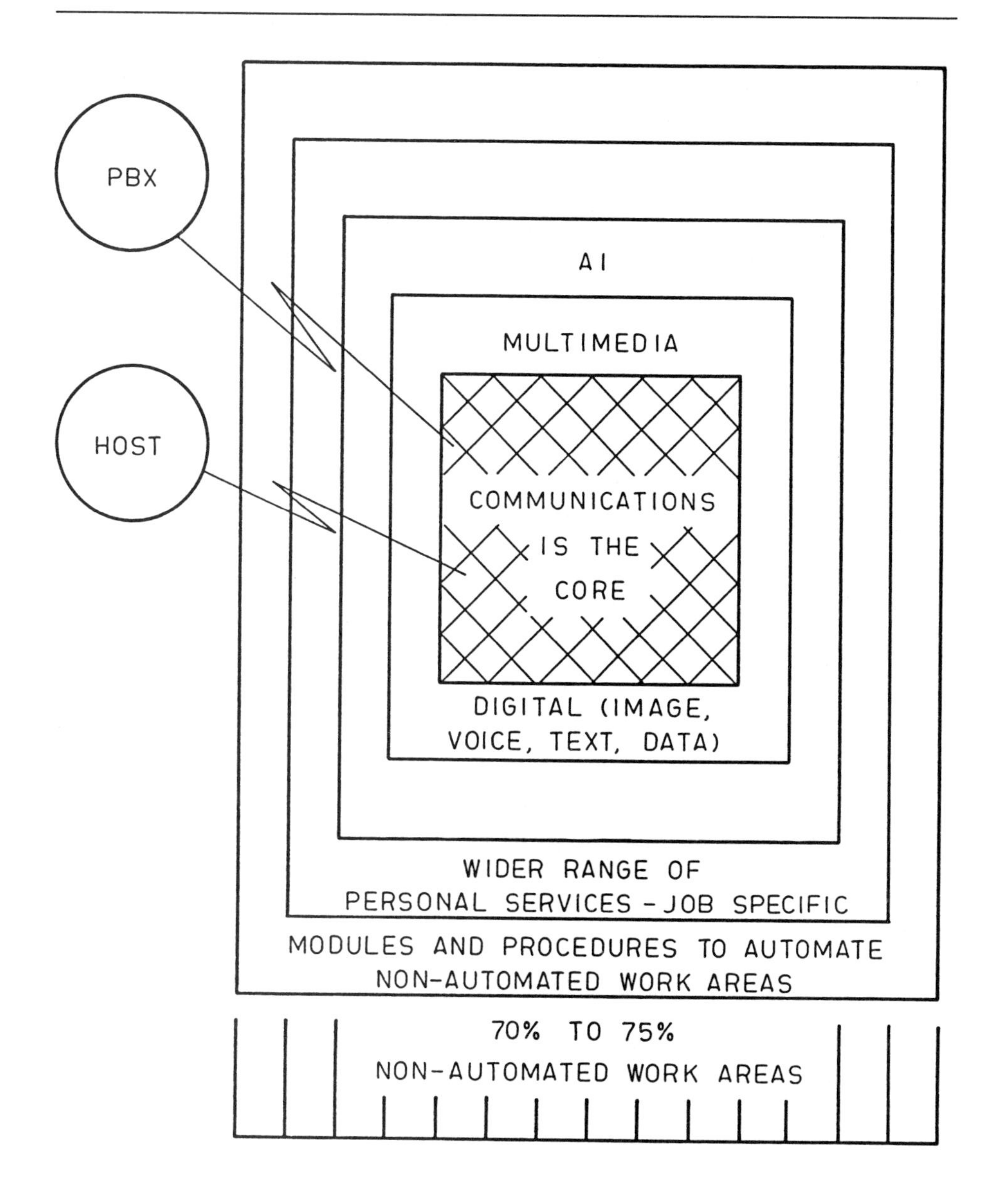

Figure 6.3 WS and AI

Significant improvements over current practices and presently-available tools have to be made in order to reach such an objective. One of the key problems in knowledge representation is interactivity. True interactivity is constrained by current DBMS. Existing query languages are too weak to capture all the essential intentional knowledge (''intention'' means rules).

Other limitations are also significant. All existing query languages being first order, they are incapable of expressing *recursion.* As a result, this type of knowledge has to be embedded in application programs, thus becoming unavailable to the DBMS for optimisation.

Logic-based AI languages, such as Prolog, do permit recursive definitions in rules, but they are not supported by the current generation of query languages, nor have they been used in DBMS development. While recursive structures can be stored extensionally in existing databases, they cannot be defined intensionally as the results of queries. However, recursive processing is required in many scheduling, routing, planning and path-finding applications.

Another shortcoming of current DBMS is the lack of facilities to process multiple information types. Most of the information types that are not satisfactorily handled by current DBMS share a common characteristic: *dimensionality.* Objects of these types are oriented in space or time. Users of computer resources live in a dimensional world. Hence, dimensional information is predominant in many applications. Maps, signals and drawings have both spatial and temporal attributes.

Building the semantics of *space* and *time* into the intelligent system serves as the basis for adding and integrating dimensional information types. As technology advances, and our applications benefit from it, we will be increasingly focusing attention on:

- *Realspace* requirements,

- Rather than the more traditional *Realtime.*

Increasingly, business and industry need to handle knowledge and information *as if* all operating units concentrating on a given subject were really at the same point in space - although, in reality, they are distributed over wide geographic areas.

Realspace can be achieved through current technology, if we design it into the system which we are building. Both General Motors and Ford see to it that their design laboratories in the USA, UK and West Germany work together on a network, as if they were in the same physical location.

As the world becomes *one* market place, financial institutions feel and face similar realspace requirements. Electronic banking operates 24-hours per day. Whether during the time when the Exchanges in New York, London, Zurich or Tokyo are open, or after hours, decisions are made in realspace - and so are the orders being given executed.

A further contribution supported by intelligent systems stems from their ability to represent and handle *complex objects.* For instance, objects which are highly structured and consisting of other objects. A machine assembly is a part hierarchy composed of subassemblies and parts. For effective, real world representation, facilities are needed to map the internal structure of complex objects. They should also:

- Permit the manipulation of the entire object as a whole,
- Capture complex constraints and interrelationships among the components of an object, *and*
- Support rules for mapping operations on the object into operations on its individual components, without violating any constraints on objects or interrelationships.

Still another crucial subject is *extendibility.* This is needed for two reasons. First, it is unfeasible to build into the DBMS detailed knowledge of all activities. As the implementation environment develops in many dimensions this becomes too cumbersome. Second, and at the same time, it is important to provide for integra-tion of different information types - which requires increasing amounts of complex knowledge.

The true integration of a system starts in its database. Therefore, a higher-up layer of sophistication, a knowledgebank engine, can help this process in a significant way. The more the knowledgebank layer is enriched with command activities, the better it will control the database. We have already mentioned KBMS.

Extendibility and integration, the two reasons that we just reviewed, are further underlined if we consider total system perspectives. A global approach easily demonstrates that an increasingly-felt requirement is to provide integrated access to all kinds of information needed by applications - and to do so with AI support.

To satisfy this need, different types of information and knowledge must be extracted from a variety of separate storage systems as well as from human experts in the specific domain to which a given project addresses itself. Over the years, this has been done mainly through manual approaches, but such processing is arduous and susceptible to errors and delays. Automated knowledge-based solutions are more efficient and much more elegant.

Applications today increasingly rely on specialised knowledge for processing new information types. Such specialised knowledge may be captured in the form of rules, special operations or semantic constraints. The software we provide must be capable of assimilating and utilising this knowledge effectively. This means AI.

The need for 5th Generation Computer approaches is compounded by the fact that new applications often strain or degrade performance requirements. Enormous volumes of data are involved, very short access times are called for, and more functionality is necessary than is provided by classical software. For more power, and greater inference capability, it is necessary to exploit parallel processing and optimisation solutions.

Evolution of Architectural Characteristics

Statistics and projections help bring into perspective an impressive growth in demand for computers and communications. In turn, this increases the pressure for new solutions, with non-von Neumann architectures and artificial intelligence at the forefront. A typical large scale system today features 10,000 to 20,000 online terminals. While the number of PC workstations is increasing, these are still in a minority. Such a system features between 200 and 400 gigabytes of online storage and 100 to 300 MIPS of installed power. Some 60-100 transactions per second (TPS) represent the current online data load.

Most of the 40 or so mainframes needed to support such an environment are in clusters of 2, 3 or 4 machines. This still leaves between 10 and 20 clusters which are not interconnected through hot switches and, therefore, not so well coordinated. Such dispersion of computer resources impacts all the way from system reliability and uptime to enduser service.

The way in which IBM foresees the future of such complex systems is that *without a radical change in technology they will be impossible to manage.* Consider this projection made for 1992-95 (only 3 to 6 years from now):

- 150,000 terminals operating online per large scale system
- The vast majority of these terminals will be PCs, posing impressive data service requirements
- 6 or more terabytes (trillion bytes) of storage will be available online
- 4 billion instructions per second (GIPS), or more, should be featured
- 1,200 transactions and messages per second will need to be handled
- A very large networking structure, including online customer access for transactions and messaging, will become a "must".

While this is a forecast, it is solid enough to be taken for projections on the dimensioning of large scale systems. It is also proper to add that a large number of the online terminals, and the corresponding transaction load, will be at client premises, thus creating an extensive network, with corresponding increased complexity in respect to system management.

No matter what sort of extension is applied to it, current computer technology cannot handle such requirements - much less respond to the networking prerequisites that they entail. Coupling 10 to 40 mainframes together is bad enough. But 4 GIPS would require 100 to 300 von Neumann type mainframes. Such a system would become unmanageable.

Similarly, in a system sense, there is no means for integration at such levels of complexity - particularly when machine diversity and incompatibility are accounted for. Hence, focused partitioning of systems power will become the dominant approach, with 5th Generation Computers assigned to communications, storage, transactions, messages and so on. Figure 6.4 presents one of the solutions being projected. The role of the message memory is dual:

- Serving as an enterprise-wide mailbox, *and*

- Isolating communications failures.

The role of the data memory is that of a database engine. More specifically, the coordination of distributed database engines which may be local or long haul - but which should be interconnected through an any-to-any network.

Such multisystem architecture presupposes at each level intelligent machines of 5GC-type endowed with appropriate expert systems. Most will be specialised to the area which they address, with the global system tuned to enhance availability, uptime and reliability. Design-wise, it will be selling by increments of availability rather than by storage-and-MIPS upgrade.

High bandwidth interconnection will be standard with such systems. The same is true of the AI constructs to assure security, journaling and recovery, as well as other specific types of protection. Architectural integration will itself be AI-based.

The projection currently made on this evolution of architectural characteristics is that, rather than computer cycle, the unit of measurement will be *System Runs*. That is, multiple instances of the computers and communications aggregate as defined by the environment the system is designed to serve. This permits the integration of machines of different operating systems and technologies - as it does not make sense to abandon or substitute today's operating system (OS) on a massive scale. Prototyping is a point of reference for new system design strategies. At all major computer manufacturers today system research focuses on:

1. An architectured system structure;

2. Interconnecting capabilities;

3. Load balancing features;

4. Failure handling routines;

5. Recovery methods;

6. Feasibility of incorporating present software within the new environment;

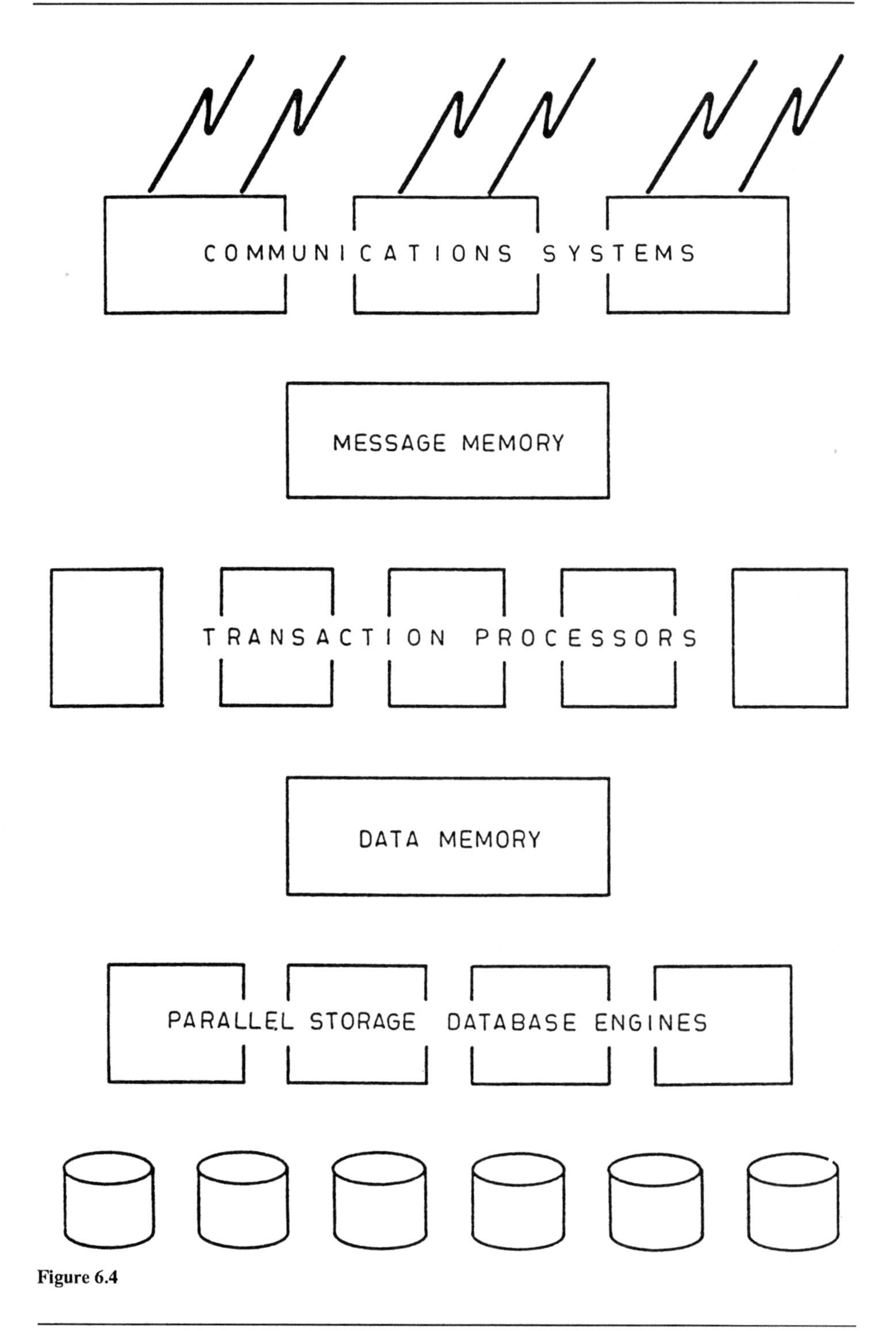
COMMUNICATIONS SYSTEMS
MESSAGE MEMORY
TRANSACTION PROCESSORS
DATA MEMORY
PARALLEL STORAGE DATABASE ENGINES

Figure 6.4

7. Use of artificial intelligence constructs to manage operations; *and*

8. Prototyping of large scale computer and communications aggregates.

Multisystems requirements are particularly emphasized to ensure that the whole aggregate appears as *one system* to endusers, programmers and running applications. Any authorised user should be accessing any resource without direct switching and the process should be independent of location. Here is where AI will make a major system level contribution.

The Japanese computers and communications manufacturers are working on a macroengineering project bearing several characteristics similar to those described above. Their approach is exemplified in Figure 6.5. The role of AI in a structural context - from VLSI to intelligent applications systems - can be particularly appreciated.

While technological developments tend to dominate, as the above references show, it is just as true that such systems will not materialise unless we are able to assure a *cultural* change in our organisation. If we do not manage change we will be overrun by the events to come - but the management of change has certain prerequisites.

Emphasis must be placed on research and development effort; training must become a vital part of our system approach; we must have the courage to preserve the developed store of knowledge and to assure an orderly transition to the new environment. Nothing short of a well thought-out plan can solve our present and future problems - guaranteeing the survival of our organisation through its own, conscious transformation.

Advancing technology increasingly implies multiple changes in our way of thinking. Education offers an example. The 1990s will see:

- The passage from collective to individual instruction, through interactive micro-processors,

- The conversion of the now passive instruction into active, because of the availability of computer supported media, and

- The implementation of far-reaching educational schemes on a distributed basis.

An integral and vital part of any architecture are the people who would be working with the systems we are designing. People-oriented drawbacks can only be overcome with imaginative approaches and people-oriented applications.

Our whole philosophy of life should evolve. Preparing for the knowledge society, we should appreciate that we are living at a time of turbulence. This makes the environment within which we work irregular and erratic, but the underlying causes can be analysed, predicted and managed. We must put a lot of thought into how problems are solved.

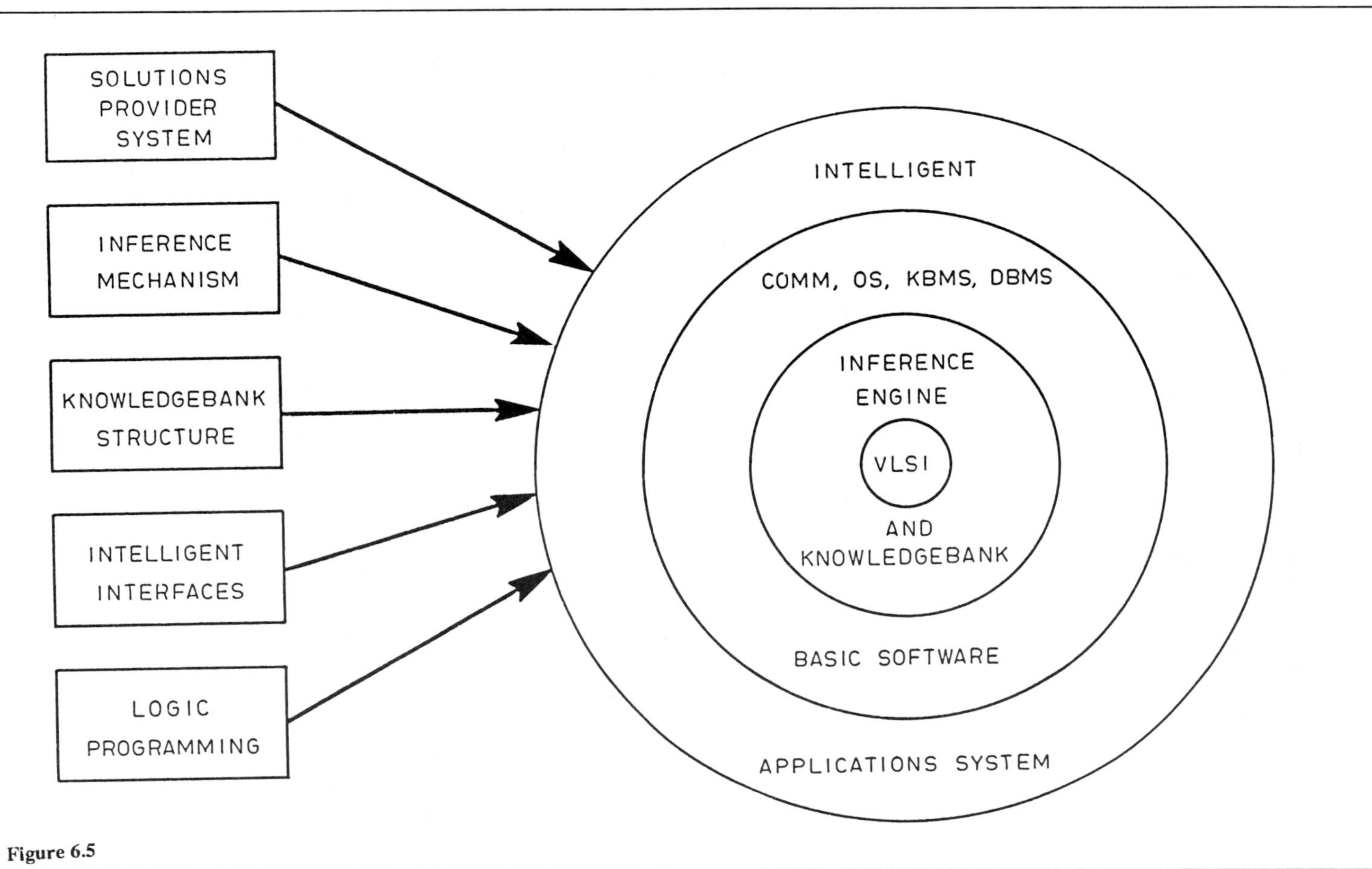
SOLUTIONS
PROVIDER
SYSTEM
INFERENCE
MECHANISM
KNOWLEDGEBANK
STRUCTURE
INTELLIGENT
INTERFACES
LOGIC
PROGRAMMING
INTELLIGENT
COMM, OS, KBMS, DBMS
INFERENCE
ENGINE
VLSI
AND
KNOWLEDGEBANK
BASIC SOFTWARE
APPLICATIONS SYSTEM

Figure 6.5

The time to ask the tough questions; to exercise radical surgery; restructure and reorient; as well as to train the human capital, is not when our organisation is in trouble, but while it is still successful. Success always means an effort of meditated abandonment in order to reach new goals. To be successful, we must:

- Isolate the skill requirements for the jobs of the future.

- Appreciate that knowledgeable people are a basic ingredient to profits.

- Identify the necessary skills to be developed through training; *and*

- Match the skills to the new job requirements.

A failure to do so lies in the background of many unsuccessful efforts. We should not be repeating the same mistakes.

For instance, when we start spreading terminals around the organisation we should know that everybody will be affected. The floor changes; the work conditions are altered; the role of the mail room is redefined; the way we do the day-to-day work is no longer what it used to be. We have to manage the people in this conversion process, and we have to plan for this need.

It is necessary to develop timetables of events; assign responsibilities; get commitments; establish feedback channels; promote team building; eliminate personality conflicts; describe (in writing) the new positions; announce, not only new titles, but also their specific responsibilities; review salary grades and ranges; and install an incentive system. We must also install a system able to implement correct action.

There is no easy way to an elegant solution. A valid architectural approach has to consider every crucial factor that enters the development equation. This consideration has to precede the project; it must not follow afterwards as a patching-up process.

The Knowledge Worker

Today, the largest part of the American workforce is made up of white collar workers. *Knowledge workers* alone, those engaged in work pertaining to information, represent nearly half the labour force; they take home about 60 percent of the nation's total wages, and see to it that 55 percent of the GNP is related to their activity.

Touch labour has shrunk to about 5 percent of total production costs in some plants. It is fast becoming just as trivial at most others. Without high labour costs to justify technology investments, manufacturers are being forced to rethink their whole approach to profit and loss management. Financial institutions will follow the same path in a few years.

Investment in high technology can be justified, not by labor savings alone, but by:

- Improved quality
- Faster product introduction
- Largely automated delivery,
- Lower inventories, *and*
- Steady cost reduction.

However, putting such goals into practice will require a new culture from managers. Tailoring better systems to manage costs, rather than simply allocating them after the fact, demands:

1. Clear vision of a company's long-range strategy, *and*

2. Well-rounded understanding of every facet of its business.

Managers do not necessarily acquire those insights in business schools. Hence, the need for intensive in-house training supporting new departures in an able manner.

The knowledge society which emerges opens new social and creative horizons. It is also demanding in terms of the quality and quantity of knowledge support and information services that it needs. To face our oncoming requirements in an able manner, we must define:

- *Who* are the information workers?

We must distinguish between "creative information workers", who need specialist education as well as attention, and "routine workers in the information sector".

- *Where* should information workers be educated?

The prerequisite is to establish a knowledge acquisition discipline and to set goals. Then, we should look after the fields of study and the criteria for success in training, as well as the steady recycling needs.

- *What* are the knowledge and information functions?

This calls for a classification flexible enough and open to future developments, but also capable of identifying not only present goals, but also their future evolution.

- *When* should the skills of the information workers be updated?

Different levels of expertise in information handling are required at different stages of professional development. We should clearly establish these stages, ensuring that they do not become monolithic, ossified or obsolete.

- *Why* should access to knowledge and information be reserved to authorised persons?

We cannot implement security measures unless we answer this query. We should also define who, when and why a person is authorised to access the text and databases -and to which degree. Particular care should be exercised not to create another class of mandarins like the EDPiers. Personal computers gave rise to a whole new category of corporate tools for efficiency. But, because of misapplication by the EDPiers they also created a whole new category of corporate waste. In a large number of cases, the inability of DP/MIS to adapt to change led to a massive redundancy of effort and a multiplication of cost, spreadsheet malpractice, report overload and an increase in, rather than the elimination of, paper.

Technology knows no frontiers, and can be friend or foe, depending on our attitude. To be competitive as a person, a company or a nation, we must feed our opportunities and starve our problems. Is this what we are doing? How keen are we in meeting the coming challenges head-on? Trouble can be predicted: we have the means for projecting on tomorrow's problems and opportunities. But do we take advantage of them?

Able, punctual answers to the questions being asked call for: a national information policy; a freedom of information act; and information privacy and protection laws. A great deal of legislation is needed on copyrights; eavesdropping; electronic funds transfer; electronic mail; and the sorting out of conflicting policies and interests. There are questions of regulations to be looked after. In the USA, the UK and Japan the communications industry is still semi-regulated. The computer industry is not. In other countries, telecommunications are fully regulated, which is not the case with computers. But where is the one industry beginning and the other ending?

In short, we need a set of principles to serve as guidelines; laws able to cope with the pervasive, often elusive and penetrating, nature of *knowledge* and of *information*. Our new legal structure should sustain the need for more knowledge work - the future of our society depends on it. And while legislation, and the physical or logical resources which it regulates are so important, we should not forget that, behind any worthwhile endeavour the key element is man. Technology or no technology, man is still the pivotal point.

We must prepare our people with orientation, practice and training: projects do not fail - people do. We should create a project environment that allows people the opportunities to "win". *Winning* means, from beginning to end, delivering *results* which are:

- On goal,

- On schedule, *and*

- Within budget.

We should orient our people to future opportunities. A human being is human only to the extent to which he or she is not concerned with himself. This may be work to do, a cause to devote oneself to, or another person. In having commitments man finds meaning in life. Technology or no technology, man's search for meaning is the basic human motivation - the one that keeps him alive and makes being alive have value. Such a man is capable of suffering when he sees reason for it, as witnessed by political revolutions throughout history. The man who has not found meaning is most likely to take his life, even amid plenty. What is more, man has within himself the basic tools for dealing with fears, frustrations and dangers: his inherent ability for reaching out, his sense of humour, and a strength -physical and emotional - beyond that which most people perceive as possible.

If we can give purpose to our human capital - a purpose consistent with the new environment in which people think, work and operate - we can magnify the human effort through computer power. Systems are going to be a lot easier to understand: the data, text and images that people use is more important than the programs we write.

The 1970s brought to the enduser text and word processing, but that is only the tip of the iceberg. In the 1980s we got into graphics, image formation and new voice capabilities. By the next decade, diverse technologies will come together, and this will require polyvalent executives and systems specialists able to handle the developing systems.

Management, both in industrial companies and financial institutions, must recognise the need to use communications and computers, not so much as an abstract service, but more as an inhouse resource. This calls for an educational function, and a coordinating process that disseminates technology for the benefit, not only of man, but also of mankind.

7

A Realspace Concept in System Integration

Introduction

The large scale systems which we discusssed in Chapter 6 will typically involve many diverse and incompatible computers, operating systems (OS), database management systems (DBMS), knowledgebank management systems (KBMS), AI constructs and other applications software. A similar reference is valid in regard to the communications devices which such systems will include.

The point often missed in system design is that, as long as key components of these systems remain incompatible, they do not form a working aggregate, but rather a hill of sand. An information system must work like a clock and present *one* consistent, logical image to its users. Hence the need for global integration.

The first prerequisite is a clear concept. Even if we have the best integration plans - and this is rarely the case - we could risk doing a half-baked job unless the goal to be reached is *clearly defined.* Describing is not defining:

- *Defining* focuses on detail

- *Describing* typically sticks to generalities.

We have to have detail at the system, subsystem, assembly and device levels, in order to do the integration job in an able manner. What do we *really* want to reach? Why do we make the *investment*? Which are the *measurable* results? *How fast* should we reach them, milestone-after-milestone? Which *resources* are we committing in:

- Knowhow

- Professional Prestige

- Money, *and*

- Software and Hardware

What is the Return on Investment (ROI) which we judge to be *acceptable*? What is the *maximum* ROI we are aiming for? Is it achieveable? What kind of resources would help us to reach our goals in a competent manner?

Like all successful enterprises, system integration - as well as any investment in computers and communications -should start with a *goal*. Classical data processing is *not* a goal. It is a process of regression, sinking deeper into the mistakes of the past. It is hardly worth pouring money into it, except for fixing jobs of the low-grade mechanic variety.

By contrast, *Realspace* can be a worthwhile goal for organisations that need it. Whether manufacturing company or financial institution, any enterprise which operates in a geographically distributed sense (national or international) needs to pull together its information resources making them available:

1. At any time

2. In any place *and*

3. When they are needed

Items No. 1 and No. 2 call for *Realspace solutions* based on networks, global databases, artificial intelligence and interactive workstations. Item No. 3 is served through *Realtime* approaches such as we have known for 25 years. All three items call for *Systems Integration.*

A successful implementation of systems integration makes it mandatory that *our* computers and communications specialists undergo a profound change in the philosophy that they apply to *architectural design.* It takes a long, orderly experience to acquire the new knowhow - a time significantly extended by the fact that people hate to forget about the old practices, and that they are uncomfortable with anything new.

To take an example, *interactive* solutions have a significant impact on the:

- Way the job should be approached

- Design of input and output formats

- Organisation and structure of the database

- Data communications requirements to be met

- Resulting decision support systems, *and*

- Enduser productivity.

All these factors taken together greatly impact on the all-important personnel costs. That is why the issue of interactivity in a Realspace sense is so vital. It reaches into every facet of computer work.

Systems integration perspectives should command steady attention. If this is not done from the beginning, and throughout the project, failures in linkages and interfaces will be haunting us for years, eventually obliging us to reproject and redo most of the work we have already done. This results in loss of time and spoilage of money. To avoid losses, any project done today should account for integration requirements, answering them in the most effective manner.

An example of problems and opportunities can be taken from very large scale integration (VLSI) at the chip level - which, as stated in Chapter 5, lately reached an inflection point in its development. Technology has increased the number of gates per chip so much that virtually an entirely electronic system can now be placed on a single chip. This is the proper type of integration at the chip level.

Such chip level integration brought benefits, but also created a dilemma for makers of complex systems based on semiconductors: computers, communications, consumer electronics, precision instrumentation, military electronics, factory robots and so on. How can they realise the cost and performance advantages of *systems-on-a-chip* without surrendering the ability to design the logic structure of the system to the chip vendor, thereby losing the ability to differentiate hardware?

Highly integrated 32-bit per word (BPW) microprocessors, as well as cheap, high-density, random access memory, have made possible microcomputer systems that at the high end pose a business and systems threat to minicomputer designs executed using standard transitor-to-transistor logic (TTL) on the printed circuit (PCB) level. Systems makers who can use very large scale integration technology have *cost and performance advantages* over competitors who cannot. Something similar can be stated at the large scale system level for business applications.

Implementing a Realspace Environment

The need for a new concept able to effectively bring together the increasingly-intelligent communicating workstations available today on every desk has been underlined in the introduction. The image of Realtime developed and applied in the 1960s in no way fills this requirement, but new approaches should definitely account for implementation perspectives.

Many advanced applications today rest on the ability to share *multimedia* database resources and integrated communications networks. Yet, even in these environments, data processing and word processing are individualised. Sometimes they come standalone; rarely are they truly integrated. Few approaches currently correspond to the developing information technology requirements. What endusers need most is *not just to compress time, but also to compress space*. This is the concept behind Realspace.

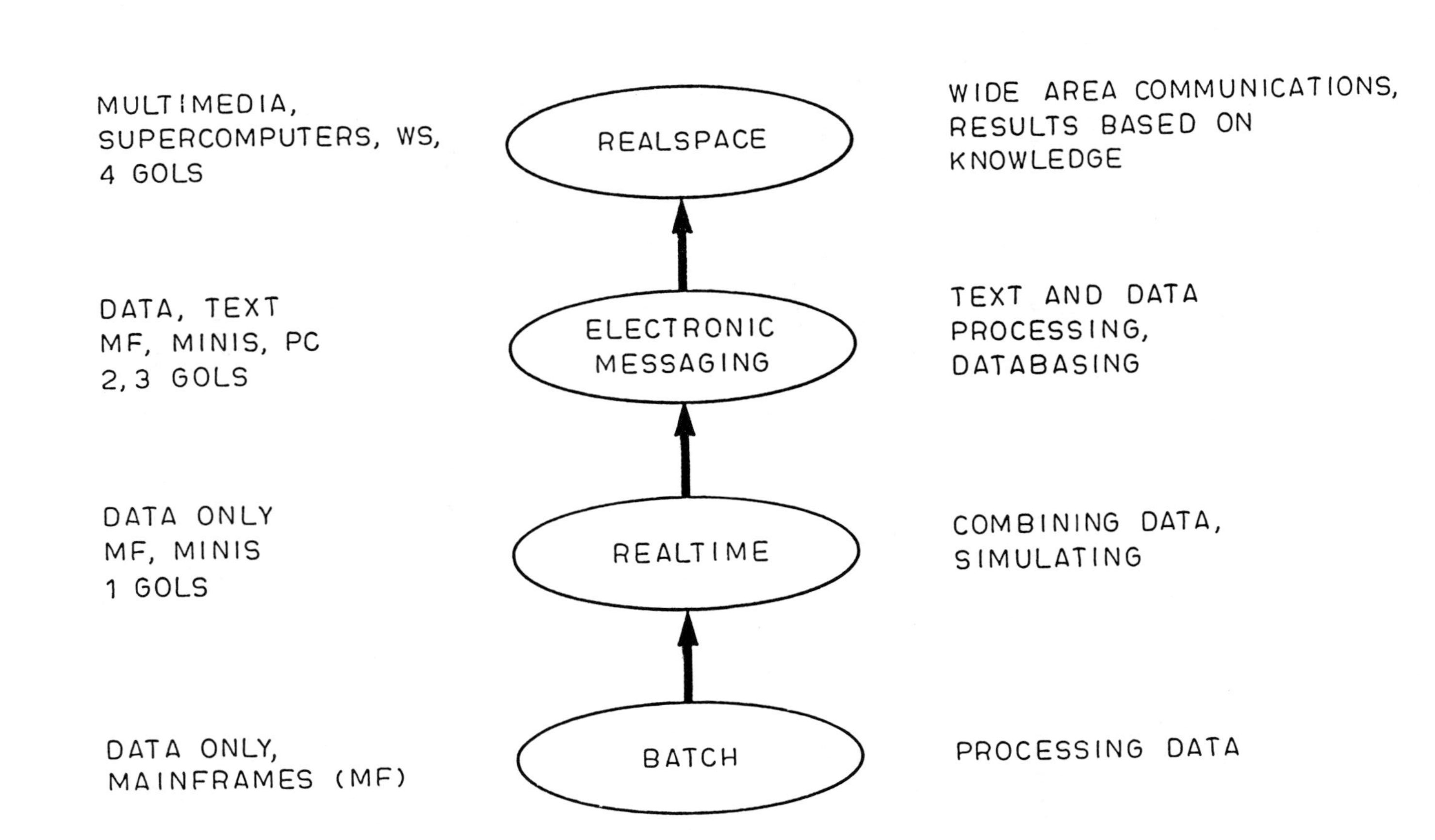

MULTIMEDIA,
SUPERCOMPUTERS, WS,
4 GOLS
REALSPACE
WIDE AREA COMMUNICATIONS,
RESULTS BASED ON
KNOWLEDGE
DATA, TEXT
MF, MINIS, PC
2,3 GOLS
ELECTRONIC MESSAGING
TEXT AND DATA
PROCESSING,
DATABASING
DATA ONLY
MF, MINIS
1 GOLS
REALTIME
COMBINING DATA,
SIMULATING
DATA ONLY,
MAINFRAMES (MF)
BATCH
PROCESSING DATA

Figure 7.1

The means at our disposition make feasible such implementation, provided that we have the appropriate culture for it. Figure 7.1 exemplifies the transition from batch to *Realspace*:

- In the 1950s, classical data processing has been Batch.

It still remains that way in the larger number of user organisations who look toward the future through the rearview mirror. These are retrograde applications, and money spent on them typically goes down the drain.

- Twenty-five years ago, *Realtime* (RT) permitted accessing and combining information online, but still addressed itself only to data - and primarily to transactions.

This is still not peak technology, though it is better than batch. Realtime features are still important, but the conversion of batch applications to RT revealed that many of them could be done quite profitably online in *Real Enough Time* (RET) - meaning immediate response but 24-hour synchronisation in terms of update.

- By the mid-1970s, *Electronic Messaging* systems made feasible the first significant computer support at management level.

Emphasising Electronic Mail (Email) capabilities and integrating a variety of older networks (such as Telex) electronic messaging opened a new era in communications. At the same time, decision support systems (DSS) constituted the other leg of computer-based assistance to management. Email and DSS are interconnected in more than one way.

- Experience, however, demonstrated that the needs of a communications-intense environment call for both *multimedia* and an incredible *shrinking of the distance* separating decision makers.

Decision makers must act together *as if* they were in the *same place* while in reality they are thousands of miles apart. This is the mission now given to *Realspace* implementation. The need is not only present, but is also quite evident - and it can be answered through available technology.

Supercomputers, which we discuss in Chapter 8, are vital to Realspace implementation. Parallelism, in the processing sense, is typically associated with the decomposition of an *information domain* (Space) and the simultaneous processing of a different information element - the fine or coarse grains formed in the decomposition. Parallelism is fundamental in the handling of distributed memory architectures, which increasingly imply a domain decomposition process that we are now starting to understand.

A money centre bank, for instance, operates 24 hours per day: from New York, to London/Zurich, to Tokyo. The sun never sets on the international bank. When the financial markets open in Europe they are just closing in Japan. When the exchanges open in New York, it is afternoon in Europe.

Dealers, traders and managers in different financial centres need to communicate among themselves *as if* they were in the same place. It is *zero space* that they are after, and technology can offer it to them - provided that systems architects are *able to conceive* this multifunctional network of information technology resources in a comprehensive, integrated sense.

High capacity networks, supercomputers and artificial intelligence constructs permit *mapping the market into the system.* This is a necessary condition for taking the whole portfolio and turning it around. Also, for *global risk management,* including the ability to:

- Track positions
- Pinpoint opportunities
- Evaluate risks
- Develop hedging strategies, *and*
- Do cross-market arbitrage

Realspace solutions to be provided should incorporate an impressive range of financial instruments and *bring together markets separated through tens of thousands of miles.* Not only *Realspace is feasible,* but it also can be achieved in a highly functional manner if we know what we are doing.

The automobile industry offers another example of the need for Realspace. Both General Motors and Ford have design laboratories in the USA, England and West Germany. Such laboratories need to work together *as if* they were in the same place. Their engineers exchange drawings, not just data. At times they may even be working on the same car.

There is still another Realspace requirement in the manufacturing industry. It is posed by the developing concept of very close coordination between design engineers, the production shop and inventory management. Realspace solutions are a prerequisite to a policy of zero (just-in-time) inventories.

Both in an organisational and in a technological sense, Realspace approaches address themselves to different types of individuals than those who can still be satisfied with realtime. Not only may the managerial layer to which they belong be different, but also the personality of the recipient of the service - including reasoning premises and structure vs. ambiguity in the decision process.

People of inductive reasoning who live and thrive in an environment characterised by ambiguity are *conceptual* in nature.* For them, Realspace is much more important

---

* See also D.N. Chorafas "Membership to the Board of Directors - The Job Top Executives Want No More", Macmillan Press, London and New York, 1988

than Realtime as it permits them to bring together in an integrated picture diverse cultures and information environments. By contrast, deductive people need more a realtime than a realspace service. This includes two classes: analytically oriented individuals, and administrators.

In terms of technology, *photonics* can be a major contributor to realspace. This reference is documented by the high bandwidth supported by optical fibres, as well as the high capacity, at low cost per bit, of optical disks. When it becomes available in practical applications, optical switching will provide another dimension to realspace implementation.

If the growing range of applications requirements is the primary reason for new departures, improvements through available technology is the second reason. There is a lot we can do with the convergence of new technologies. The rapid growth and application of *imagebanks,* as contrasted to classical databases, is an example. *Multimedia* storage facilities and communications links are increasingly important as:

- Graphics, image processing, video and voice

- Join the current large memory systems of text and data, in an increasingly integrated fashion.

No doubt, the coming years will see radical changes in the structure of the realspace market - from communications systems to storage solutions and human windows. While, for instance, the handling of imagebases is presently divided up mainly between the generation and processing of images, the consensus opinion is that in the next few years all that will change, with management of the imagebases becoming the focal point of interest.

Integration perspectives will be applicable at different levels: (1) Optical disk for central, departmental and WS level multimedia resources; (2) Megastream and gigastream channel capacity from long haul to departmental, with 24-hour per day availability; (3) Very high resolution on video, for compound electronic document reasons. Each area, and all three together, are in need of new concepts; fully knowledgeable, steadily trained personnel; and first class software, enriched through AI.

The Business of System Integration

Realspace approaches can be fruitful in many environments. Although they primarily address themselves to professional and managerial levels, to be supported through increasingly powerful workstations under a properly thought-out architectural solution, once these applications have been realised and justified, they will also spread to lower organisational levels.

Because of lifecycle requirements, the architectural approaches which we adopt must look further into the future, both in terms of the competent use of technology and of its profitability to our organisation. Increasingly, a company's infrastructure is being based on:

- System architecture
- Sophisticated networks
- Multimedia databases, *and*
- Artificial intelligence

These are the fields where, in the late 1990s, return on investment will be the highest. We must position our organisation against the forces of the 1990s. To do so, we must capitalise on the fact that computers and communications have made feasible a realspace environment, as information technology eliminates geographic distance.

The emphasis on projected lifecycle is fundamental. When a designer develops a new computer and communications system, the first question to be asked is: "Will the system survive the test of time?". Survival, in that sense, means coming out with a solution that is compatible with users' existing equipment and planning strategies while, at the same time, it has a demonstrated ability to grow as operations grow. Only flexible systems, designed along a layered concept, can transit through time in a valid manner.

Systems integration is not simply a technical term reserved for the specialist. Instead, it defines *a philosophy* applicable to every business. Better than any other solution, an integrated system identifies the organisation advantage which will evolve out of the need to adapt to the fast developing technology. An integrated system will maintain its value through organisational commitment to systems quality, compatibility and support.

The concept of integration itself transits through time, and as it transits it grows more sophisticated. During the 35 years of computer systems implementa- tion, the *toughest job* has successfully moved through levels of skill and expertise definition:

- Throughout the 1950s it was *programming.*
- In the early 1960s *systems analysis* became the challenger.
- The focal point in the late 1960s and early 1970s shifted to *system design.*

This change in emphasis, in terms of the most important job to be done, was itself a reflection of the transition which took place in data processing as reflected in the first row of Figure 7.2: From batch to classical realtime (point-to-point networks); then distributed data processing with minicomputers in a semi-peer setting.

However, the development of concepts did not freeze at the level of the latter reference:

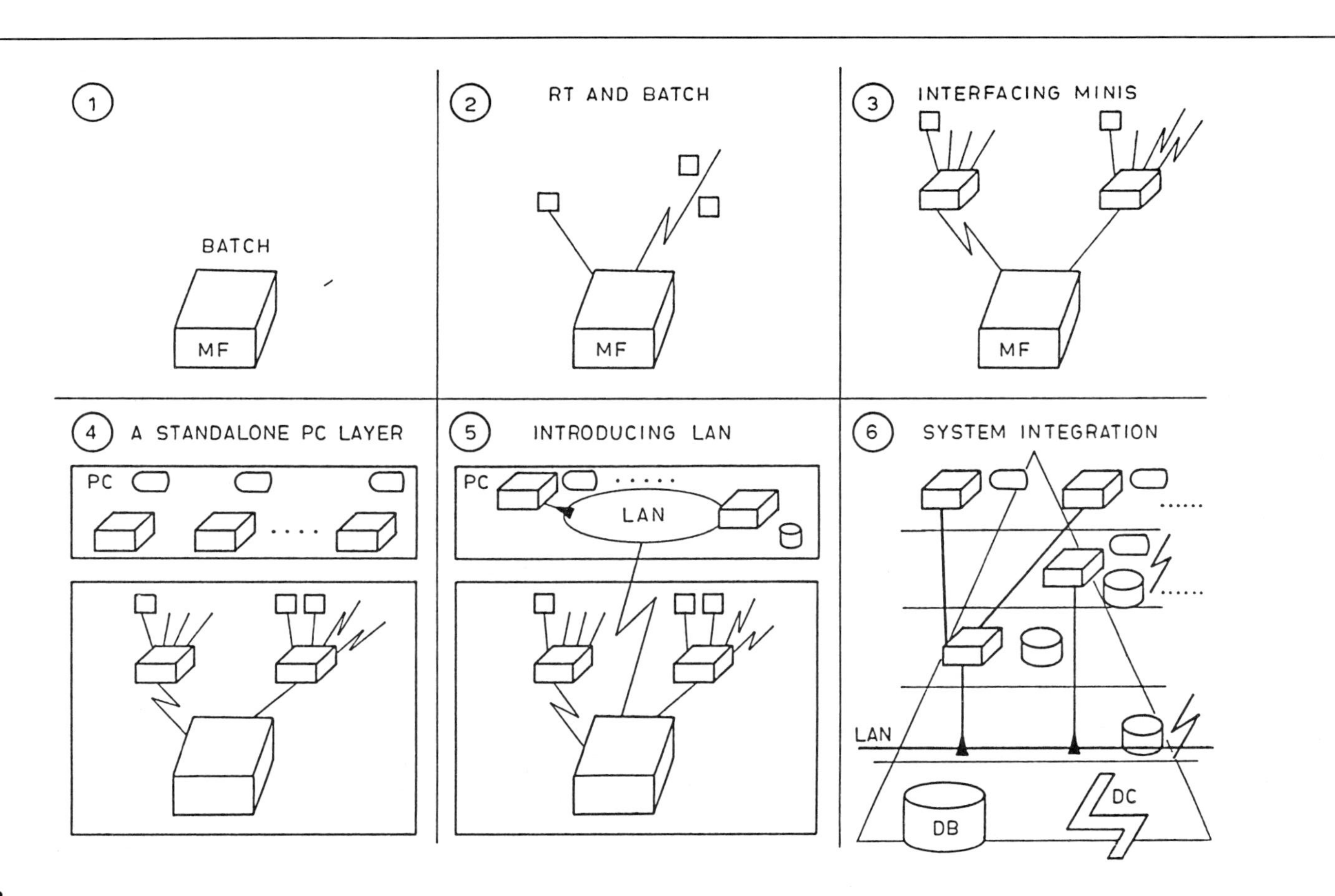
1
BATCH
MF
2
RT AND BATCH
MF
3
INTERFACING MINIS
MF
4
A STANDALONE PC LAYER
PC
5
INTRODUCING LAN
PC
LAN
6
SYSTEM INTEGRATION
LAN
DB
DC

Figure 7.2

- In the late 1970s came the standalone personal computers (PC) bringing the *enduser* into the picture.

- By the mid 1980s, local area networks (LAN) interconnected these PCs among themselves, and to the other computer resources, giving rise to the concept of an *inhouse system.*

- In the late 1980s, *system integration* took the spotlight as interconnected systems had to work as one aggregate.

The key person for this task is the system architect, who should work under the authority of the Chief Technology Officer (CTO) of the organisation, and who should include in his duties and goals architectural design for system integration.

Not only do the daily advances in computers and communications demand a new type of experience, but also the integration of the endusers into *one logical any-to-any system* implies a different skill from that which has so far been necessary. The emphasis is on *aggregate functionality,* enhanced through the best approaches technology permits, from AI to security/protection.

Sensitive information elements (IE) must be isolated and carefully watched. Procedures must be established to prevent IE which are being sent out to other networks from becoming exposed. Databases must be protected; access to them being limited to authorised personnel, identified through the appropriate authentication and authorisation keys. Keys must be managed.

System architecture and system integration and security perspectives are vital. But, while principles are necessary,* let us always recall that it is the knowledgeable men in our employment who turn the wheels of fortune and get things done. With computers and communications, the Chief Technology Officer and his people must be integration specialists able to:

1. Bundle microcomputers;
2. Integrate micros and mainframes;
3. Develop distributed databases;
4. Provide gateways for efficient communication;
5. Work out the new wave of office automation perspectives (global office);
6. Design effective workstations for user-based computing;
7. Apply 4th and 5th generation programming languages;
8. Choose a limited number of OS and DBMS;
9. Teach the users at all levels of the organization; and
10. Ensure that the company keeps up-to-date with high technology.

Historical precedents press in this direction. The failure to keep up with technology largely caused the shrinkage of the railroads, and brought major woes to the steel and

* See also D.N. Chorafas: "System Integration, System Architecture and the Role of Artificial Intelligence" McGraw-Hill, New York, 1989

automobile industries. Likewise, computer manufacturers who kept on focusing on mainframes lost large chunks of their market share, while those who made networks their strategy have been performing very well -DEC being a practical example.

The backbone of the modern information system is the *network* that links its subsystems in an able manner, connecting them to the external world. Networking, databasing, and information handling at large, is dependent on microcomputers, distributed processors and telecommunications.

However, the emphasis in communications is shifting towards digital networks. These interconnect, not only voice traffic and data transmission facilities, but also intelligent devices endowed with AI and able to provide services that were previously impossible. And, while we talk of integrated networks, differentiation is also necessary. Data traffic differs from voice traffic. It:

- May be queued for delivery;
- Can be stored, then forwarded;
- Calls for three orders of magnitude lower bit error rates (BER);
- Can be transmitted discontinuously in blocks or packets;
- Has different holding times; *and*
- For efficient services, it requires full duplex facilities.

Unlike people, machines can transmit and receive simultaneously. They can also process, in parallel, different types of jobs.

Integrated networks should be designed to support a wide variety of services, from teleconferencing and electronic mail to document handling, over and above voice requirements. Imaginative developments are called for as the solutions currently existing, and most of those projected for the near future, are not as radical as they need to be.

The question today is not simply how do we improve upon the telephone as we have known it for over 100 years. The new perspectives of telecommunications have outgrown, not only the classical telephone set, but also Bell's vision of a "grand system". Both men and machines require systematic upgrading within a perspective which evolves over time through major steps.

One of the keys to the developing capabilities lies in the use of electronic switching and transmission technologies under software control. Another is the integration of traffic. We have spoken of multimedia. Voice, text, data and image must be carried in a general purpose network whose:

● Speed

● Quality of service

● Cost *and*

● Mode of operation

should reflect future, not past, requirements and technological capabilities.

The coming decade is crucial, not because of the number of networks to be developed and implemented, but because companies, both large and small, must repeatedly make some *major* capital investment decisions. These will have to take into account the most probable shape of communications and computer facilities, as well as the way in which they will evolve during the next 10, 15 or 20 years.

All these references are part and parcel of the business of system integration, which itself is one of the important directives for the future. The way in which we approach the integration job must be practical, but should also be based on solid theoretical foundations. The latter are necessary to enhance the life cycle of the solution that we would implement.

Applying System Integration. Transactions and Messages

Implementing systems integration means tackling a large-scale job. The task goes beyond purchasing and putting together all of the necessary pieces of a computers and communications aggregate. It requires a global perspective, specific goals to go after, standards to be observed and well-defined connectivity requirements.

The premise is that a properly done system integration can greatly reduce the cost of all forms of communications: voice, image, graphics, text and data. Among the benefits from system integration is the logical compatibility of attached hardware and software devices in terms of: protocols, operating systems, data formats, data rates and transmission media being used.

Properly designed and implemented, system integration makes it feasible to handle multiple vendors. Other benefits come from:

1. Cost reduction,

2. Greater system effectiveness,

3. Simpler maintainability, *and*

4. Improved reliability

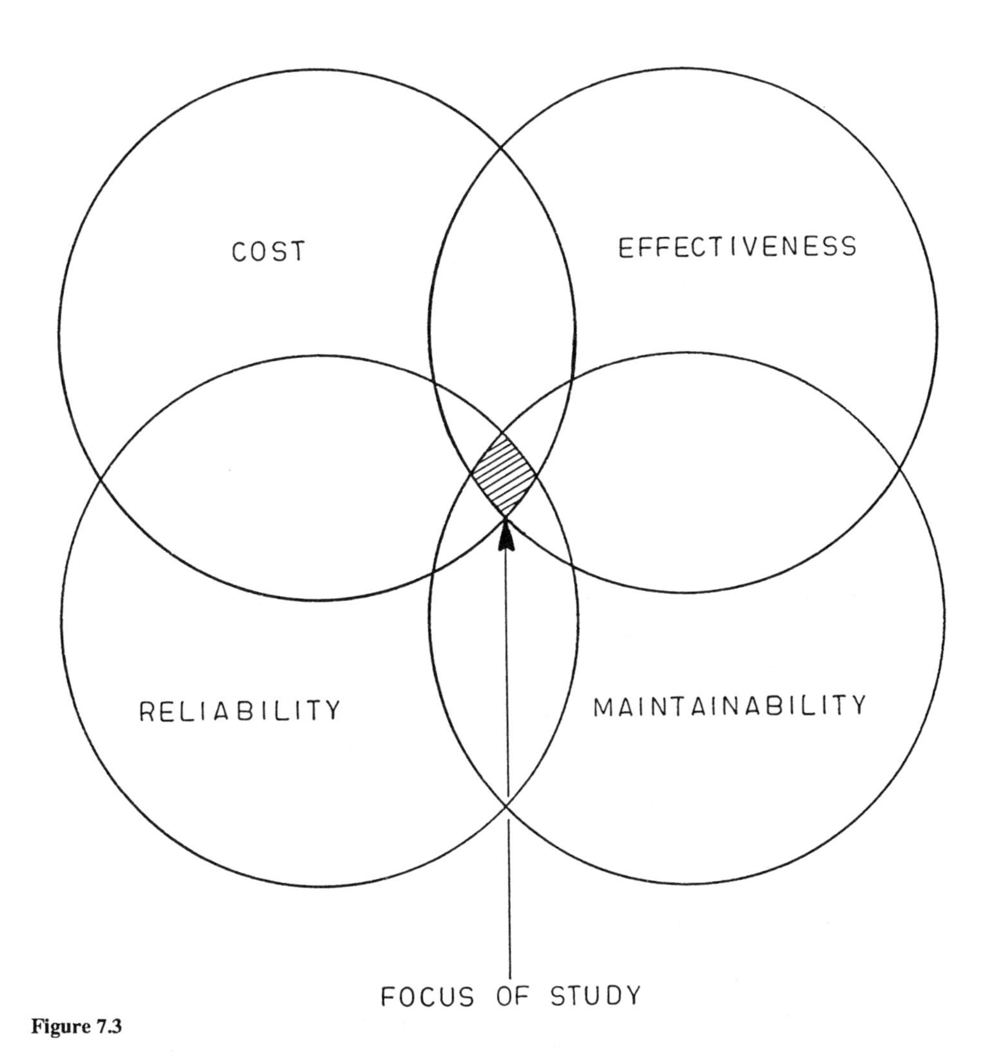

Figure 7.3

As Figure 7.3 suggests, these four areas of interest are interrelated. They have in common a core which should be the focal point of our study.

We develop networks in order to become, and remain, low-cost producers and distributors of products and services. Therefore, we simply cannot disregard the costs which are involved in the network itself. We must both plan them and control them.

Along this line of reasoning, for instance, EDS has been ripping out much of AT&T's existing equipment, lines, telephones and Centrex systems at General Motors, replacing them with more modern equipment, including a high-capacity digital PBX. Moreover, it is gradually taking control of the pipelines that tie all of these systems together.

By circumventing the Bell System's land lines with microwave and satellite communications, EDS plans to reduce the cost of GM's communications by about 20% to 30% per transmission. Leading American banks and insurance companies are also active in this sort of effort.

From banking and insurance to mechanical manufacturing, every labour intensive industry has been, for at least 10 to 15 years, actively seeking to automate its operations. But, most often this has been done by automating discreet ''islands'', with the result that many installed systems do not talk to each other. Now the emphasis has shifted to integration.

As the preceding sections of this chapter underlined, the able implementation of system integration should be achieved in a distributed sense. This brings into perspective the need for *base technologies* permitting us to organise and manage distributed databases.

A major part of the implementation reference regards the creation of database structures able to support increasing volumes of text, data and images that need to be stored, retrieved, processed and presented to the enduser. Another key part is linking them into one comprehensive aggregate. Ultimately, the database is where the components of an organisation's information system converge.

Perhaps more than any other topic, database integration requires conceptual capabilities. It also calls for organisation and a structure able to ensure that the database levels in a distributed environment fulfill two requirements which superficially seem to contradict one-another:

1. Being distinct in their enduser orientation, *and*

2. Creating among themselves a uniform aggregate.

The integration job should cover from personal, to departmental and central resources. If a *microfile* (hard disk at every workstation) solution is adopted - as it should be done - then an integration pyramid will exhibit at least four layers of databasing to be interconnected:

1. The microfile at the WS level
2. The LAN- or maxicomputer-based layer, at departmental level
3. The central information files (CIF) at the data centre(s) on mainframes.
4. The public databases perspective.

At each of these levels, databases will behave in a dynamic manner. Therefore, the whole layering concept should be worked out as an interactive system, from central resource to the enduser level of implementation. In terms of usage, the keywords are:

- *Transactions* and

- *Messages*

Transactions are short requests to the databases that generate a rapid response. They trigger authentication/authorisation and updating procedures. Transaction handling through distributed local databases offers:

- lower communications costs;

- more organisational flexibility; *and*

- potentially greater reliability than the practice of holding the entire database in a single, large computer.

A distributed processing of transactions permits users to exploit the power of personal computers by sharing the processing load among many processors, provided that our system operates as one integrated engine. Parallelism is achieved as transactions are processed simultaneously through each of the intelligent workstations. By contrast, in a centralised system resources are strained by a sharp increase in transaction numbers, not only because of the extra power that is necessary, but also for the reason of shared information files.

Messages are typically addressed workstation-to-workstation. Hence, they have even less of a reason to transit through central resources. As a result, they gain twice by being pipelined in a fully distributed sense:

- Direct access to destination

- Simpler, more cost/effective system.

Such benefits are real, provided that our system integration approach has been successful - prerequisites being that our work in the multimedia domain has *insight* and *foresight.*

The existing heterogeneity of media (computers and communications) will not be overcome through brute force. Key design prerequisites are to predict the direction of user requirements, capitalise on developing technology, invest to take advantage of lower cost solutions, and operate to enhance user satisfaction -and, therefore, corporate profitability.

The new and rational way in designing a system is to think of computers as peripherals. We must start with the network. Mainframes, departmental machines and personal computers are elements of the system. So is software. But, the *commanding authority is the network.*

At the same time, particular attention must be paid to the delicate balance to be maintained between information delivery and information processing. While many companies are looking for an equilibrium between the processing and delivery of information, thanks to any-to-any networks, the delivery side of services has become: simpler, more rapid, very accurate and of lower cost than it used to be.

No doubt, the choices to be made during the 1990s will provoke significant changes to our firm, as well as to its competition. Technical solutions should be fully dynamic on account of the fact that about every six months something quite significant happens that changes the way we look at the workplace and the communications means at our disposal. As far as we can currently project, this will continue to happen in some measure over the next 10 years.

Levels and Rules of Architectural Design

Having defined the primary levels of reference in a distributed information environment supported through personal computer power, we can now think in terms of rules to guide system integration. Such rules will typically apply at specific levels of architectural design.

Well thought-out architectural levels will themselves be part of a grand design. It is now generally agreed that failure to integrate the new technologies flooding the workplace has dire management consequences. Without:

- A long-term architectural goal, *and*
- Careful planning for integration, carried out immediately,

organisations risk building unmanageably complex aggregates. At the same time, management can no longer look to vendors to integrate systems. Every company has got to do it by itself - or it will be, and remain, at the mercy of different vendors and the half-baked approaches which have become the order of the day.

In terms of professional capability, a true integrator understands what information is going where, and how it can be used to solve problems. The Chief Technology Officer must be able to see the whole picture and put the parts together:

- text/data
- image
- voice
- software
- hardware
- interfaces.

The CTO must be almost a wizard in integrating components into cohesive information systems. For his part, the integration specialist must be keen to design and coordinate the proliferation of user-oriented computer tools, helping to create practical, usable aggregates. But, he must also provide a smooth, *comprehensive transition* path from the current structure to the new one so that the workflow remains integrated, and of practical use.

One of the focal points is the rules to be observed for system integration. Once established, they should be a subject of strict observance within *our* organisation. There are no universal rules yet available on how to develop and maintain a systems architecture. For this reason, based on experience, I am writing the following fundamental principles for distributed information systems:

Rule 1: We do not deal with 1 but with 4 networks: data communications (DC); databasing (DB); data processing (DP); and enduser functions (EUF). Of these, DP is the best known - although often misused -resource. The three foremost responsibilities of a network administrator (NA), database administrator (DBA), and enduser functions manager are shown in Figure 7.4. The same diagram also outlines how some of these responsibilities can be interconnected.

Rule 2: Two frames of reference must be observed in each one of the aforementioned architectural/network levels: *logical* and *physical*. One example of the distinction to be made is the *pel* and *pixel* in softcopy presentation at workstations. Of the two, the logical frame will be changing the fastest - yet presentation to the enduser must maintain consistency.

Rule 3: A message is a file in the database (pipe), and every file must be designed as a message. This becomes a general conscience: New languages incorporate pipe management commands and mailboxes. Addressing is increasingly packet-oriented, emulating what is learned in communications.

Rule 4: Local storage should be used for local operations. The traffic in the network should be kept at the necessary minimum. Preference must be given to Real Enough Time (RET) solutions over realtime alternatives, while those information elements subject to immediate response should be stored at microfile level.

Rule 5: Microfiles and local databases should be updated in realtime. Consistency with central databases should be assured in 24-hour intervals (RET), and a primary image of the information elements (IE) must be defined for that purpose.

Rule 6: Programs should, by preference, be stored at microfile level of the WS. Programs are more local than data. Therefore, *security* should be enhanced all the way: from central resources to departmental and microfiles. Both programs and data should be treated as a corporate resource.

Rule 7: Data entry should be done *only once*, and the data entered used many times. This one entry/many uses principle can save a lot of money, and it is relatively easy

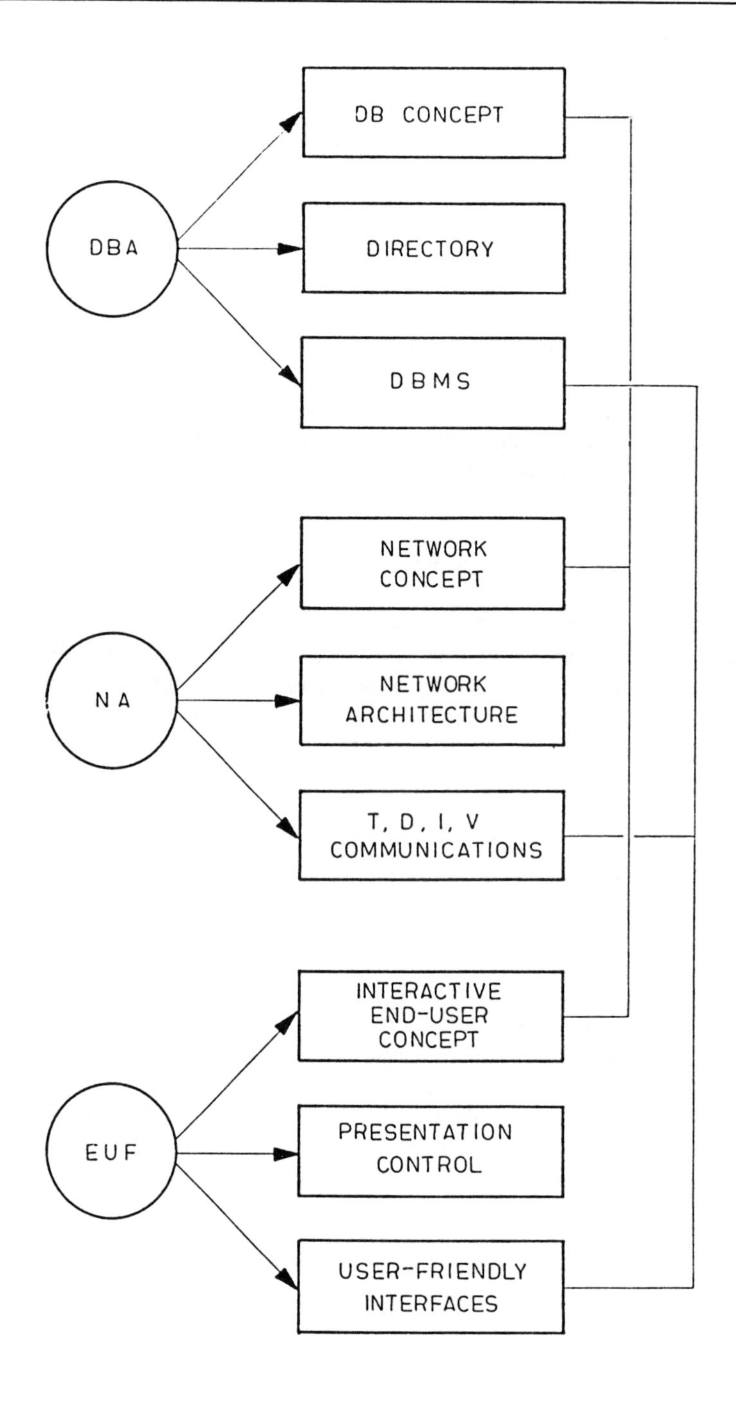
DB CONCEPT
DBA
DIRECTORY
DBMS
NETWORK CONCEPT
NA
NETWORK ARCHITECTURE
T, D, I, V COMMUNICATIONS
INTERACTIVE END-USER CONCEPT
EUF
PRESENTATION CONTROL
USER-FRIENDLY INTERFACES

Figure 7.4

to implement if our system has been properly integrated. Another benefit is much lower error rates, and the weeding out of file design incompatibilities.

In fact, *our* organisation has an interest to see to it that data entry takes place at the office of its clients and of its suppliers. Subsequently it should be transmitted online:

● Realtime, if urgent, and the corresponding cost is justified.

● Real Enough Time otherwise.

Rule 8: Computers are *not* meant to be programmed. First locate the software (in packages), then test this software and negotiate its price. When satisfactory software has been properly identified, then comes the time to locate, test and negotiate hardware on which this software runs.

Rule 9: Systems planning should lay stress on: portability (machine independence); reliability (for high systems availability); restart/recovery capability (including journaling and statistics); online auditing; and security/protection. These are integral parts of an architectural solution which should also present the other characteristics we have been discussing in this chapter.

Rule 10: Simplicity in design should be a steady guideline. Only what is simple is valuable. Keep the computers and communications systems simple, small, and able to be used by a "stupid" user. The better is the enemy of the good.

While the observance of these rules would not automatically assure success, failure to follow them opens a Pandora's box to many other failures. Experience has always been a good consultant, and these rules are based on experience gained from practical implementations. Experience also suggests that, although software and hardware would reflect the current level of technology, our design should have a longer lifecycle expectancy, to remain valid for the next "X" years.

Unavoidably, any design will present tradeoffs. For a network architecture, one of the biggest tradeoff issues is the choice of switching technology. Switching choices essentially mean a decision on how to share the communications facility and, through it, the database resources.

With point-to-point operations, the alternatives have been multiplexing and switching. With multipoint (multidrop) approaches the available choice was polling/selecting, polling proper and contention. They all involve line switching.

There are three basic disciplines available today for line switching: circuit, message and packet. Circuit switching is the oldest, and the more widely used. Packet switching is the newer, offering broader facilities. Relatively speaking, it is the more efficient. But, for design decisions, we also have to pay attention to:

● The topology over which our network will span;

- The use of polyvalent multifunctional endstations;
- Transmission proper characteristics (media, capacity);
- Basic electrical characteristics (noise, conditioning, synchronisation, regeneration, frequency spectrum);
- Protocols (at each layer of the layered architecture), *and*
- Effective interfaces (between successive layers).

The last two references are logical in nature. They presuppose the adoption of corporate-wide system architecture - which follows a l*ayered approach.* End-to-end digital connectivity must be based as far as possible on standards. These are background factors supporting the more sophisticated part of the network.

Other considerations, too, need to be brought to the forefront. By and large, they have to do with pipe capacity (bit streams), bit error rates (BER), and - for inhouse installations - the topology of the cabling plant. Together with reliability, such factors will define a significant part of the visibility reaching the enduser level:

1. High *network availability:* 99.9% or better;

2. Network *response time:* 1 to 2 seconds, or less;

3. Network *data integrity,* doing away with corrupting, losing or misdirecting messages and transactions;

4. *Network integration i*n the global service sense which we should be promoting.

A careful examination of the foregoing factors helps to illustrate the fact that systems integration requires that the communications choices regarding architectural design are properly made. As section 4 identified, able design choices are one of the prerequisites to database distribution. The latter should aim both at high functionality and at the lowest possible cost. Therefore, it will involve tradeoffs between microfiles, local storage, and central storage, on the one hand; and communications costs on the other.

In the realm of databasing and communications solutions we will also be explicitly choosing uptime, avail-ability and maintainability. These three factors should be constant preoccupations. The same is true of the ability to handle fluctuations in data flow, including the way in which they affect operations.

Let us recapitulate. To do a clean job in system integration we must start with the following objectives: choose the transmission facilities; establish the switching methods; project the layers of databasing; elaborate the topological layout; select the WS equipment; control the central resources; settle on the protocols; and establish the interfacing. This means choosing from various alternatives and documenting all the choices being made.

The multifunctional aspect of these technical references, and their breadth, easily identifies the fact that the job of the system integrator exceeds anything we have projected in the past in terms of computer system analysis and design. We must now consider formatting, text/data presentation, connection control, addressing schemes, maximum packet sizes, routing techniques, contention, collisions, timeout, error recovery and status reporting.

Error Detection and Correction

Superficially, it might seem that error detection and correction is a separate subject from system integration. In reality, exactly the opposite is true.

Realspace environments must be both reliable and dependent. System *reliability* relates to uptime. Reliability is not an ability. It is the probability that a well-defined system will operate without failure over a specified period of time, within a determined environment.

As defined in this text, *dependability* relates to the freedom from error. As I have had the opportunity to explain in several cases, in the general case an error is no malicious or utterly unwarranted event. It is the means which provides us with the opportunity to correct the behaviour of our system providing:

- a feedback, *or*
- a feedforward mechanism.

Quite definitely, we must use this opportunity. Taking bit error level as the frame of reference, the following section explains the methods available to effect improvements in bit error rates (BER).

In communications systems redundant information is added to messages so that transmission errors can be detected. Such redundant information is generated according to a prearranged function, whose domain is the bit or byte string being transmitted.

The sender applies a protocol-established function to the data, adding the resulting redundant information to the message. The receiver uses the same function to recreate the redundant information from the bit or byte string.

- If the generated redundant information does not match that received from the sending station, the receiver knows that it has an error in the text.
- If a sufficiently large amount of redundant information was added to the message, it would be possible for the receiver to correct, as well as detect, errors.

For many years, it has been cheaper to build terminals which only detect errors, and which they request the generating station to retransmit the unrecognised message until it is recognised. The two most commonly-used methods of adding redundant information are:

1. A vertical redundancy checking bit (or bits)

2. A block check character (or characters) added to groups of transmitted characters.

Parity (vertical redundancy) bits are added to each character to cause the characters always to contain, say, an even number of 1 bits in even-parity systems. In odd-parity systems, each character is adjusted to contain an odd number of 1 bits.

The receiver checks each character to ensure that it contains an even (or odd) number of 1 bits. Vertical redundancy checks are not the most efficient method of adding redundant information to transmitted data, primarily because of the nature of errors in actual communication facilities.

Errors tend to occur in groups, or *bursts,* rather than to be uniformly distributed. This is especially true in high speed transmissions, where a line error of a given time-duration will affect a larger number of bits than would be the case if the line were operated at a lower speed. For this purpose, the block check character is used, either alone, or in addition to vertical redundancy checking. Block check characters are generated using all the information contained in the message.

With packet switching we are using a two-byte long cyclic redundancy check (CRC). This is a polynominal employed by the receiver to actually detect errors. It is always transmitted at the end of the data and covers the whole length of the packet.

These are the classical approaches to error correcting in datacomms. Their development was a matter which occupied communications specialists in the 1960s. Subsequently, new solutions have been put in place, intended to improve significantly the bit error rate (BER) in telephone transmission lines.

The new CCITT V.32 recommended standard incorporates forward error correction (FEC) into what will probably be a better generation of modems. Interest in its implementation is enhanced by the fact that chip-level error detection and correction modules are getting cheaper, while their capabilities are being enhanced.

With plain old telephone service (POTS) the goal is to be able to improve the bit error rate performance of a transmission channel by two to three orders of magnitude, without having to effect changes to the actual circuit. Through forward error correction, the receiver may be able to reconstruct a data block (where, for instance, error bursts have altered a hundred consecutive bits), without requiring a retransmission from the sender.

FEC works by processing a bit stream through a series of algorithms at the sender level prior to transmission:

- The bit sequence is rearranged,
- Sometimes the binary values are changed, *and*
- There are extra bits added to the original data block

At the receiver end of the circuit, another FEC processor uses decoding algorithms after receiving the message. Control bits inserted at the transmitting end are employed to determine if the user block was received correctly; if not, they are used to correct the errors.

Through this approach, a remarkably distorted data transmission can still be reconstructed at the receiving end. This is accomplished without having to retransmit any part of the original data.

Let us recall that, with classical coding, each bit of a user data stream is compared with one or more of the bits sent in a parity control group of bits. With FEC, the value of each bit which may be changed by the coder is tied to the value of the other bits. Then a redundant bit is added for every group of bits handled in this manner.

When a bit is compared only with the bit that preceded it, the number of redundant bits required to assure decoding at the receiver end is high, though the complexity of processing at the other end is minimised. On the contrary, when the bit is compared with a large number of previously transmitted bits (the constraint length in FEC), the number of redundant bits is minimised, but the complexity of processing at both ends is higher.

Bandwidth has to be used to account for error incurred in transmission. With the control codeword for requesting retransmission (ARQ), which is the most popular approach today, any frame or block received in error is sent again. This error correction overhead can be viewed as part of an expanded bandwidth requirement.

Just because part of the bandwidth has been classically allocated to data assurance, FEC is an improved discipline. With forward error correction - as with CRC - the bandwidth expansion consists of the redundant bits added to the user data.

Compared to the CRC used in ARQ schemes, the FEC overhead, as a percentage of total bits sent, is often higher - but, as stated, it saves the need for retransmission. For the time being, the force working against FEC is not in communications overheads but in the cost of modules to perform forward error correction.

This is changing. The cost of such equipment over the last 10 years has dropped by a factor of 10. As a result, applications that could have used FEC to improve channel performance, but which did not warrant the added cost, are now finding it affordable. For instance, FEC has been included as an integral part of the CCITT's new dialup modem specification.

Along with bit error rate improvements, the network must assure the proper housekeeping facilities. Software should provide:

1. *Outage* incidents detail by application line, and connected station, giving a specific description of the actual incident that caused the outage.

2. *Circuit* incidents detail, listing problems encountered by circuit number referenced to the line and modem.

Both reports should show the incident duration, and the total relevant outage encountered by that circuit.

3. *Pending incident analysis*, showing all unresolved network problems.

4. *Detail by location*, exhibiting all incidents for a geographic area in a closed or pending status for a specified timeframe.

Data analysers are one of the means for network control. They are applied in any one of three modes: *active*, *semi-active*, or *passive*. In the passive mode, they simply capture and store all online data. Keeping their buffers continuously full, they retain the most recent data and discard what came earlier.

A semiactive data analyser can be programmed to search for a specified character or bit stream. If the search is successful, it traps the specified sequence and alerts the operator, while also capturing the subsequent data.

In the active mode, the data analyser records/evaluates, but also transmits, data, selected commands and responses, into the network. With this, it becomes possible to emulate any intelligent data terminal equipment or data circuit terminating equipment.

In either mode the data analyser can be used for both troubleshooting and development purposes. The operator, or another machine, exercises the analyser's various options:

- Once the initial transmit actions are taken, the operator waits for a valid response message.

- When it is received, an appropriate transmit message is generated while the analyser enters the next operational state.

But there is a limitation with these solutions, and this has to do with channel capacity. The data rates an analyser handles are typically up to 19.2 KBPS. Some analyser vendors provide additional interfaces for printer output and remote access. Remote access minimises human intervention at remote sites.

Most data analysers can interpret the standard codes of ASCII, EBCDIC and hexadecimal identification of the frame architecture. The operator no longer needs to interpret the "1" and "O" in control fields of bit-oriented protocols to determine the type of frame or packet.

This is accomplished by condensing the information of several fields, contained within the protocol's frame architecture, into a readable format. A split screen can be used to display source data from a data terminal and data circuit terminating equipment simultaneously.

Table 1 presents a generalised approach to error handling and recovery procedures along the three phases discussed in the preceding section: Association, Data Transfer and Dissociation. The outlined procedure can be significantly improved through the use of equipment discussed in this section.

Since our future increasingly depends on communications, and we know that the current plant is ageing - hence, faulty - error detection and correction become an integral part of the system architect's frame of attention. He may not be a specialist in every field, but he should use specialist assistance to ensure that the system he is designing is both reliable and dependable in the functions which it supports.

Table 1: Error Handling and Recovery Procedures

Error in:	*Error Handling:*
Association Phase	Break attempt to establish a connection, or cancel connection at remote host.
Data Transfer Phase	Available alternatives are: ● termination of session after finishing local and remote file actions ● retransmission of information elements ● checkpointing and restart
Dissociation Phase	Terminating of session after finishing local and remote file actions without performing additional functions.

8

The Era of Supercomputers

Introduction

A ''strategic product'' is one on which our organization depends for its survival. That is also the meaning of a ''strategic solution''. In the larger sense of computers and communications, from central resources to personal workstations, system solutions have become of strategic importance.

Organisations which cannot manage their information resources enter a crisis and lose their competitiveness. But, proper management does not come as a matter of course. First, we should choose our goals. Second, focus on our salient problems. Third, change our:

1. Culture

2. Policies

3. Methods *and*

4. Tools

Having taken these steps, for the next ten years managing computers and communications would involve much more than chores performed so far. Enduser computing, direct access to global databases, wide use of electronic message services, an increasing amount of electronic document handling, expert systems, a growing range of other AI applications, such as pattern recognition and language translation, as well as the need to drive these developing large scale systems, will all greatly impact on computer power requirements.

Power requirements are expected to be massive and, therefore, cannot be met by step-wise increases in mainframe-supported MIPS. This is a fact already reflected through current estimates on the projected growth of the computer market (Table 1). The fastest growing sector is supercomputer hardware, closely followed by supercomputer software.

A 1988 study by *Datamation* indicated that of 600 readers who responded:

A. 12 percent use supercomputers, while

B. 30 percent employ high-performance computing.

Class B includes: minisupercomputers, mainframes with attached vector processors or supercomputer timesharing services. Quite significantly, 10 percent of the total population who did not have supercomputers planned to purchase one.

Supercomputers seem to be a rewarding market but, as 35 years of experience has shown, at the bottom line the limiting factor is software - not hardware.

Table 1: Compound 5-Year Growth of the Computer Industry in the American Market

Estimated 1987 to 1992 Performance (Optimistic Estimate in all Classes)

	Percent Growth in 1987/1992	
	Compound Growth	***Est. Total USA Market Value in 1992****
Supercomputers	257%	$1.8 Billion
Mainframes	66%	$15.0 Billion
Superminis	122%	$31.8 Billion
PCs	160%	$58.0 Billion
Software for Supercomputers	250%	$2.2 Billion

Unless we change our methods of developing and testing software, the era of supercomputers will pose staggering requirements to which it will be difficult to respond. Current experience serves to document this reference.

Dr. David L. Parnas was a member of the Consultative Council for computer usage of SDI. His thesis has been that, under current technology, it is impossible to guarantee *system reliability in software*.

Dr. Parnas takes the operating system 360/370 as an example. When OS/360 was first developed, 5,000 errors were found in it during the testing period. Never has a large scale software program been built without errors, and errors cannot really be weeded out without long tests under real operating conditions - and real operating conditions can only be established through actual usage.

* Only the professional PC's price is projected to increase due to added functionality. All other prices will drop. Nevertheless, the growth in dollar value is impressive.

The Parnas argument is not necessarily 100% solid, as operating conditions can be simulated, and errors weeded out. This is bound to be a long and tedious process, but it is not impossible. What is impossible is to foresee all local events and environmental interactions which may be happening in the territory covered by the large scale software system. Only AI can be of help in such a task - and this necessarily means new types of software.

But fully failure-free operations are another matter. On more than one occasion, the computers of the North-American Air Defense System (NORAD) interpreted flocks of birds as missile firings. In 1975, the fire in a Russian gas pipeline was first interpreted by computers in America as missile firing. Other examples also weigh heavily on the risk factor.

For reasonably failure-free operations, supercomputers *require* a new type of software. Great improvements in software are embedded in artificial intelligence,* but AI is still in its infancy. The software hurdle which currently exists is not expected to be solved before the year 2000 - and to solve it we need very highspeed parallel processing.

Parallel processing and artificial intelligence would work in synergy, accelerating the rate of change towards generations of computers helping to create new forms and solutions previously unknown and, at this point, hard to visualise. The future involves an increasing amount of parallel processing, as both scientific and business problems are layered with parallelism, and increasingly require an impressive number of giga instructions per second (GIPS).

To appreciate where we are going, it is wise to turn back to the first generation of computers, in a rapid follow-up of the successive steps in computer generations. This can be instrumental in exemplifying design philosophies, and in explaining the goals to be reached through 5th generation computer engines.

The first generation of computer hardware featured about 10,000 operations per second. Hence, the unit of measurement was KIPS. These machines were basically designed for numerical calculations and they were programmed in machine language. Physically, they were built with vacuum tubes (1953-1958) and they were working in a sequential manner.

By substituting vacuum tubes with transistors, the machine size with second generation computers was substantially reduced. Their reliability increased so that they could run for days without failure. In terms of power, they featured a few hundred thousand instructions per second (1958-1963).

However, a different way of looking at this matter of transition in the history of computers is that the development between 1953 and 1963 made their weight felt on

---

* See also Chapter 4.

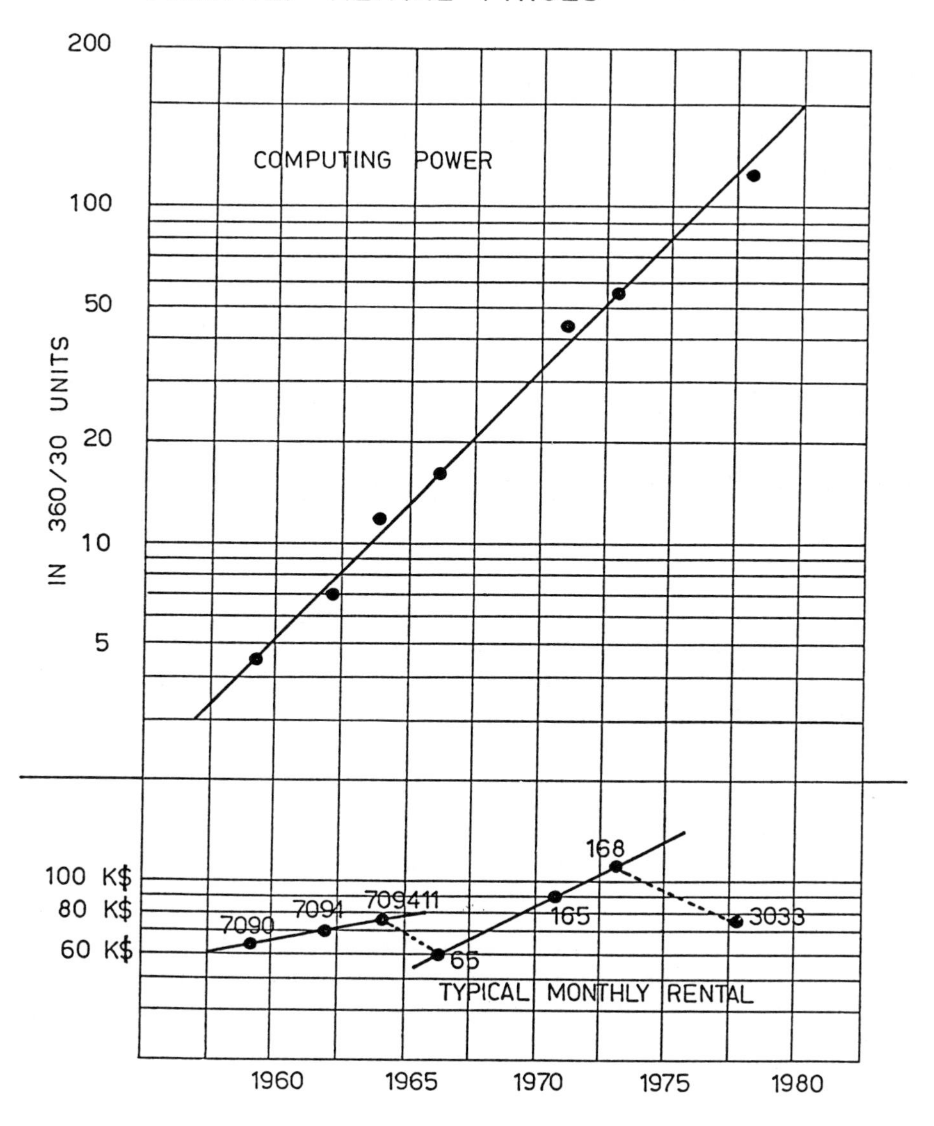
HISTORICAL TRENDS IN IBM LARGE SCALE COMPUTER SYSTEMS AND MONTHLY RENTAL PRICES
200
100
50
20
10
5
IN 360/30 UNITS
COMPUTING POWER
100 K$
80 K$
60 K$
7090
7091
709411
65
165
168
3033
TYPICAL MONTHLY RENTAL
1960
1965
1970
1975
1980

Figure 8.1

the third computer generation (1963 to 1972), which was essentially distinguished by an effort towards upward compatibility, as exemplified by IBM's 360 series. But:

- Unlike their predecessors, third generation computers were memory-centreed, rather than arithmetic and logical unit (ALU) centreed.

This laid the foundations for a significant change in the way that we look at computers and computing in a networked sense.

- As technology advanced and computing power increased, the costs start dropping, thus leading to a truly affordable computer power (Figure 8.1).

The challenge with really different and more cost/effective computer design concepts started in the early 1970s, with the business implementation of the minicomputer, and in the late 1970s with the PC. The motto became: cheaper and cheaper. It stayed this way until about 1985, with the goal being one of making a von Neumann-type computer, but cheaper than ever before.

Computer size changed, and so did prices, but the structure practically remained the same until a different goal dominated design: newer and newer. This is a post-1985 objective, and it can be expected to last until the end of this century. The focal point is *parallelism,* one of the pillars of 5th generation computers (5GC).

Problem Processing through Parallel Systems

For all leading computer manufacturers, *parallel computer systems* are strategic products. This direction is taken in reflection of the fact that everyone understands very clearly the limitations of current technology. But, parallel systems and 5th generation computers are not necessarily synonymous terms.

Parallel (or highly parallel) stands for tightly coupled systems. They emphasise synchronisation on a byte-by-byte basis. However, the number of processors included in the architecture does not need to match the byte-by-byte level. Furthermore, in the general case, parallel systems have no AI orientation - while AI is one of the pillars of 5GC.

With this definition in mind, we can look into the types of parallel systems and their classification. Both von Neumann and non-von Neumann machines are included in this definition, as some parallel computers still follow a von Neumann architecture, while 5GC is a totally different departure.

- MIMD, stands for multiple instructions, multiple data stream.

With some exceptions, this is basically a von Neumann type machine, but of improved characteristics. Its design is along control flow lines and, therefore, it partly overlaps with Data Flow computers.

- Data Flow design is usually based on a packet communication type which influences machine organisation.

Such organisation consists of a circular instruction execution pipeline of available resources. In it, processors, communications switches and memories are interspersed with a pool of work elements.

- SIMD, single instruction, multiple data stream

With MIMD machines each node can perform independent tasks asynchronously. By contrast, SIMD are lock-step engines where, on each clock cycle, each node operates identical instructions. This requires a fine network to connect nodes to memory in the shared memory case, or nodes to other nodes in the *distributed memory* case. Of these two approaches, the distributed memory is by far the more interesting for the coming decade.

For many years shared memory SIMD machines have been commercially available, such as array and vector processors. However, it is only during the 1985/86 period that general purpose distributed memory SIMD machines have become available. Many of these engines are *hypercube-type* multiprocessors, which consist of a number (a power of 2) of loosely coupled processors in a binary n-cube configuration.

Known as *Boolean Hypercube* or *Cosmic Cube*, this architecture has been invented at the California Institute of Technology (Caltech) in the Parallel Processing Project, whose goal was to develop the capability of concurrent processors. At Caltech, essentially all *software* for the hypercube has been written from scratch. This new software is polyvalent and can usually be run on either parallel or sequential machines.

The available experience with MIMD and SIMD machines worldwide is not so great as to lead to firm statements. Available indicators lead to the tentative conclusion that MIMD architectures offer somewhat greater flexibility, though possibly lower peak performance, than SIMD machines. Table 2 correlates memory structure (shared vs. distributed) and grain size. It also gives examples of equipment currently available, or at an advanced development stage.

The references given in Table 2 help document the fact that a wealth of research can be done in parallel processing, where there exists a variety of solutions. Computers of several architectures have shown good performance in many fields, though machines of conventional von Neumann architectures still dominate MXMP, Modest Coarse Shared the commercial market. So far, highly parallel machines with over a hundred processors have largely been confined to university and industry research organisations, but the trend is now towards broader use in business and industry. Dow Jones for instance, has installed two Connection Machines.

We said that a variety of approaches is followed in computer design. An SIMD design can have variations such as SISD (single instruction, single data stream). This

Table 2: Memory Structure and Grain Size of MIMD and SIMD Computers

Control Type	*Grain Size*	*Memory Structure*	*Examples*
SIMD Many Processors	Fine	Distributed	ConnectionMachine (CM-1, CM-2), Goodyear MPP, ICL CAP, AMT, Mini DAP
SIMD	Coarse	Shared-Distributed	IBM Yorktown GF11, Columbia QCD Machine, APE (Rome)
SIMD Modest Number of Processors	Coarse	Shared	Alliant, CRAY-CRAY 2, ELXSI, ETA-10, IBM 3090 VF, Sequent, Flex, Encore
MIMD	Coarse	Distributed	iPSC, Ametek, FPS, NCube, Transputer Arrays (CSA, Inmos, Meiko), Suprenum (GMD)
MIMD	Coarse	Shared	BBN Butterfly, CEDAR, IBM RP3

is basically a von Neumann computer solution. All the same, MIMD can have variations, as this design tends to merge with data flow.

As we know from everyday practice, definitions in basic computer design are not fool-proof as new technologies expand in a variety of ways. Since communications and, therefore, interconnection is the critical element, the way to connectivity conditions the infrastructure of the machine. Still, we can distinguish a von Neumann from a non-von Neumann architecture, and this is shown in Figure 8.2.

- The von Neumann architecture is basically composed of five parts: input, output, memory, ALU and control.

- In prevailing non-von Neumann architectures: input, output, and control are on one side; memory and ALU on the other. The latter can be massively parallel.

There are also fine detail differences in design, one of the most important relating to switching. We can, for instance, differentiate between *switching chip* interconnections for high bandwidth vs. local area network (LAN) type approaches. Often, switching circuit interconnections are mesh-type, but the more modern type is hypercube connectivity.

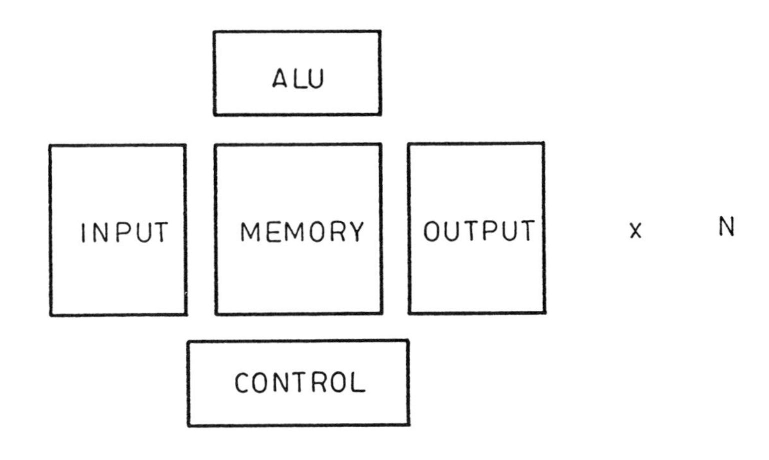
A. VON NEUMANN ARCHITECTURE
ALU
INPUT
MEMORY
OUTPUT
X N
CONTROL
A TIGHTLY COUPLED SYSTEM

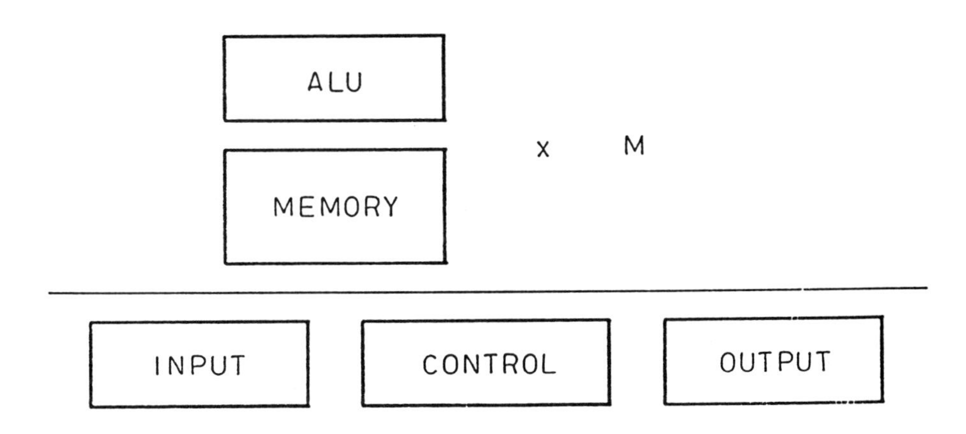
B. NON-VON NEUMANN ARCHITECTURE
ALU
X M
MEMORY
INPUT
CONTROL
OUTPUT

Figure 8.2

Vitally important basic design features include: high bandwidth, alternative pathing, enhanced availability, high capacity and add-on capability. The competent choice of key parallel processing characteristics is instrumental in the construction of cost/effective systems. These may be:

1. General purpose computers, designed in a modular manner;

2. Specialised processors for applications ranging from design automation to refinery control;

3. Artificial intelligence engines, addressed to a pallet of applications;

4. Signal and image processors, which are recently becoming of increasing importance; and

5. Dedicated database computers for management of large text, data, voice and image warehouses.

In all these fields, and more, current research focuses on system structure and performance - including load balancing, offloading, separation of functions, elimination of bottlenecks, and recovery methods. A key point regards the feasibility of evolving today's software to a status able to provide very efficient interconnection of hardware. Here, too, AI has a role to play.

In a networking sense, data transport instances in the switching structure map into session control and route control. They include transmission line control, and see to it that movements of data transport instances between processors remain transparent to the network.

Diagnostics are typically run through an input/output analysis protocol. Solutions for continuous operation necessarily involve dynamic address space reconstruction for changes in underlying processor hardware/software, including the application of maintenance to the control program. Techniques are necessary for online module replacement, and for reconfiguration without process interruption.

Parallelism and 5GC concepts in machine design help provide new approaches to existing services - as well as the ground for new types of services. This may mean expert systems assistance in decision making, and new product lines which are tightly knit with computers and communications.

Technology permits polyvalent approaches, but to run such an environment in an able manner we must manage change. Managing change means a new culture, not just investments. It also calls for skills. Long term perspectives, and the able use of human capital, should definitely be part of the picture.

To a very substantial degree, in business and industry today such multipurpose systems are not yet well understood. And systems which are not well understood can neither be used efficiently nor become popular. While we do appreciate that large problems are inherently parallel, we also understand that machine cycles alone will not make the difference. We also need first class software and a valid methodology.

Do We Need 5th Generation Computers?

Faster, bigger and more reliable computers have been the designers' goal for over three decades. But, over the last five years the race towards bigger computers has taken a different turn. As Dr. Grace Hopper (the inventor of symbolic programming languages) was to underline in a conference, "when we need more computer power, the answer is not to get a bigger computer". That is what the synergy of 5th generation computing and AI is all about.

Fifth generation computers and intelligent information networks are the answer for projects that require:

- Huge amounts of computing power

- New online solutions, *and*

- Very fast access database engines.

While the era of photonics will, most likely, make itself felt by the mid-to-late 1990s, 5th generation computers and parallel processing are making their impact today.

Figure 8.3 exemplifies this transition *from electronics to photonics.* It also positions microprocessor-using hypercube architectures at the end of electronics era -that is, the remaining decade of this century.

1. In von Neumann type architectures, reflected in the first 4 computer generations, memory and processors (ALU) were tightly coupled 1:1

2. Though of different types and designs, multiprocessors share the memory resource among different processors.

3. Hypercube architectures return to the tight memory and processor coupling, but support massive parallelism through a network structure.

In this sense, hundreds and thousands of processor-and-memory (P+M) units are the nodes of the network. They communicate among themselves very effectively through hypercube links.* Transmission speeds reach the range of 2 to 3 gigabits per second (GBPS).

The Caltech Hypercube has been used in applications covering an impressive range: computer graphics; event driven simulation; computer aided design (CAD); finite element analysis of earthquake engineering;

---

* See also CHORAFAS and STEINMANN: "Supercomputers. New Generation Computer Technology and the Impact of AI," McGraw-Hill, New York, 1989

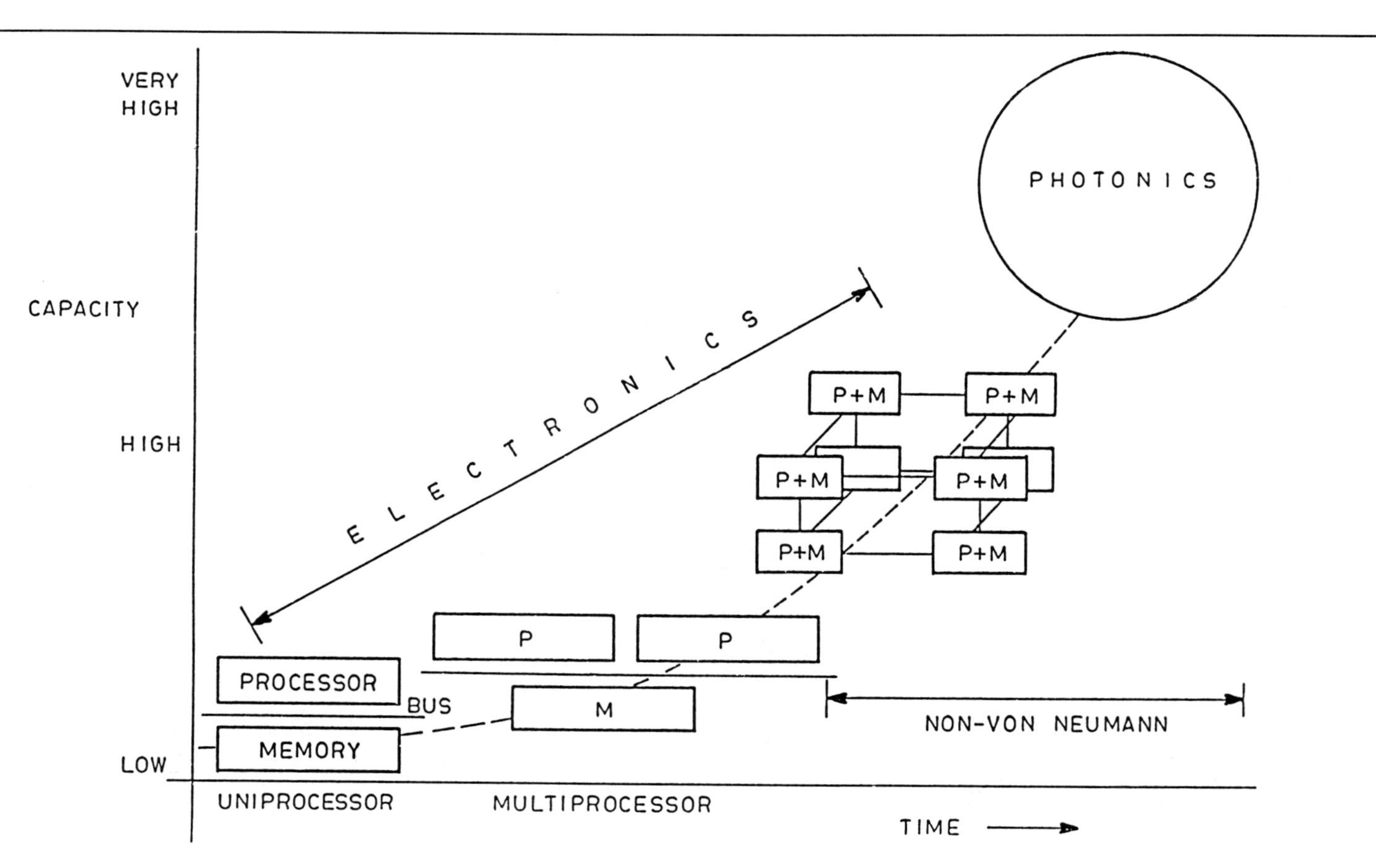
VERY HIGH
CAPACITY
HIGH
LOW
PHOTONICS
ELECTRONICS
P+M
P+M
P+M
P+M
P+M
P+M
P
P
M
PROCESSOR
BUS
MEMORY
NON-VON NEUMANN
UNIPROCESSOR
MULTIPROCESSOR
TIME

Figure 8.3

matrix algorithms; load balancing algorithms; optimisation; parallel shooting; multigrid adaptive meshes; study of antenna field distributions; mathematics and logic; modelling of cortex and applied neural networks; protein dynamics; chemical reaction dynamics; plasma physics; molecular gas dynamics; convolution decoding; and robotics. It has also been used in financial and insurance applications.

Hypercubes have not been the first major departure from classical computer design. We already said that in the mid-1960s, with the 3rd generation, memory became the pivotal point, replacing the ALU in that role. An important part of the 4th generation have been *array processor* architectures, which permitted the first supercomputers to be built. The products of Cray Research are among the best known array processors.

In an architectural sense, hypercubes are an improvement upon the capabilities offered by array processors. There are also other architectural approaches with 5GC. Utilising its experience with long haul packet switched networks, Bolt Beranek Newman (BBN) has designed a 5th generation computer based on network system concepts, with nodes and links.

Though a network type design of packet switching orientation and the hypercube are by no means the same, they do share similarities. They:

- Efficiently use a communications discipline in their basic design, and

- Represent peak technology able to deliver large amounts of computer power.

The second release of the Connection Machine, for instance, delivers 10,000 MIPS peak power and 2,500 MIPS steady power. This is an improvement over its own record with the first release, whose instruction rate was 7,500 and 1,000 MIPS respectively; and the second, much more powerful, release came to the market only 18 months after the first.

Hypercube architectures are in full evolution. By contrast, array type architectures have reached significant maturity, the proof of this being they have become miniaturised. For instance, the Alliant FX Series of compatible, expandable mini-supercomputers is based on a vector architecture. With up to four vector processors, the FX/40 delivers 94.4 MFLOPS* peak computational power, making its facilities accessible to workgroups.

As step No. 2 in Figure 8.3 indicates, with such architectures all processors in the system maintain a common view of global memory. Design-wise they are sharing physical storage through a coherent cache. The system's Concentric operating system is based on Unix, with extensions for parallel computing. This brings into perspective the software issue.

---

* There exists no direct correspondence between MIPS and MFLOPS. They are different metrics. However, based on benchmarks some organisations tend to think that 1 MFLOPS = 10 MIPS. If this was true then the peak power of FX/40 would roughly be 950 MIPS.

One of the key reasons behind the new thrust in hardware advances is the need to simplify software, as current software products have become too complex, unwiedly, difficult to maintain and of ridiculously large dimensions. In this sense, it is hoped that changes in architecture will reduce complexity. The focal point for this change is *parallelism.*

Therefore, to answer in a documented manner the query: ''Do we need 5th Generation Computers'', we should look more carefully into the evolution of software. A historical perspective tells that the first and second computer generation had different programming languages. The former featured machine language code and, by mid-range, assembler. The latter benefitted from improved assembler constructs, macrooperations and compilers. But all operations were *sequential.*

Back in the 1960s, with the third computer generation, larger mainframes featured computer power to permit an effort towards integration of programming instructions with data storage functionality. This came in two successive steps:

- The development of an operating system (OS) in 1964, *and*
- The appearance of the first database management systems (DBMS) during 1965-1966.

The fact that the third computer generation expanded vertically into data communications and databasing made necessary the competent use of both OS and DBMS. By 1968 the OS was enriched with teleprocessing routines (TPR). Yet, while such developments were significant, the whole operation remained sequential.

1. There was only one processor and this meant that it had to be shared between data communications, data processing and databasing activities.

2. The concept of handling in a parallel fashion not being available, every operation was stealing machine cycles from the other which waited in a linear queue (stack).

Some sort of parallelism was first achieved by interconnecting central processing units (CPU) and central memory (CM). At the time, we called it multiprocessing. This did remain a von Neumann type architecture, featuring n times the basic block: input, output, memory, ALU and control. But, such architecture has its limits in terms of:

- System reliability
- Maximum power *and*
- The cost of possible solutions

While the price of mainframes has been dropping over the years, it is still much too high for the power that they deliver. Figure 8.4 reflects a study made by the Swedish PK Banken. The cost per MIPS, with 3 MB of central memory, is $4,000 for an Altos type supermicro, but $171,000 for a large mainframe. This is a ratio of 1:43.

COST PER MIPS WITH 3 MBY/CM*

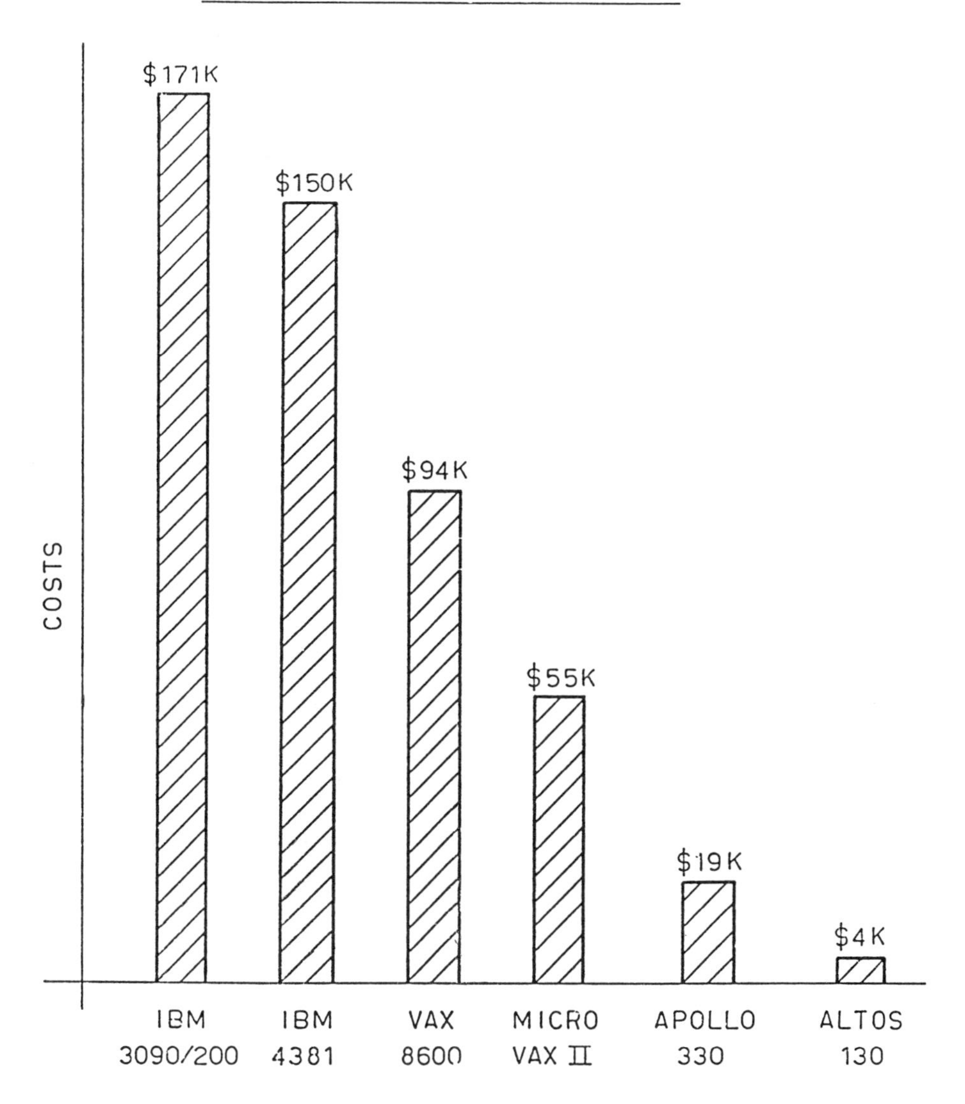

THAT IS AN INVERSE GROSCH LAW.

---

* BY PK BANKEN

Figure 8.4

Comparing 5GC to a mainframe, a similar benchmark gave the ratio of 1:40 in favour of a coarse grain hypercube; and 1:100 when a fine grain hypercube was contrasted to the mainframe. Two orders of magnitude go a long way in making the mainframe obsolete - as well as in answering the query: "Do we *really* need 5GC?".

Those Who are Left Behind: The EDPiers

It is surprising, but nevertheless true, that not everybody understands the significance of such large differences in cost/effectiveness. Ironically, those who find it particularly difficult to penetrate the subject are the EDPiers. As explained in an earlier chapter, back in the 1950s, "EDP" was an abbreviation for "Electronic Data Processing". Now it means "Emotionally Disturbed Persons" - the destabilised mass of supposedly computer professionals who:

- Still think of classical mainframes
- Program in Cobol;
- Have not yet discovered the utility of 4th and 5th Generation Languages (4GL, 5GL);
- Depend on secrecy about what they are doing to enhance their status; *and*
- Are part of an amorphous do-nothing-bureaucracy whose sole objective is to defend their turf.

What EDPiers do not understand is that the communications and computers business has radically changed. They choose to live in the past either because of low MIPS, vested interests or ossification. Yet, enduser requirements for communications and computers are no longer what they used to be, and never will be the same again. Not only patchwork-batchwork, but also old-hat realtime - the slow-going hierarchical system with the multidrop lines - no longer fits the bill. Paul Strassman made a study which proved that:

1. There is no correlation between spending big money on information technology and business performance, and

2. Productivity is flat both in the service and in the manufacturing industries.

There is no wonder that both of these conditions still apply as long as the EDPiers continue doing things which are totally irrelevant to the enduser. Yet, these same EDPiers consume large sums of money from *our* company's resources. An OECD study demonstrated that (excluding real estate) 1/3 of capital expenditures today are allocated to computers and communications.

We have spoken of an improvement equal to between 1 1/2 and 2 orders of magnitude that can be obtained by using high technology instead of old DP approaches. A much greater ratio in cost/effectiveness prevails if we compare a hypercube architecture to the now classical and obsolete mainframe line - even to its newest announcements. No corporate executive or computer system specialist can forego one and two orders of magnitude in difference. If he does so, he is not worth his salt -nor is he worth the salary which he is earning.

It is not only cost/effectiveness which demands parallel computing. It is also the changing nature of the problems that we are faced with. We said that large problems are inherently parallel. To be solved, they need new types of software and hardware where speed-up and performance are often impressive.

Code must be scalable to larger concurrent machines in order to meet future requirements. This can be more effectively done if the EDPiers are able to appreciate that there exist parallel processing misconceptions:

1. Parallel computers require new parallel languages.

This is not necessarily so. Depending on the problem standard languages, such as C, Lisp or Ada, can do the job.

2. A brand new code is absolutely needed.

Though new code specifically written for parallel processors will be more effective, it is also feasible to take existing code and port to one single node of say, a hypercube, and then proceed through the process of program decomposition.

3. The use of 5GC involves confusing parallel programming extensions.

This is another misunderstanding, and the answer to such a claim is "No!". A few simple system calls suffice.

4. Parallel processing is very difficult to master.

Here the answer is straightforward: it is *not harder;* it is simply a different approach - provided that one has the necessary GIPS to do the job.

The key to the new environment is: *Think in parallel.* It is a new culture opening up great professional perspectives. Experience shows that, once the parallel culture is acquired, moving so-far-serial applications to concurrent architecture represents a relatively modest effort.

Within the next 5 to 7 years, the world of communications and computers is going to be parallel. Therefore, it is better to make the change now. The milestones of the past are fading away. The old milestones involved:

- 32-bit machines which created the supermini and maxi classes
- Desktop computers, which were originally resisted by the EDPiers but which featured dramatic cost reductions, thus allowing mass delivery and creating a major social impact;

- Logic machines, specifically designed and implemented for AI, but without major communications capabilities and no clear aperture for parallelism.

Contrasted to these characteristics, the 5th generation contains thousands of high-speed computers that perform parallel processing. It will eventually feature several billion operations per second. Most importantly, it can already manipulate abstract symbols to simulate human thought.

This is the perspective under which should be seen 5th generation computers. Of all the references made in relation to their characteristics, the most important is *data level parallelism.* It is the key to achieving high computation rates, and also new types of logical processing.

Data level parallelism is particularly appropriate for databases, natural language constructs, and image processing operations. Also for finite element modelling, graphics, and the simulation of physical processes. It permits high resolution graphics, image synthesis, animation and image enhancement. These are the milestones of the future.

A New Generation Engine: The Hypercube

Reference to the Hypercube has been made in the preceding sections. A computer designed along networking lines, as is the case with the Hypercube, presents a number of significant advantages. These include modular and expandable design with no fundamental bottlenecks; fast concurrent access to mass storage devices; and direct parallel channels into processing nodes from standard peripherals.

The overall structure results in a flexible system for combining I/O, memory processing, and customised functions. Since the basic computer design is itself network oriented, its implementation can underline distributed capabilities ideal for multisite transaction and messaging systems.

As a development engine, the 5GC can be effectively used as a prototyping vehicle and concurrent debugging facility. These functions are enhanced by a flexible configuration but also the relative ease of implementation of AI constructs. Concurrent Lisp, concurrent Prolog, expert shells, and other S/W development tools see to it that knowledgebases are incorporated into usable and profitable applications.

Sections 2 and 3 have explained that new generation computers particularly emphasise two versions which are sometimes combined in the same machine:

1. In *vector,* or lock step processing, relatively simple arithmetic operations are performed simultaneously on a set of data.

2. In *concurrent* systems, dozens, hundreds, or thousands, of individual processors work on different aspects of a problem at the same time.

The latter approach is suitable to complex problems, while at the same time parallel database technology makes feasible:

- Large storage capacity scalable to application requirements;
- High-speed concurrent access;
- Parallel searches from multiple processes; *and*
- Application support for parallel database operations.

Multichannel communications permit concurrent, independent flexible linkage with many different systems and integration with existing systems. Speech handling, graphics presentation and interactive applications at large call for an impressive array of value added features.

An example of a 5GC equipment which can offer such facilities is the new release of Intel's Personal Super Computer (iPSC/2). This system consists of three major components:

1. *The Hypercube*, featuring 32 to 128 computa-tional nodes, each built around Intel's 32-bit 80386 as node processor, coupled with an 80387 floating point accelerator. Memory capacity is 1,4,8 or 16 MB per node. In the not-too-distant future the 486 will be replacing the 386.

Two numerical performance enhancements are available. Scalar Extension (SX) roughly triples the scalar performance of the basic iPSC system. Vector Extension (VX) increases vector performance over that of the iPSC by about one order of magnitude.

2. *The Routing Network* connects all of the iPSC/2 processing nodes together.

Although nodes are physically linked in the sense of a hypercube, an iPSC/2 facility known as Direct Connect Routing offers greater performance by functionally interconnecting every node to every other node.

3. *The Concurrent Workbench.* This is a systems programming support package that provides a variety of languages, a simulator, vector development tools and concurrent debugger.

The associated software development environment is hosted on the System Resource Manager. The latter is accessible from a network of workstations and serves as the administrative console and gateway to the hypercube for network users. The System Resource Manager is also an 80386-based system, and features 8 MB of memory, a 140 MB hard disc, and an Ethernet connection.

Design-wise, the System Resource Manager can be used as an information and intelligence filter, making feasible the independent processing of multiple information streams from a wide variety of independent sources. It can also be employed to facilitate expert system interaction with transaction and large online data sources.

One of the uses of the hypercube is in an interfacing role promoting large-capacity database interaction with a variety of workstations, local and long-distance communications, and the concurrent routing of realtime data between different systems - or different tasks. By combining multiple independent communications channels it offers the basis for *realspace* applications.

Furthermore, the nature of a hypercube offers itself to prototyping new information exchange standards. It can be used to implement scalable system resources configurable to meet specific processing, symbolic or mass storage loads. A significant degree of compatibility can also be obtained between distributed intelligent terminals and central or departmental 5th generation computers.

In terms of database technology, for example, iPSC/2 can feature up to 67 GByte storage with devoted multi-processor I/O and distributed file managers operating at 8 to 64 MB per second. DBMS processing and mass storage capacity can be scaled with specific application requirements. Concurrent data access ensures low message latency and fast user response

Another hypercube architecture, the Connection Machine, works fully in parallel, featuring distributed arithmetic/logical operations and distributed memory. It is precisely this memory distribution which offers the most logical programming model, one which is easy to debug, but which is also cost/effective and flexible, with open-ended scaleability.

Both with the Connection Machine and iPSC, this cell of memory with ALU is not storing information in any permanent fashion. It is only handling it while in processing. This is very important in understanding machine structure and the strategy to follow in its utilisation. It is also the reason why applications are fully distributed within the system.

Compared to iPSC, the Connection Machine features a network of 65,536 interconnected individual processors handling an equal number of messages in a random pattern. Each node has a computer, and each computer is enriched through 4,096 bits of memory. The Connection Machine also comes in a reduced size of 16,384 nodes.

These 16,384 or 65,536 processor-and-memory nodes are interconnected through a massive communications system. The *Router* operates at 500 MBPS, but the physical line capacity is at the 2 to 3 GBPS level. The machine's operating system fully supports networking, with central system services assuring locking, journaling and recovery.

Also based on networking principles, but not a hypercube architecture, BBN's Butterfly is a self-contained and expandable parallel processor. Machine configuration can be increased by adding processing nodes up to the maximum of 256.* Thus the user organisation can start with a relatively small system and add what is needed as processing requirements increase.

There are many interesting applications to which hypercube and network type architectures can be put. In terms of future perspectives, one of the most important is *idea databases*,** an implementation able to incorporate a significant degree of vagueness and uncertainty in query and search - in contrast to the crisp database accesses we have had so far. For this reason, the concept of *intentional* and *extensional* databases is being promoted as one of the most promising domains of AI applications.

But AI implementation is not the only reason for using new generation computer technology. Some industries have been traditionally in the foreground in the employment of supercomputers. For instance, defence (some 40 percent of installed units); oil companies; universities; nuclear research; space research; meteorology; and the computer makers themselves. Now the need for supercomputer power is spreading throughout business and industry, including banks and insurance companies.

Fifth generation computers are not just parallel processors which come in different versions. What really counts is that the whole implementation environment is radically changing. But is *our* organization ready to capitalise on this new tidal wave?

The 5th Generation Computer Project in Japan

Information systems engineering presupposes the creation of a working environment composed of concepts, methods, norms, tools and metrics. All five references find an answer within the context of the 5th Generation Computer effort in Japan which has been brought to its culmination in November 1988.

The way the Japanese look at the subject, artificial intelligence is a broad, *cognition science* field. Hence, it is much larger than 5th Generation Computing, which is an object, a tool fitting within this AI perspective. The foremost objectives of the Japanese effort are:

1. Development of a *knowledge information processing system* by using AI;

2. Improvement of *software productivity* through new software engineering technology; and

3. Development of a very high-performance parallel machine.

---

* BBN Advanced Computers has been awarded a contract from the Naval Research Laboratory to upgrade and expand its Butterfly. The upgrade will increase the system to 320 MIPS.

* * See also Chapter 9

One of the leading applications is *delivery planning*, including routing and resource allocation with distributed cooperative problem and constraint solving. Another example of using AI constructs on 5GC equipment is a diagnostic system for an electronic exchange facility. This diagnoses undetectable faults by internal testing and uses hypothetical reasoning (dependency directed backtracking) as well as deep knowledge.

While the applications perspectives range widely, what one appreciates most in the Japanese 5GC effort is the orderly and timely progression towards goals, as well as the setting of the right priorities in terms of implementation. As a general trend:

- *Managerial and financial subjects* account for the 45 percent of all AI applications;
- Specific focal areas are: *analysis, planning, control* - in that order;
- Emphasis is placed on *agile, user-friendly* man-machine interfaces, made feasible through AI.

With the now classical computer applications, the dialogue between the man and the information stored in the machine has been one of the difficult points. Solutions must be very user-friendly, but the so-far-available technology does not provide efficient interfacing. AI can do the job. Hence, the emphasis placed by the Japanese on language translation and on voice input.

A master in this transition has been the Institute of New Generation Computer Technology (ICOT). Funded by the Ministry of Foreign Trade and Industry (MITI), it was projected to cover three major phases:

- The first three years (1982-1984) were devoted to the development of the personal inference engine (PSI) and the extended self-contained Prolog (ESP) language;
- The next four years (1985-1988) have produced a number of improved PSI engines and competent networking solutions;
- The last, 3-year long, phase (1989-1991) was originally scheduled to focus on highly competitive parallel inference engines, designed to give Japanese companies dominance in world markets.

Japan's 5th Generation project aims to leapfrog the rest of the world in artificial intelligence and high power parallel processing. The first commercial product to emerge from this program was the personal sequential inference (PSI) machine. The next product has been Delta - an intelligent database engine.

- PSI deals with logical inferences,
- Delta with knowledgebanks and large databases.

Immediately after the first release of PSI in 1984 the Japanese set a goal of a tenfold boost of its power. This has been achieved. The next goal was the development of intelligent networked databases by ICOT.

Contrary to the confused signals which reached the western countries, PSI has been one of the major products so far made - and this from the initial stage of the project. It also was the first Prolog workstation to be developed, and featured a large main memory: up to 320 MB (PSI-II).

Software-wise, PSI's extended self-contained Prolog (ESP) has essentially the features of logic programming. It is a Prolog-based development of an object-oriented approach, and serves as a system description language for SIMPOS. The latter is the programming and operating system for PSI, supporting bit-map windows, editors, librarian, debugger, file control system, networking and so on.

The ICOT prototype of late 1984 has been immediately followed by new, more powerful, developments (PSI-II by Mitsubishi Electric, CHI by NEC, PSI-X by NTT,

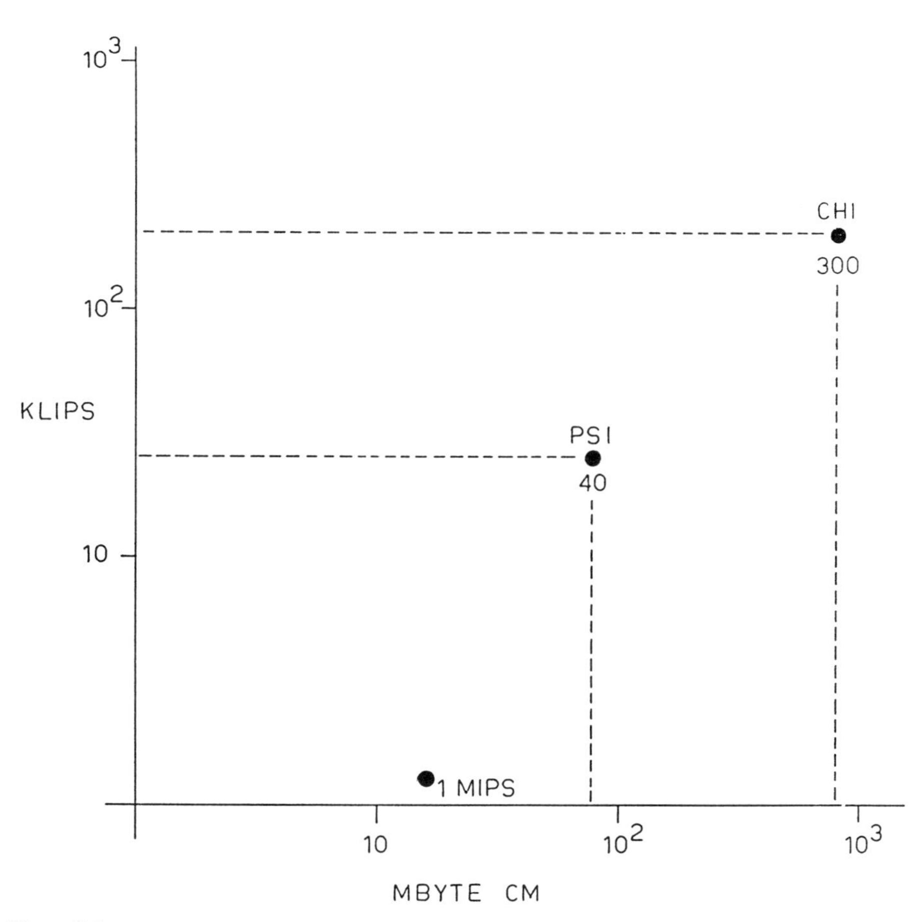

Figure 8.5

etc). Figure 8.5 compares the original release of two of the available models to a reference 1 MIPS power. (Remember that the "P" in PSI stands for *Personal.)* Two criteria are used in the classification:

- Megabytes in central memory (CM), and

- Thousands of logical instructions per second (KLIPS).

Logical instructions per second is the metric used with inference engines. As in the case of MIPS and MFLOPS, there is no direct correspondence between MIPS and KLIPS. Some companies use as rule of thumb 1 LIPS = 100 IPS (or 1 KLIPS = 0.1 MIPS). This is, however, very approximate and at times inaccurate.

The original personal sequential inference engine (PSI) has been a 5th Generation Computer operating at 30 to 40 KLIPS. The PSI-II effort aimed at 100 to 400 KLIPS; it surpassed it. Other projects such as CHI (by NEC) also developed a personal sequential inference engine of significant performance (well over 300 KLIPS).

CHI stands for cooperative high-performance sequential inference machine. It is a commercial product. Like other inference engines in Japan, CHI is an outgrowth of the ICOT project, providing:

- a sophisticated, intelligent processing environment

- for a wide range of applications and development tasks.

Compact in size, CHI executes five to ten times faster than the major conventional large-scale mainframes. According to some accounts, the upper range of CHI executes twice faster than the 3090/S600. It also offers significant flexibility in developing and processing applications. Other critical comments concerning current computers include the fact that:

1. High-level control mechanisms, such as intelligent backtracking, are not readily incorporated - although they should be;

2. The use of underlying languages (external calls) is complex and must be made much simpler;

3. Higher-levels of problem-solving techniques, such as constraint solving and handling of hypothetical knowledge, are still beyond the scope of current computers - but they are part of PSI and other 5GC projects;

4. There is no methodology or guideline for building expert systems using current tools, yet such methodology is most essential.

What supercomputers and optical computers are expected to do for Japan's hardware skills and products, AI is likely to do in terms of software knowhow. The new generation of machines have the ability to *infer*. Instead of merely manipulating data or words, they process knowledge; as we have already seen, AI and 5GC are highly related topics.

Quite importantly, 5th generation computers will be more tolerant of people, accepting ill-defined problems in the form of spoken language. They will present the answers in a manner that humans can easily understand. We will return to this issue in the following sections.

Compared with similar efforts in Western Europe, not only have the Japanese had a head start, but they are also keen in marketing the products which they develop. They do not let them languish in the laboratories, and they do not load them with unnecessary costs so that they are not able to take-off in the international market.

MITI thinks that 5th Generation Computer technology will significantly improve the competitiveness of Japanese industry - and this on a worldwide basis. Not only is parallel processing the perfect vehicle for artificial intelligence, but also AI constructs have entered the practice of Japanese business, strongly promoted by the local computer manufacturers.

Ensuring that there is enough power for AI implementation on a large scale helps provide a competitive edge. It also breaks the present-day stalemate underlined in preceding chapters. Let us always recall that about 70 percent to 75 percent of essential management functions have not been automated so far - and cannot be automated through classical computer approaches.

The timing is impressive. Instead of waiting for theoretical developments to be followed by engineering, and with a time lag in manufacturing, and then in a further-out phase by applications, the Japanese do these three major phases in parallel. There is a similarity between the solution currently followed and that adopted in the early 1940s for the Manhattan Project - the making of the atomic bomb in America. And there are basic reasons for the rush.

The prevailing concept in Japan is that classical data processing could bring automation up to a point, but no further. *New departures* are necessary if we are going to increase:

- Our level of sophistication in the implementation of computers and communications; *and*

- The quality of services which we can offer to our clients - hence our competitiveness in the market place.

If they succeed with the data level parallelism on which they are currently working - that is with the parallel PSI - the Japanese may overtake the Americans in the computer field. However, the game is not yet over. The Japanese have still to prove that their big national and corporate teams can do better than the American start-ups led by ingenious individuals and small groups.

Table 3: Share of Current Microprocessor Market (about $4.5 billion in sales)

United States	43%
Japan	34%
Western Europe	18%
Others	5%

The Japanese make many other American microprocessors under license, often cheaper and of better quality than those made by their USA inventors.

As a result, NEC, Hitachi and Fujitsu, have taken over the leadership from Motorola, Intel and Texas Instruments.

However, even if it does not get the lead on a worldwide basis, the New Generation Computer project has helped Japan in improving its software production, which has long been a weakness. For the present time, the USA lead in software remains strong. The question is: for how long?

Designing software is creative work that lends itself well to the individualistic American approach. On the other hand, the world thought that the Japanese are handicapped by their relative lack of innovation in microprocessors. As Table 3 demonstrates, they have made up for the lag.

Throughout the 1990s and well into the 21st Century, we can expect Japan to produce a torrent of new technology. Already the PSI engine, of which we spoke earlier, is a relatively small device. This is shown in Figure 8.6. Both in terms of:

- Thousands of logical instructions per second (KLIPS), *and*
- Megabytes of central memory

As we have already seen, CHI, one of the second generation of PSI, is more performant than the original PSI by more than an order of magnitude. Yet, American hypercube developments - particularly the new Connection Machine (CM-2) - exceeds that capacity by a comfortable margin. Let alone the fact that CM-2 is by far the most powerful computer in the American arsenal, once again the question is raised: for how long will it surpass the best that the Japanese can offer?

European Efforts with Supercomputers and AI

America, Japan, Western Europe, and even Eastern Europe, are all very active in 5th generation projects. There is, of course, a reason for this: *more power*. The story of more power is closely linked to what man has done on earth. This is a principle known in politics. Less understood is the fact that science and industry, too, are subject to the pull and push power provides.

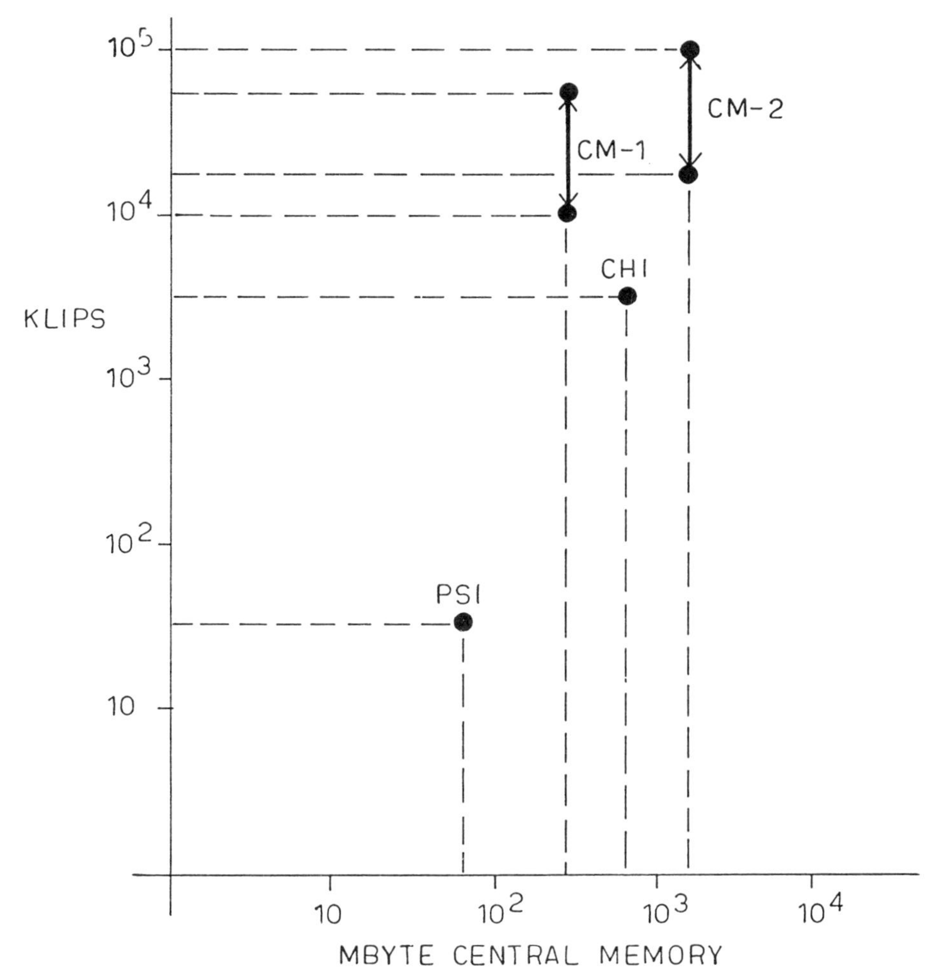

Figure 8.6

When the Hoover Dam was made in 1936, it supplied 97 percent of the electric power needed in Los Angeles. Fifty years later, in 1986, Hoover Dam (whose output has basically remained unchanged) supplies only 3 percent of the power needed in Los Angeles; a telling example on power consumption and power increases in the electrical engineering field.

The same happens with computers. At 2 to 4 million instructions per second (MIPS), and at a cost of $4,000 to $8,000, a personal computer today offers twice the power of a 1976 mainframe (large central computer) whose cost was $2,000,000. Although few companies may end up removing their mainframes, this desk-level revolution has given new choices about how a computing job should be done. It also called for greater computer power at the centre.

To drive the thousands of installed personal computers companies now need immense central resources. At the same time, advances in artificial intelligence - such as language translation and artificial vision - require two to three orders of magnitude more powerful computers than the biggest that we now have available. They need a different type of machine, with logical facilities, rather than just computing.

Quite evidently, it is not only the Japanese and the Americans who appreciate these facts. In Europe, too, this subject has been of interest to the mainstream of industrial thinking. *The battle of the 21st Century* will be fought with:

- very high speed communications;
- superfast computers;
- large, but easy-to-access databases; *and*
- more efficient human interfaces.

A *human window* is the term describing an efficient, user friendly interface between the user of a computer and the information stored in the machine. Expert systems help in implementing human windows which are simple to actuate, employ natural language, and are forgiving the user's mistakes. All this needs computer power to be put into effect.

Statistics still favour the Japanese and the American side of the worldwide competition. According to some estimates, 12 to 15 percent of USA households today include a family member who uses, or used, a computer at work. This percentage is slightly higher in Japan, but is much lower in Western Europe, which practically means that more than *90 percent of people in the industrial countries are not yet computer literate*. Making computers easier to use should attract many of that 90 percent.

This is precisely where massive national efforts should be oriented: towards turning the current 90 percent rate of computer illiteracy into a 90 percent, or better, rate of computer literacy. Unfortunately for Europe, this is not duly appreciated. Instead of having the right focus, national supercomputer efforts - where they exist - are primarily hardware oriented. While necessary, this is not enough. Let us look at the facts behind this statement.

Largely assisted through Government money, European, American and Japanese companies work in 5GC and AI. Many companies are interested in this subject. Everywhere the stakes are high. As a result, steady work on 5th Generation Computers is going on. There are four high technology computer projects worth mentioning in Europe:

1. The European Strategic Project for Information Technology (Esprit). Based in Brussels, Belgium, it finances projects throughout the Common Market;

2. Eureka, a joint effort of Western European governments, in which countries other than the 12 of the Common Market also participate. Fund allocation is performed by a council of ministers.

Eureka is virtually free of government control, even though it gets half its money from its members' governments. Unlike previous European efforts to pool technological talents, a project must originate with an industrial firm in another country. Universities and government laboratories can be brought in, but industry is the initiator and the home base.

3. Project Alvey, London, England. This is a British government initiative, particularly focusing on British industry's needs - but the government does not seem enthusiastic about investing more money, as it should do.

Technology transfer has been the focal point in Alvey, which is financed by the British government. This is done through the *Clubs,* one per industrial sector: banks, insurance, mechanical engineering, electrical engineering, and so on. Alvey has also taken a good initiative in bringing universities and industries together. It is a little late for doing so, but better now than never.

One of the several problems that Alvey faces is the appropriate budget for infrastructure. American projects in high technology typically allocate 20 percent to 25 percent for project management. Alvey keeps that budget to under 10 percent. This simply is not enough for properly planning and controlling very advanced projects and, at the same time, turning the technology transfer wheels. But, the greatest shortcoming of Alvey is the failure to focus on *marketing* the products that sponsored research helps to create.

Work on supercomputing products is being done by researchers at the Universities of Edinburgh and Southampton, as well as by emerging companies such as Sension and Meiko. An early British lead with transputer technology (high density VLSI processor chips lending themselves to parallel computing) has been harnessed by Floating Point Systems of Beaverton, Oregon, rather than by Inmos, its UK developer.

European governments (and European companies) are slow to understand that *the only valid test of a new product is in the marketplace*. This is a pity, as good products are being designed, consuming significant financial and human resources, but then they are left in the wall closet for eternity.

4. GMD (Gesellschaft für Mathematik und Datenverarbeitung), in St. Augustin, West Germany, financed by the German government.

The German GMD Project has three major divisions: one addresses itself to the theoretical research, the other to applied research, and the third to applications, including expert systems. There are also two minor divisions, primarily concerned with education and technology transfer.

GMD is financed by the West German government to bring the German industry into the 21st Century. However, my findings are that the technology transfer effort is being found wanting, and the research projects themselves are not well tuned to *face market requirements.* They are primarily theoretical efforts with little or no perspective for financial success.

Just as importantly, the timetables are not fast moving. They take too long to produce too little. Surely, they are no match for the Japanese and the American 5GC projects.

Today in West Germany, two companies are making supercomputer systems: Bonn-based Suprenum (a spin-off from GMD), and Integrated Parallel Systems of Karlsruhe. The latter is a spin-off from the Karlsruhe University. As is the case with British supercomputer developments, originally the technology has been competitive - but the marketing is not very effective, and *without aggressive marketing the technology decays.*

The story of Esprit helps exemplify what is right and wrong with European-based efforts in high technology. Its strength and major weakness rests on the concept of *pre-competitive research,* far from market orientation - as Esprit has been looking after developments in high technology, particularly in 5th Generation Computers and artificial intelligence, in the Common Market countries.

In the positive side, this pre-competitive effort forged a Common Market-wide research spirit particularly in the expert systems domain. Race, a parallel project, focuses on high technology in communications. But, on both counts, pre-competitive research handicapped the developing products' marketability. They got up to a half-baked point and stayed there.

Major phases can be distinguished in connection with the European Strategic Project for Information Technology. *Esprit I* set itself the following noble objectives:

1. An increase in functionality of European information technology projects by providing more tools and promoting their implementa-tion;

2. Research on, and development of, models for office systems, including their requirements and specifications;

3. A level of integration both in hardware and software, extending into interfaces necessary for a common environment to be developed; and

4. Prototypes and some emerging standards to be brought forward in a subsequent phase.

Not all of these objectives have been reached. By contrast, some issues initially considered byproducts became the real output of Esprit I. The best reference is cooperation among Europeans. Another one is cooperation between universities and industries.

In the footsteps of Esprit I followed *Esprit II*. Its first significant new direction is the change from office systems to integrated applications. The second, and even more important, element is the emphasis which starts being placed on *intellectual* breakthroughs rather than tools for pushing paper around the office or out of it.

The concepts of integrated applications and expert systems led to an emphasis on multimedia. Notice has been taken of the fact that most computers today do not utilise multimedia objects. Technology makes this feasible, but our concepts did not yet adjust to the task.

Esprit II promises to focus on this lack of knowledge, and also on the possible development of a unique knowledgebase rather than knowhow fragmented into pieces which are often inconsistent. However, evidence has not yet been given on how this goal will be achieved.

Also, it cannot be repeated too often that the key limitation of this project is the fact that, until recently, funding has been limited on pre-competitive research. While originally a correct decision, given the multinational and multicompany nature of research funded by Esprit, this subsequently became a *constraint*. Only products which can compete successfully in the market are worth pursuing, but such a guideline has not yet been issued.

Responding to calls from the leaders of the 12 nations, the European Common Market Commission also proposed another plan for greater technological cooperation: the Framework Program, which is expected to offer an extra dimension of Community cooperation. It includes computer technology, biotechnology, telecommunications, marine research and nuclear energy.

Framework is described as the management blueprint for all of the EC's basic research projects, such as Esprit, Race and Brite. But, it acts against invested interests. Some EC member countries are said to worry that increased research spending will force reductions in farm outlays.

What a great medieval mentality! Western Europe is investing where it has a surplus and does not know what to do with it - not where funding is most needed to keep a leading position in the Battle of the 21st Century. That is precisely what politicians deprived of any imagination (and associated brains) can do.

Instead of feeding our weaknesses and strangling our strengths, we should focus our attention where it pays dividends. That is, the vast domain in need of urgent fixing: *marketing*. European scientists tend to fall in love with their project, and like to see it in the laboratory for a long time. But, this is not the way that high technology works.

The product, if successful, *has to move fast out of the laboratory and into the market*. It has to become known and accepted. This calls for fine-grade sales skills. Americans are wizards in the sales effort. The business of America is business. The Japanese are following the same business steps; indeed, they are improving upon them.

In Eastern Europe, the Russians, too, have launched a 5th generation project of their own called *Start*. It is being led by a consortium of computer institutes in Moscow and Novosibirsk. The Russian and other East European researchers are trying to build a massive parallel computer, code-named *Mars*. If the Western Europeans cannot put their act together, the East Europeans cannot do it either.

While Russian scientists are very good in theoretical subjects, the lack of adequate computer resources sees to it that software is not an art this is flourishing in the Soviet Union. Some of the basic software being used is Western, with minor modifications. One source of evidence points to Western programs which are copied illegally in Bulgaria, where the accompanying manuals are also translated into Russian.

For instance, Total. This is an eighteen-year-old, and by now obsolete, database management system (DBMS) by Cincom Systems, a company based in Cincinati. The Russians (and their Bulgarian underlings) fail to understand that the prospects of such copying policies can range from counterproductive to disastrous. In terms of results they are plainly limited, but in the longer run they divert skill which might have been otherwise allocated to original developments.

An ironic fact about the wider-range Russian problems with computers is that they have been viewed as an indispensable tool of central planning. In 1971 the Communist party essentially endorsed this proposition: The Ninth Five-Year Plan should create a computer network that would link organisations at every level of the economy. The Twelfth Five-Year Plan in 1986 repeated this premise - which, by all available evidence, is still a premise and may stay so until the 21st century.

The planner's dream is that timely data would flow up from the factories to the ministries, to incorporate data into the planning goals and to provide feedback. Party congresses have endorsed this concept, while not specifying the date at which it should be implemented. But, at the same time, the same planners and their ossified planning mechanism do their utmost to keep the Russian computer industry non-competitive. Worse than EDPiers!

Fifth Generation Computing in America

We said that large central computing resources are needed to drive the distributed information systems which get more powerful all the time. User organisations require central databases accessible in an easy, but secure, way from the periphery. Online systems call for a consistent image of transactions and messages supported in a global sense.

All this consumes large amounts of computer power. In America today, a large scale system will typically have 10,000 to 80,000 terminals attached to it; will feature some 100 to 250 MIPS on multiple data processors; and access from 250 gigabytes (GB) to nearly 1 terabyte (TB) online. All this will probably operate well beyond the 100 transactions-per-second rate, which itself increases as the applications expand.

Established American computer manufacturers are in a race to develop and produce more powerful machines -among themselves and against the Japanese. Three references are necessary to describe the high technology effort going on in America today:

1. The most brilliant are the start-ups.

From massively parallel computers to artificial intelligence, new materials and biotechnology, the products which they feature have their origin in research, usually financed by the Department of Defense. Once a breakthrough has been achieved, the researchers leave and, with venture capital backing, set up their own shop.

That is the American way of hitting gold. The spirit of the old West. Most of these newborn companies will die. But some will become giants. Thinking Machines is a start-up. Both Intel and BBN are established companies, but their computer ventures are the nearest thing to a start-up.

2. The most solid is IBM's work, largely concentrated at the Thomas Watson Scientific Laboratories, at Yorktown Heights.

The largest computer company in the world is investing substantial sums of money in advanced research. These projects are progressing smoothly. Some are impressive, but it is not certain that they will overtake the Japanese in the longer run. IBM cannot compete with the startups in terms of timing, but when it will come to the market with a product it will do the utmost to master that market. In the meantime, it can finance and co-sponsor research at some of the start-ups, which is precisely what it is doing.

For instance, Steve S. Chen, the former chief designer of Cray Research and his Supercomputer Systems Inc. (SSI) of Eau Clair, Wisconsin, have a joint project with IBM aiming to design, bottom-up, a supercomputer with 64 processors operating in parallel.

In fact, SSI's is one of several efforts under way to break new ground in supercomputers. What is unique about it is that the venture has stirred unusual attention in the industry because of Dr. Chen's reputation and IBM's muscle. One particularly valuable IBM treasure SSI will probably appropriate is the technology for connecting microchips. Hooking together complex arrays of microcircuitry, and making sure that they do not overheat, is one of the most exacting and expensive feats in high technology industries.

IBM invested in SSI by forming a research and develop-ment partnership in which both companies are equal partners. A new round of investors are being offered limited partnerships. SSI's plan is to convert its research partners into equity investors once it starts building its first product, which it has dubbed the SS-1. Lining up customers as investors is a technique that several computer companies have used to get both cash and market research in one stroke. The strategy also carries a big risk: alienating prospective customers who compete with the investing companies.

At the same time, as far as IBM is concerned, software will pose a design dilemma. IBM's mainframe computers have very different software standards from present-day supercomputers, and the future machines projected to be increasingly used by scientists and engineers. It will be difficult, if not impossible, to design a computer generation that does not short-change one of these two approaches and the need is to make up one's mind. *Should the new product line look toward the 21st century or the 1960s?*

3. The organisation whose projects may have the broadest impact on American society is the Microelectronics and Computer Development Corporation (MCC), of Austin, Texas.

Set up in 1982 as a cooperative effort of several computers and communications manufacturers, MCC makes the product of its applied research available to all of them. The effort is commendable. The problem is the time the obtained results will take in each member firm's development pipeline to become marketable commodities.

The MCC effort is oriented toward obtaining breakthroughs in seven areas of attainable computer technology:

1. New semiconductor packaging to increase the density, and decrease the size, of systems containing integrated circuits (IC);

2. More efficient IC design through computer-aided approaches;

3. More efficient software design by large software teams;

4. Parallel processing machines for faster computation;

5. Artificial intelligence and knowledge-based systems;

6. Large, complex databases to store the mass of information required by knowledge-based systems; *and*

7. Better interfaces between computers and people, so that users can work more effectively.

In the beginning, MCC shareholders* joined one or more of four distinct research programs: IC packaging, software technology, computer-aided design (CAD), and advanced computer architectures. These programs were to last 6,7,8 and 10 years, respectively. But, since the beginning in 1983, there has been an interesting development in goals.

The program in advanced computer architectures, for example, has been split into four full programs: parallel processing, artificial intelligence and knowledge-based

* Originally, there were 10 shareholders, mainly computer manufacturers in the USA. There are now 22, plus government funding.

systems, large databases, and human factors technology. These four programs remain linked into a systems integration concept. There is also an integration group composed of up to nine researchers. It is to function as the fifth element in computer architectures.

The systems integration staff monitors the results from the four computer programs, acquires data from trends outside MCC, and begins to design the kind of computer system or systems that might evolve from architecture programs. Eventually, the group is to be responsible for integrating all five computer architecture programs, but systems may evolve towards being specific to classes of applications.

This is in line with MCC's ultimate goal, articulated by Admiral Robert Inman, its first President: "In the 1990s we are not talking about a computer industry, a semiconductor industry, or a telecommunications industry. We are talking about an information-handling industry. MCC ought to evolve in programs that move toward a systems approach to information handling". This, too, is a 5th Generation Computing goal.

The key to its success is MCC's database program. Eugene Lowenthal, who is in charge of database systems at MCC, had hoped a single database structure could satisfy the needs. It now appears that two different types must be developed: a logical database, and an object-oriented one.

Some 75 percent of the program's effort is oriented to the *logical database,* with Prolog-type assertions used to manipulate it. The *object-oriented* database will make use of the Lisp-type symbolic processing statements, to give the programmer better control over the computation. At the same time, a knowledgebank is being built, under Douglas Lenat, in the AI project. It includes a broad, but shallow, base of knowledge that humans consider common sense.

The industry in every industrialised country depends on cooperative laboratories like MCC to help avoid redundant effort and still obtain the basis necessary for moving into the 21st Century environment. This is particularly true of America, where a significant part of research in high technology is oriented to military goals. With much less resources placed in this race, the Russians are subject to the same constraints which, for instance, the Swiss and Japanese do not have.

While there is a good deal of fallout from military projects into civilian accomplishments, there are also deadends. Hence, a significant amount of waste. The price paid by the USA and Russia for *parading as superpowers* undermines the economies of both nations and bends their ability to effectively control civilian markets. Meanwhile, the Japanese are busily preparing to dominate the global marketplace.

9

Global Databases and Memory-Based Reasoning

Introduction

While impressive technological advances are today available for implementation, the challenge remains as to how do we employ them best. The correct use of technology can give a competitive edge. It can also get us into new lines of business faster. But, given the investments being made, the wrong use of technology can get us out of business - also faster.

Dr. J. Alvey of British Telecom, whose study initiated Britain's 5th Generation Computer and AI program, mentioned in his keynote speech at ICCC'84: "And while perhaps we have yet to demonstrate more than 2 giga-bytes through a fibre, that speed of transmission still dumps the Encyclopaedia Britannica in your in-tray in a very few seconds. Which poses the next question of what do you do with all that data - the intelligent computer helps us digest it and use it?".

Dr. Alvey has pinpointed the implementation area which is the closest to our goals for the 1990s: most of high technology's advantages can be obtained only in combination with intelligent machines and knowledge processing. This is as true of superconductors as it is of networks and global databases. Since the size and complexity of databases is increasing at an impressive pace every year, several leading manufacturers are working on parallel, dedicated database machines. Teradata of Los Angeles, California, has been the first fully-parallel-database computer to hit the market. IBM has a project in San Jose, CA, which concentrates on hardware approaches to a dedicated database engine. This is not to be confused with Project R* which is a software solution resident on mainframe.

Teradata's DBC/1012 achieves signigicant performance advantages through parallel processing, and has been in the market since 1984. It is a coarse grain 5GC which can have from 3 to 1,024 processors running in parallel:

● Interface processors (IFP) do request processing

● Access mode processors (AMP) perform relational database management

Each of the access modules supervise one of two large column Winchester disk drives (DSU). The AMPs substitute, to a large measure, the role classically played by DBMS, a software program. The IFP provide the interface to the mainframe and its applications routines, thus avoiding the need to rewrite them.

Among themselves, IFP, AMP and DSU establish the size of the database computer: 4 x 14 x 24 means 4 IFP, 14 AMP, 24 DSU. If the data load increases, redimensioning the IFP and AMP helps keep response time constant, which is a major plus in any implementation. The DBC applies parallel processing to business needs, and while it is a coarse grain 5th Generation Computer it is perfectly able to work online with 4th generation machines.

For any company above medium size, the correct policy is to look at dedicated database machines to face future growth in database access. To do so successfully, we must properly establish the criteria. As the preceding two chapters documented, these are:

● global database approaches;

● overall efficiency;

● able interfaces;

● increasing throughput; *and*

● better performance, capitalising on the low cost of microprocessors.

Studies for efficient database solutions are typically based on real life transaction processing and message exchange, using time segments of computer resources and connecting online different installations. As we know from experience, it is not enough to provide the power. We must also examine the infrastructure, global addressing, the effort to be done on partitioning, the image (shadow) copies for realtime recovery, and database integration at large.

If we look at the Teradata experience, we get the message that, with existing microprocessors, we can make powerful database engines. But the environment also impacts the results. Hence, the wisdom for proper preparation and for benchmarking, which no vendor can provide. Every user organisation must do its homework by itself.

When the Chicago Board Options Exchange (CBOE), a supervisory arm of the Securities and Exchange Commission, decided to go for 5th Generation database computers, first it rethought its mission. This has been defined as:

1. Surveillance of trading floor activities;

2. Responsibility for maintaining an orderly and honest marketplace; *and*

3. Rules enforcement and market regulation.

Then it identified the bottlenecks it faced with the computer systems it then had in operation. Difficulty of accessing data over time, was the first of them. Data storage media incompatibilities, the second. Reference data not readily available, the third.

Sounds familiar? Every major organisation today has problems which exceed the complexity that von Neumann computers can tackle. But the solution is not simply changing equipment. A structural and organisational study is necessary. The same is true about use of the best available software, particularly relational DBMS and utilities endowed with artificial intelligence.

Relational DBMS for Better Performance

To understand the changing landscape in database management, it is proper to follow what the market had to offer over a period of two decades: from the middle 1960s when IMS (hierarchical) and IDS (networking) structures became available, to the early 1980s when relational DBMS started becoming popular.

Early implementations of database management systems did not provide the benefits of relational technology. They had high performance overheads due to the record-at-a-time nature of their processing, causing an access to the database for every processed record. And they were fairly inflexible.

Recent relational database management systems overcome the channel and interface problems of these early approaches. They use set-level processing inherent in relational data models. With this, many records can be processed in a single access to the database.

Today, relational DBMS technology is mid-range of the typical 20 to 25 year life cycle of any new commercial computer system concept. Apart the fact that we now only begin to see an increasing acceptance of relational solutions, the concept itself is in evolution in terms of:

1. The constellation of linguistic tools based on relational DBMS functionality;

2. The further evolution of this functionality to cover distributed databases and very large databases; and

3. The necessary coordination with knowledge-banks managed by intentional KBMS connected to extensional DBMS.

Fourth and fifth generation languages (4GL, 5GL) are necessary for two purposes: to significantly increase the productivity of the professional programmer, and to open up the population of endusers in a self-service approach to machine programming. As 40 percent or more of the work to be done in the future will be database connected, the proper approach is to employ 4GL and 5GL tools closely connected, correspondingly, to the DBMS and KBMS.

Distributed databases impose further perspectives which can be best served through AI-based media. Three topmost requirements are: security, distributed access and concurrency control. Concurrency control of distributed databases (DDB) includes:

- Locking methodology;
- Timestamp ordering;
- Validation methods; *and*
- Hybrid approaches.

Valid solutions feature: a concept, basic algorithms, heuristic schemes, one or more versions which can be implemented, and approaches avoiding performance failures. System reliability is crucial to DDB solutions.

At the current state-of-the-art, relational databases provide improved price performance over their networking (owner, member) and hierarchical predecessors. They do not, however, necessarily feature parallel processing to execute set-level application requests. Yet, for an array of managerial and technical reasons, parallel processing solutions are becoming vital.

Competent memory management solutions are an integral and vital part of a computer architecture. As Figure 9.1 indicates, the latter includes four axes of reference; (1) the programming facility to be supported; (2) the input/output structure; (3) operating system perspectives; and (4) memory management. This makes a full circle as memory management and programming faculties are highly interrelated.

The coordinated implementation of database computers, hence hardware solutions, with relational DBMS software to manage a distributed environment is the best approach currently available. This is exemplified in the ICOT solution where:

- Delta acts as a database computer;
- The DBMS for the extentional database is Ingres; *and*
- The KBMS for the intentional database (developed in Japan) is Kaiser.

In America, Relational Technologies (the maker of Ingres) offers Ingresnet as the solution for distributed databases. Though the Japanese and American approaches address the subject of larger databases at different orientation levels, *performance* is in the background of both of them.

Ever since the early days of DBMS the primary concern of customers aware of the real problems that they should face has been performance. Enriched with parallel processing capabilities, relational DBMS require new implementation perspectives. By the same token, the facilities they offer attract a new set of users with more sophisticated types of applications.

While today we know much more than we ever did regarding database management (or perhaps just because of this fact) database solutions are in full evolution. At the initial stage, and over a period of several years, software approaches have been the way to go. As system functionality settles, hardware approaches have advantages. Parallel processing through database computers means *hardware solutions designed to satisfy large database needs*. This is particularly important to a:

- Broad range of transaction processing applications, *and*

- The servicing of information centre requirements for message handling and decision support systems; *and*

- An intensified exchange of electronic messages - from electronic mail to voice mail.

Both for decision support and for message exchange, the information centre environment addresses itself to ad hoc enduser processing. With this style, the primary consideration is the functionality provided by the database product.

Integration also imposes requirements. Until quite recently, these two different types of processing: transactional and messaging, were kept entirely separate. Nevertheless, the advent of relational DBMS has given the capability to bridge the gap between the two ends of the databasing spectrum, as available products are capable of supporting both types of information systems needs.

But not everything can be contributed by the hardware or software side. Much rests on the organisational effort. Well managed databases establish - then update - *sunset clauses* for all information elements contained in them, wherever they may be stored in a distributed sense.

The competent administration of the database through a steady database administrator (DBA) responsibility is another major challenge. DBA responsibility does not end with the choice of a database management system. This is only one of the milestones to the satisfactory execution of the job which needs to be done.

Having examined performance, user organisations typically move on to the subject of data portability -which also is part of the DBA job. The third foremost issue is the observance of standards. For instance, of SQL ANSI as the linguistic interface, including the portability of SQL-based applications.

IBM was to remark that the SQL compiler knows a lot about where data is and can skip across, assisting parallelism in terms of access. Coordinated with the

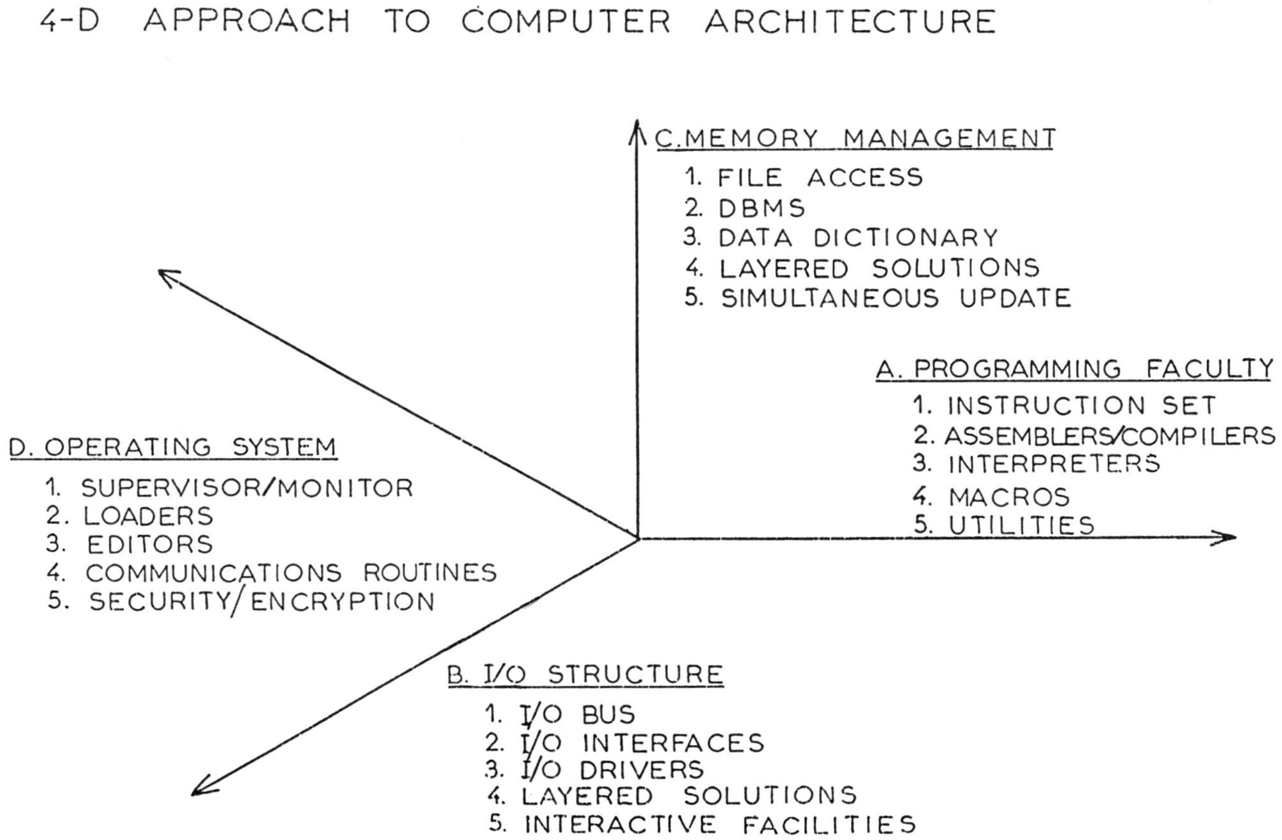

4-D APPROACH TO COMPUTER ARCHITECTURE

C. MEMORY MANAGEMENT

1. FILE ACCESS
2. DBMS
3. DATA DICTIONARY
4. LAYERED SOLUTIONS
5. SIMULTANEOUS UPDATE

A. PROGRAMMING FACULTY

1. INSTRUCTION SET
2. ASSEMBLERS/COMPILERS
3. INTERPRETERS
4. MACROS
5. UTILITIES

D. OPERATING SYSTEM

1. SUPERVISOR/MONITOR
2. LOADERS
3. EDITORS
4. COMMUNICATIONS ROUTINES
5. SECURITY/ENCRYPTION

B. I/O STRUCTURE

1. I/O BUS
2. I/O INTERFACES
3. I/O DRIVERS
4. LAYERED SOLUTIONS
5. INTERACTIVE FACILITIES

Figure 9.1

functionality provided by parallel databases engines, this helps make a better performance possible. Still, this is a general statement to be tested in each specific implementation environment, not to be accepted at face value.

A fourth key subject is that of solutions supporting PC databases (microfiles) able to exchange data with mainframe databases. Up to a certain extent, portability is conditioned by how the applications software was written. This last reference also impacts on:

- user access, *and*

- response time

though response time can be flattened through the able usage of dedicated database machines by redimensioning the number of attached microprocessors.

For all those reasons which we have seen, *emphasis is now increasingly placed on a duality of physical and logical solutions.* A logical database processor becomes the centre of the system. It is responsible for tasks relating to the execution of queries after translation from an acceptable query language.

- After translation, the query will exist in some internal form (such as a tree representation of an algebraic expression);

- An optimiser will apply transformation rules to yield an equivalent expression cheaper to evaluate; *and*

- An interpreter will carry out the computations indicated in the expression by invoking operators of the physical database processor.

The physical database processor administers the information elements - both data and metadata. All system data come under the control of this processor. Many of the database operators are also handled at this level.

Logical parallelism is the cornerstone of competent solutions. It will be assisted through physical parallelism, for instance, the fact that many more arms are moving per gigabyte. But more arms call for greater coordination, hence the corresponding increase in AMP in the Teradata engine. ''In going from 20 to 40, and from 40 to 80 (AMP), you cut the access time in half'', a Bank of America executive was to suggest.

This can be said in conclusion: The transition from any DBMS to something new is *not* to ''any relational DBMS''. The right transition is to a *controlled environment* which features both hardware and software assistance to the relational solution. Within a short while it will also incorporate artificial intelligence.

Contribution of Artificial Intelligence

During the conference at the Yorktown Heights Laboratory, IBM emphasised that, if used in an able manner, artificial intelligence can play the data more intelligently. An object-oriented DBMS manages collections of objects of various types without consideration of:

- how those objects are represented, *or*

- how operations on them are implemented

at the query language level. Extensibility permits the incorporation of specialised processors, providing parallel access capabilities.

IBM has experience with AI usage in operating systems as the Yorktown Expert System (YES/MVS) demonstrates. According to the IBM book, AI implementation in database operations must achieve two goals, which I consider to some extent to be contradictory:

1. Be very efficient, *and*

2. Follow the SQL line

Yorktown Heights seems to be working on this dual issue. If successful, it will create a logic structure able to endow database management with AI capabilities.

One of the topics under investigation, for instance, is the dual image which leads to optimisation. The AI construct under development capitalises on the fact that we read more frequently than we write. Through this and similar observations on how we use our tools, optimisa-tion can give important throughput performance advantages.

Let us, however, be sure of our goals and the way to achieve them. In database management, as in any other field, AI approaches work where there is human expertise that we can capture. If today we are using experts to perform the correct placement of information elements, a better solution is the use of an AI construct, and initial results demonstrate that this can be done.

According to the Japanese, the conventional concepts of program development and database usage will be integrated into one higher-level concept called *knowledge*. It will be manipulated uniformly, leading to the retrieval of information elements from fully distributed environments, and the computation of programs being integrated into a broader structure of problem- solving. Said a system specialists during the ICOT meeting:

''Programs and data differ in representation rather than in semantics. Programs and data are respectively regarded as intentional and extensional representations of the same knowledge''.

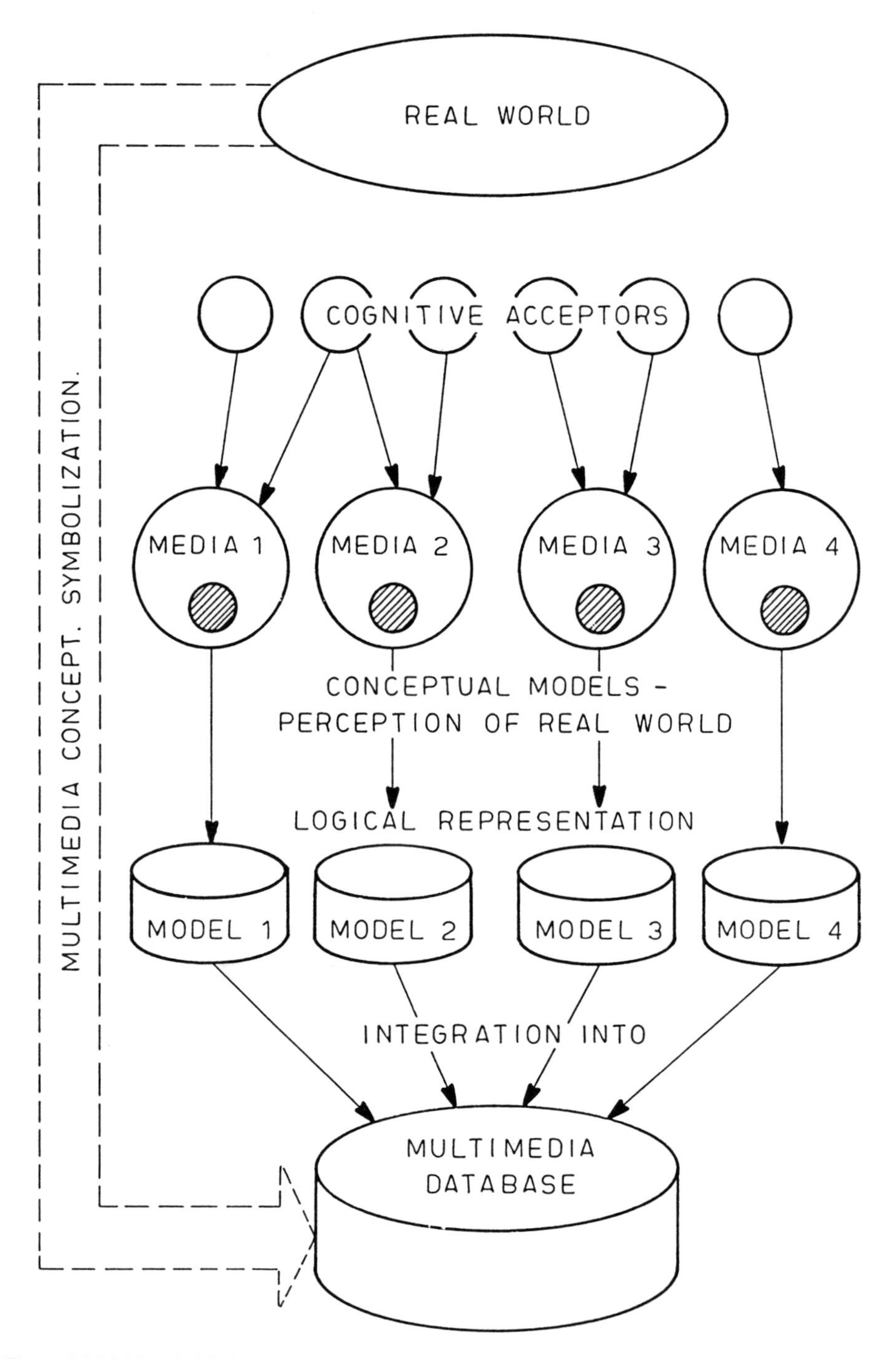

Figure 9.2 Multimedia Model

How AI constructs can integrate into and support a multimedia database is shown in Figure 9.2. They can act at three levels: (1) Cognitive acceptors; (2) conceptual model facilitators; and (3) logical representation structures.

Among other companies, General Electric also works very actively on these subjects. The goal of one of its AI constructs is to store and retrieve events from memory. It acts as an intelligent document retrieval system. Here are the problems, constraints and opportunities as seen by GE.

1. Even in the best organised companies endusers do not know exactly what they are looking for in memory.

They are particularly interested in bringing up from memory to workstation schemas which they need, but they do not have an enterprise-wide view beyond the external schema. That is not the enduser's job anyway. Therefore, say GE, Hitachi and Mitsubishi - in different instances, but in converging terms - this is an area where *AI may be of significant assistance.*

During the Tokyo meetings, for example, an evaluation has been made relative to the three major models for database management systems: hierarchical, network and relational. The conclusion was that the utility of this taxonomy diminishes as systems evolve towards a three-schema architecture with fundamental distinctions between:

- internal, physical schema
- external, user-viewpoint schema, *and*
- conceptual, enterprise-wide schema.

In the USA, the three-schema architecture of ANSI is an abstraction of the structure of existing systems rather than a new departure. So far, the three levels: internal, external, and conceptual, are not as carefully distinguished as they ought to be. But work done in this direction is promising.

As indicated by GE, the second area of implementation where AI advantages may be significant is:

2. Retrieval with partial and close matching of input features.

This includes both direct and reverse order of objects and events requested by the user.

Such reference is significant in the sense that it identifies precisely the work which can be accomplished through *memory-based reasoning* (MBR), a subject which we will discuss in Section 5. The challenge is to find AI solutions doing the MBR job without consuming an immense amount of computer power.

This is the direction in which many avant-garde companies currently work. Some database/AI projects are based on *events:* a conceptual representation of all features of something that happened in the world:

- Event-based information is *episodic*.

- Semantic information is *taxonomic*.

Semantic information, however, facilitates generalisation, abstraction and retrieval. As such, it complements the episodic approaches.

In the context of a global database solution, events are represented as a collection of features present in this event. All features are tagged with the same tag. Retrieval is guided by the conceptual representation reflected in the AI construct.

Semantic representation can effectively handle instances of an event. Figure 9.3 reflects the semantics of the database/AI research at a leading manufacturing company. Both input and output are indicated. The input leads to an *event integrator*.

Input events activate related events useful to the parser for the information to be retrieved. The process rests on the premise that instances of features have been integrated in a semantic hierarchy. This hierarchy is generic and, therefore, independent of the need to define unknown events posed by the query.

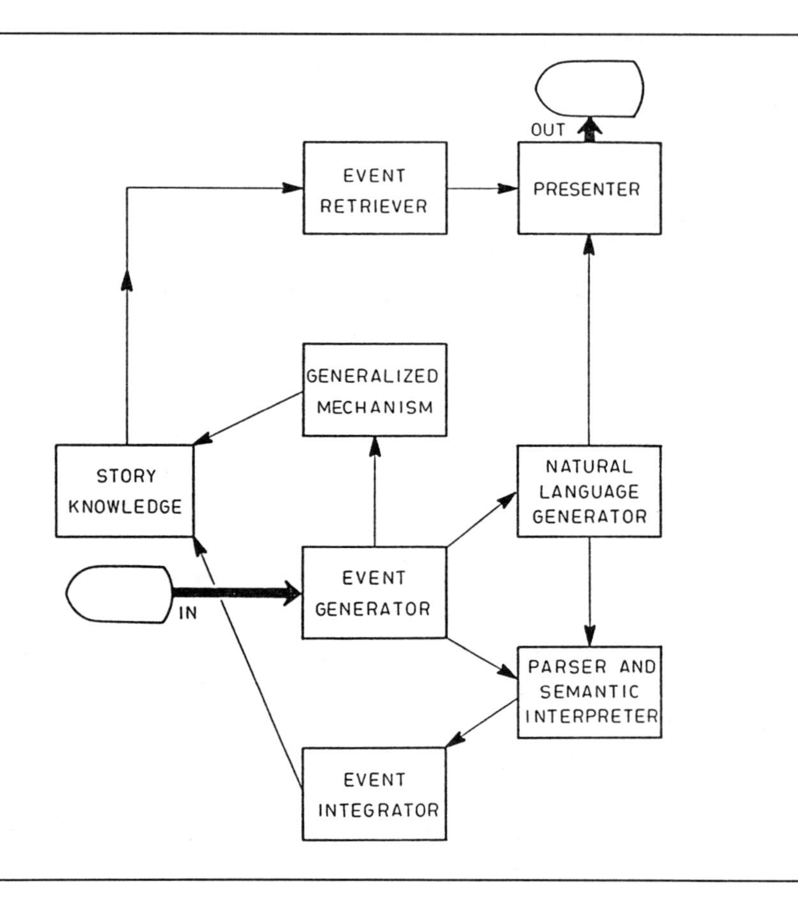

Figure 9.3

Even retrieval based on past solutions presents the problem of having to specify when to perform fresh retrieval, and what this means. Instead of this approach, GE has chosen an automated parsing from input features. Intersection is achieved at event level when activation exceeds theshold (candidates). The use of a filter mechanism performs structural match in a two-level approach. Total match is feasible under certain conditions.

In this sense, there is a multiple contribution of artificial intelligencene with significant potential. It can help define database organisation concepts; permit examples to be stored in multiple locations; produce more realistic generalisations, and so on.

One of the basic contributions expected by AI is in the actual retrieval process. The key questions here are: How do we evaluate the system? How do we ensure that the search operation is done in the right way? The answer is by building systems able to exhibit intelligent behaviour. We are still learning on this job, but the prospects are promising.

Taxonomic Reasoning and Classification

We said that there are two main approaches to the efficient exploitation of databases through artificial intelligence constructs. The one is event-based information, applying the concept of *episodic* memory. It can be very efficient provided that we have available fine grain parallel 5th Generation Computers, featuring 1,000 or more MIPS.

The alternative is a *taxonomical* approach, with or without semantic information. This is the foundation on which rest relational databases. It is a solution much more economical in MIPS, but requiring a significant amount of preparatory work, which needs to be both:

- horizontal, in the sense of classification work for surface organisation; and

- vertical or deep, for the classification proper.

The uses to which a deep model will be put are shown in Figure 9.4. It reflects the work I did some years ago for a major industrial firm. The figure also identifies the vertical exploitation of available information. This is done in successive layers, each generating information elements. Functionality ranges from sales forecasts (1st generation) to profit and loss estimates (4th generation).

The horizontal (surface) organisation is done through a classification scheme which follows taxonomical reasons. Here again, AI tools can be valuable both during development work and in the subsequent exploitation of the model by endusers.

Classification approaches have predicate value. Both pragmatic and theoretical grounds are behind this statement. Taxonomic reasoning is a deductive process. It is based on the traversal of a lattice made of descriptions. Formalisms rely upon the structure built by the classification, which should precede the automatic reasoning phase.

The organisational part is most important. Taxonomically organised knowledge must be arranged as a structure of descriptions exploited as:

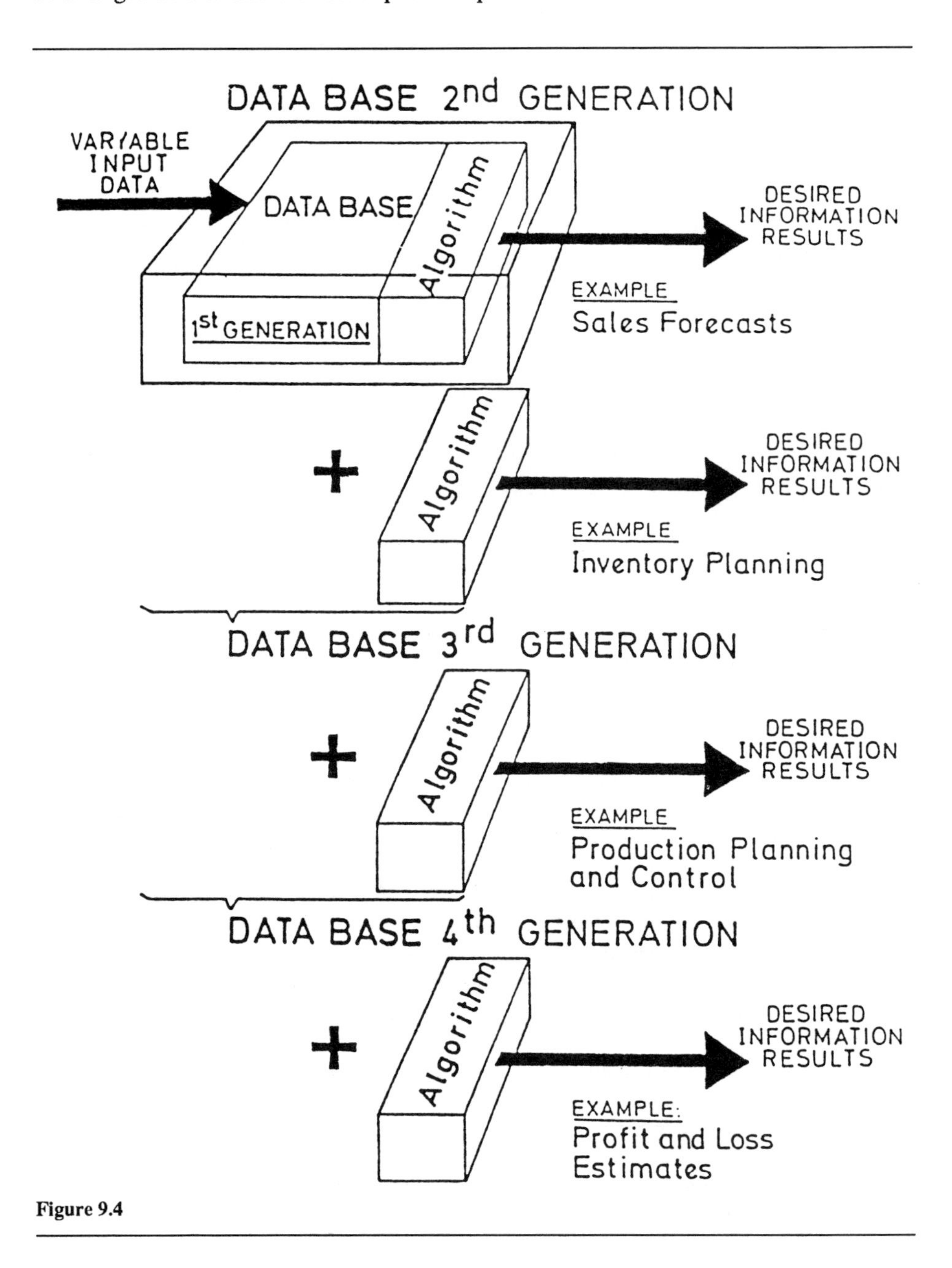

Figure 9.4

1. An efficient and consistent access path;

2. A mechanism for research and reasoning;

3. A selector of statements applicable during deduction; *and*

4. An infrastructure able to accept new descriptions.

A number of diverse or complementary operations can be performed by exploiting the same structure. In an implementation sense, the taxonomy should provide a means to attach to the lattice objects which are external to the system. This permits interfacing to distributed databases, or to an underlying linguistic support.

Taxonomies represent membership relations between classes of objects. They are the basis of an inheritance mechanism. In this sense, taxonomic reasoning has two views:

- A deduction process from taxonomical classification; *and*

- Inference about the taxonomies themselves.

An instance description can have attributions attached to it. This serves in specialising a class by describing some of its properties. Such an approach is valid when all knowledge is represented in a single lattice. Precisely, from actual knowledge, to general rules, dependencies and constraints.

When a description needs to be accessed to answer a problem, all relevant and related facts and assertions might be found directly connected to it. Also helpful are market propagation algorithms. They assist in establishing when a solution has been reached. They also determine when the search has run into a cyclic path.

Once the classification scheme has been properly done, deductions can be performed traversing the lattice, controlled by user-specified strategies. The latter determine which part of the taxonomical hierarchy to consider, and how to move around it.

Supporting tools are at a premium. While traversing the lattice, descriptions may need to be marked. Statements with variables can be instantiated. The process can be controlled by strategies defined at the metalevel, and evolving both metaknowledge and metadata.

Descriptions, and the structuring mechanisms embedded in the logic (inheritance, attributions), can be exploited to build the expert system's knowledgebank. This will be organised as a network where each node corresponds to a description and links to *its* relationship. These are the perspectives of new database management, and are way ahead of anything we have known in the past.

The inclusion in the lattice of statements with variables enables us to overcome the limitations of traditional semantic networks. The latter can only represent specific facts and cannot act as general type approaches. Taxonomical solutions are much more generic, and they are needed to solve tomorrow's problems in an able manner.

With the new generation of databases and knowledgebanks, all deductions that are allowed within a valid classification scheme can be performed by a process of lattice traversal. This concept holds good provided there is a suitable representation enabling statements with variables to be embedded in the lattice.

Figure 9.5 gives an example of the implementation of taxonomical principles in the manufacturing industry. Icons were chosen to help in classification. Similar concepts can be used in the financial industry and in a wide range of other applications.

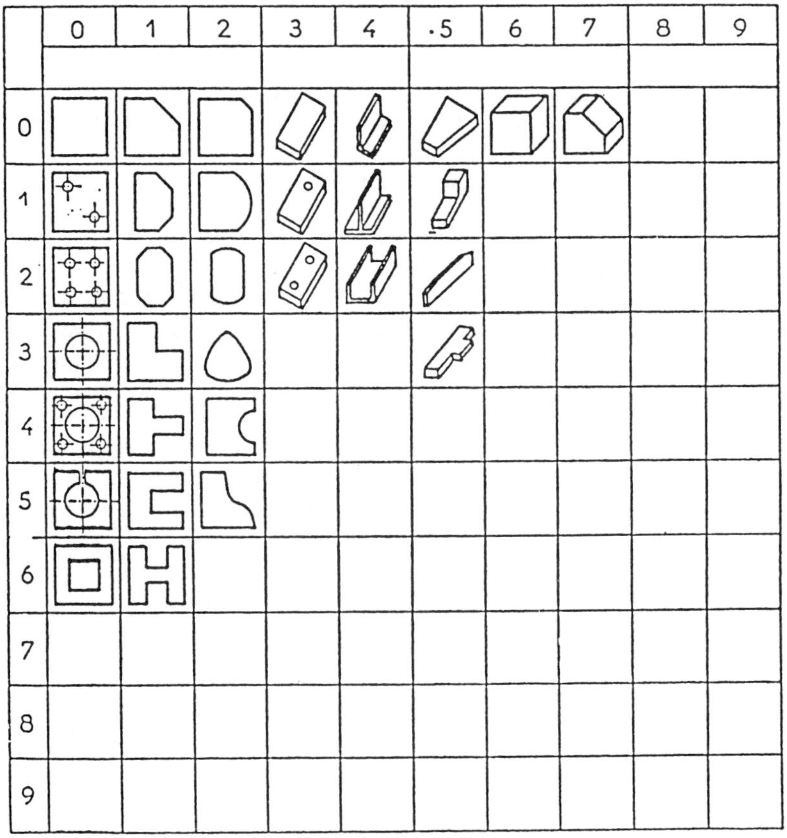

Figure 9.5

We must properly appreciate the fact that the next office revolution will be the *personal visual computer*. Computers are not only shedding their traditional jobs of sorting business records and doing number crunching, they are also taking on an exciting new chore: *image processing* being an example.* That does for images what word processing is doing for the typed word.

* See also Chapter 10

Within a few years, thanks to the sharp decline in the price of computer power, millions of professionals and other office workers will be able to capture, store, change, merge, send, receive and display:

- pictures,
- graphs, *and*
- illustrations

and just about any other image, in ways unknown only a short time ago. In virtually every field, *digital processing of pictures* is changing the way that work is done, and how much it costs.

Enriched with AI constructs, we can now capitalise on the massive work done for 3 1/2 decades in data processing, exploiting the fact that billions of printed words now pass through a computer at least once. During the 1990s the same will soon be true for photographs and illustrations.

An example of the potential which exists in manipulating images is given by what is taking place in the film and video industries. Today, special effects are almost all computer-created. It is only a matter of time, insiders say, before someone programs a computer to create an entire full-length feature film.

Carry this concept into communications and the gains will be dramatic. Affordable personal computers and cheap laser printers have already slashed the cost of producing newsletters, complete with four-colour illustrations in some cases. Meanwhile, emerging technology is enabling images to be shared among numerous electronic media, including video, computers and facsimile machines.

But, while in terms of cycles computer graphics have the potential to soak up a great deal of computer power and storage, *it is in new types of software - for instance AI - that the leadership game will be played.* Among other prerequisites, a valid classification structure would be instrumental in building an access mechanism whenever a new description is entered into the knowledgebank:

- The access formalism would properly connect the new description to other descriptions related by inheritance;
- At the frontend,the user's desk, the image processing will edit, zoom, shrink, alter or integrate the picture; *and*
- To connect to the other elements in the database, the classification algorithm will be invoked when a new description has to be inserted in the lattice.

Such classification formalism helps in restricting the attention to that part of the relation which is explicitly represented by links and paths. The relationship is induced by a few structural properties of descriptions, the total being subject to expert systems support.

Taxonomical methodology drives the problem-solving task, by issuing queries to the knowledgebank, acting and interacting with the external world, applying rules to find answers. *Taxonomy provides fundamental reasoning capabilities.* On top of it, more sophisticated strategies can be programmed to handle complex problem solving situations - from the more classical to advanced image handling.

Memory-Based Reasoning

As more of the information we need to do our daily work is locked in computer memory, the ability to access it quickly and use it become critical. Parallel processing can be applied to databases to improve the speed of the access mechanism, as well as its dependability.

This complements the notion of very fast computing that we have just discussed, and it makes use of inference capabilities in which, as stated, 5th Generation Computers are strong. In principle, we gain speed if we:

- Reduce the number of instruction accesses, *and*

- Eliminate most of the data accesses.

Logic programming and database machines help in cutting both of these down by about two thirds. Furthermore, we improve program quality if errors are recognised and corrected by the machine. Logic programming does so through a process known as equilibration.

Data level parallelism helps to significantly reduce the time needed for access, while it improves the level of sophistication with which this operation is done. Data level parallelism features a single control sequence, hence a single program, and executes it on all data simultaneously, operating on the whole data set at once.

An example of a 5th Generation Computer exhibiting such features is the Connection Machine which has a distinct processor for each data element, each computer with its own 4,096 bits of memory, but interconnected through a massive communications system. *Memory-based reasoning* has been developed by Dr. David Waltz of Brandeis University, one of the designers of the Connection Machine. It involves two phases:

1. Hypothesis formation

The system selects from a large database part situations similar to the current situation that we wish to understand. For good performance, it is important that the database is large.

2. Examination of the hypothesis suggested by the returned instances.

On the Connection Machine, this is done at a rate 200 times faster than on the largest available mainframe. Ten Gigabyte memory can be managed through fresh, complete comparison every minute.

The hypercube structure of the Connection Machine is not the only one to serve MBT ends. However, parallel processing *is* necessary to obtain high performance -which means processors and memories integrated in a networking sense.

In a memory-based reasoning implementation, dedicated hardware performs associative searches of knowledge. Retrieval and update efficiency is improved through metaphors and the use of a knowledge compiler with a knowledge server and/or predicate server - which may be either hardware or software constructs.

This is a different way of stating that, in terms of processing power, a new approach to AI can be implemented through MBR addressing an episodic memory structure where events and objects are stored as episodes, rather than being subjected to formal classification/identification as discussed in section 4. The price for this approach is very high machine power, typically more than 1,000 MIPS, or equivalent metrics.

One of the advantages of a memory-based reasoning solution is that no rules are necessary to drive the system. Neither is there a need for a knowledge acquisition phase. Another major advantage is the fast retrieval of information *based on concepts* rather than keywords.

In this sense, artificial intelligence opens the gates for new departures. The implementation of artificial intelligence constructs follows any one of three ways, from the older to the newer:

1. Rule-based systems;

2. Domain knowledge; *or*

3. Episodic memory.

In the general case, rule-based systems are deep knowledge oriented - specific in the area for which they have been designed. As explained in the preceding paragraphs, episodic memory is event oriented, and leads to a different type of reasoning: the intensive use of storage to recall specific *episodes*.

This contrasts with the traditional assumption in AI that most expert knowledge is encoded in the form of rules. It also leads to the concept of *idea databases*. An idea database (or idea processor) is *free-form*. Free-form means that information is viewed in a semi-organised manner in line with real-life folders and piles of papers, rather than the artificially regimented structure of current database and file managers.

Idea databases have much to do with messaging and communications, that is, with the future of computing. A networking structure directs the propagation of information through a web of nodes. Each link joins two nodes, one antecedent, the other consequent. Connections between nodes are specified by experts based on their domain knowledge - or a free form.

''I tend to avoid equations as much as possible'', says Dr. Stephen W. Hawking. ''I simply cannot manage very complicated equations, so I have developed geometrical ways of thinking instead. I choose to concentrate on problems that can be given a geometrical, diagrammatic interpretation. I can manage equations so long as they do not involve too many terms''. This is the leading trend in computer-assisted technology. What you see is what you get.

At the applications end, the raw speed of the parallel machine and its database support lead to much faster decisions than those possible with other technologies. In a highly competitive marketplace, this gives its user a critical jump over competitors who do not possess such tools.

There are parallels to what took place in 1981/82, on a much smaller scale. At that time the novelty was the personal computer. Horizontal software brought the micro onto the executive desk, along with a rich library of ready-made vertical (applications) software oriented to commodity operating systems and commodity micro-processors. Fast program development through Fourth Generation Languages (4GL) opened up the productivity challenge.

Times have changed. Leapfrogging competition is no longer done through the PC, but it can be done by means of a valid implementation of 5th Generation Computing - provided that we have a clear image of what we wish to accomplish, and how to go about it.

Today, valid implementation means *symbolic processing*. That is, manipulation by computers of information and knowledge represented as symbols, analogous to the human reason with knowledge that they possess. Symbols are used for real-world objects and properties associated with them. They can be linked together, using structures such as networks or graphs. Here again, emphasis is on image processing.

This short paragraph summarises an event of significant impact, from man-machine communications to the development of software to run the machines which we construct. Research has determined that classical programming languages, which were developed for repetitive operations on numbers, are inefficient and inadequate for use in decision making applications. Yet, these applications are the focal point of the future.

There is, of course, a close connection between this reference and AI, although an often-heard argument is that man lacks skill in effectively processing symbolic and logical problems. The counter argument is that we did not know about the eye when we made a camera. Chances are that intelligent man-made machines will be quite different from the human brain, just like flying machines work quite differently from the way in which birds fly.

But we can learn lessons from the proper observation of natural processes. Jean Piaget felt that children learn fundamentals by building their own intellectual structures. The basic concepts of mathematics can be comprehended by young children through a discovery method, with intellectual experiences related to total cognitive development. A similar approach is necessary in the domain of our main discussion.

Researchers in machine learning maintain that we must *break with past practices* of thinking about computers. This is especially true in the area of computer applications to the total learning process. It is hard to renew our way of working with computers if we project onto them the properties and limitations that we think we know today. New departures are needed. Are we ready for them?

Data Integrity for Distributed Databases

Data integrity refers to the accuracy or correctness of data in the database. Rules for checking integrity tend to be one of two types: *domain and relational.* Both must be brought into perspective.

In a relational database, each of the attributes has a domain which includes a pool of permissible values for the attribute. In most cases, such a domain is specified upon creation of the relation. Domain constraints are usually required for each attribute in each relation.

To insert or modify the value of an attribute, the system must check it against its domain. Such a check is handled automatically by the system as the concept of domain integrity is basically an extension of the concept of data typing in a programming language.

In addition to domain restrictions, a number of systems also include range checking. For instance: not full, unique, set, max/min; also the control of format. AI tools can be used to check for specific format(s), as well as for information elements within a format, a concept which, though clear, I find difficult to imprint on the mind of computer professionals.

In the extentional database sense, relational systems must assure integrity. *Relational integrity* refers to constraints which specify that two or more values in the database are required to be in agreement in some way. This is typically handled through user-specified procedures. Syntax helps in delineating relational constraints.

These concepts are most important within a global database construct. Not only is the issue of integrity both absolutely necessary and difficult to implement, but also global solutions should support logical data independence, automatically mapping queries and other actions on the integrated, global schema.

Just as vital is physical data independence, addressing parameters that involve distribution, replication and organization of the stored data without affecting the user's view. This calls for mutual consistency, managing the propagation of any changes.

Intelligent security mechanisms must assure continued operation in spite of site failures, even if some copies are temporarily inaccessible. Recovering sites must be automatically integrated into the system without interrupting ongoing activities. The same is true of dynamic integration of new sites, as well as of reconfiguring.

As ICOT underlined in the course of the Tokyo meetings, the risk is always present that distributed interconnected databases may be incomplete and inconsistent relative to each other. Incompatibilities may include differences in naming conventions, or in underlying data structures, representations or scales. To operate properly, it may be necessary to:

- supply missing data values;
- aggregate data across databases; *and*
- provide a consistent view of the data to the enduser.

Traditionally, the burden of treating incompatibilities has fallen entirely on patches created through interface programs. AI takes a major step in shifting this burden to the system.

With intelligent information management structures, the database administrator provides rules for how incompa-tibilities should be handled. For each query in which an incompatibility arises, an intelligent system can automatically apply rules to resolve such incompatibility - thus creating a virtually single, integrated database.

Concurrency control and reliability mechanisms perform through the intelligent structure which is sensitive to data access patterns, as well as to the assignment of data and programs to sites. While it is well known that the problem of distributed database design is closely related to concurrency control and reliability, classical solutions have not permitted us to address it in an adequate manner. Significant improvements must take place before truly distributed databases can come of age.

Working meetings with leading American computer manufacturers have added to this subject that of *portability*. They underlined its importance in a dual sense:

1. User activated within a specific applications environment; *and*

2. Due to the fact that the whole industry moves towards connectivity of heterogenous systems.

''One of the big issues is compatibility with data in a distributed environment'', said a senior executive. ''Compatibility calls for able OS interfaces. Not so much for work at the language level''. In fact, global compatibility requires homogeneity in:

- system calls;
- data models; *and*
- display protocols.

The goal is to provide a single, high level of database management for uniformly handling data in the multiple databases. Also, to assure a *single view* of the available data to endusers, while retaining local autonomy for organising and updating databases. Just as important is the provision for continued validity of all existing application programs.

Global retrieval requests must be expressed in intentional structures without knowledge of how data is distributed among databases, or which fast access paths are available where, for locating data. An optimisation strategy also needs to be implemented for mapping global requests into efficient sequences of local operations.

Interfacing permits us to address the problem in a modular way, each module kept at a relatively simple level. "If you try to cover the whole range of functionality, you need to write a full OS as complex as the old", suggested one of the leading experts in the field.

But there are also negatives in excess interfacing. The remark was made that, as experience with interfaces documents, the DBMS behind shows through. The same is true of firmware. And interfacing adds another layer to the structure, therefore affecting performance. Inter-facing, however, can be made much more effective if expert systems are written to make it selective and focused.

While no single development tips the balances towards a new environment, a sequence of well planned steps does so. As this and the preceding chapter help document, the 5th Generation Computer effort, global databasing perspectives and AI approaches are not single events. They are polyvalent solutions for industrial survival, closely knit with what a knowledge society requires.

10

Photonics Media in Information Storage

Introduction

The diffusion of CD-ROM players for personal computer users has opened a wider market place to imagebase systems that could not have existed otherwise. Because of it, it became reasonable to capitalise on the low cost of optical disk technology which joined the other available types, both as an alternative and as a further extension of what could be done in the past with other media.

As we know from experience, when it comes to the storage of information three types of solutions are in common use, (other than paper which we definitely wish to eliminate although we do not always succeed in doing so):

- *Photographic,* including microfilm, microfiche, and computer output to microform (COM).
- *Magnetic,* which became available in chronological order on tape, drum and disc, and
- *Optical* storage, using laser readers/writers, feasible on different options.
- Figure 10.1 identifies the different options that we have available in terms of optical disk systems. The three main classes are:

1. *Compact Disks* for the WS level, as well as for entertainment.
2. *Laser Video Disks* mainly for entertainment, *and*
3. *Digital Optical Disks* for central and distributed computer resources.

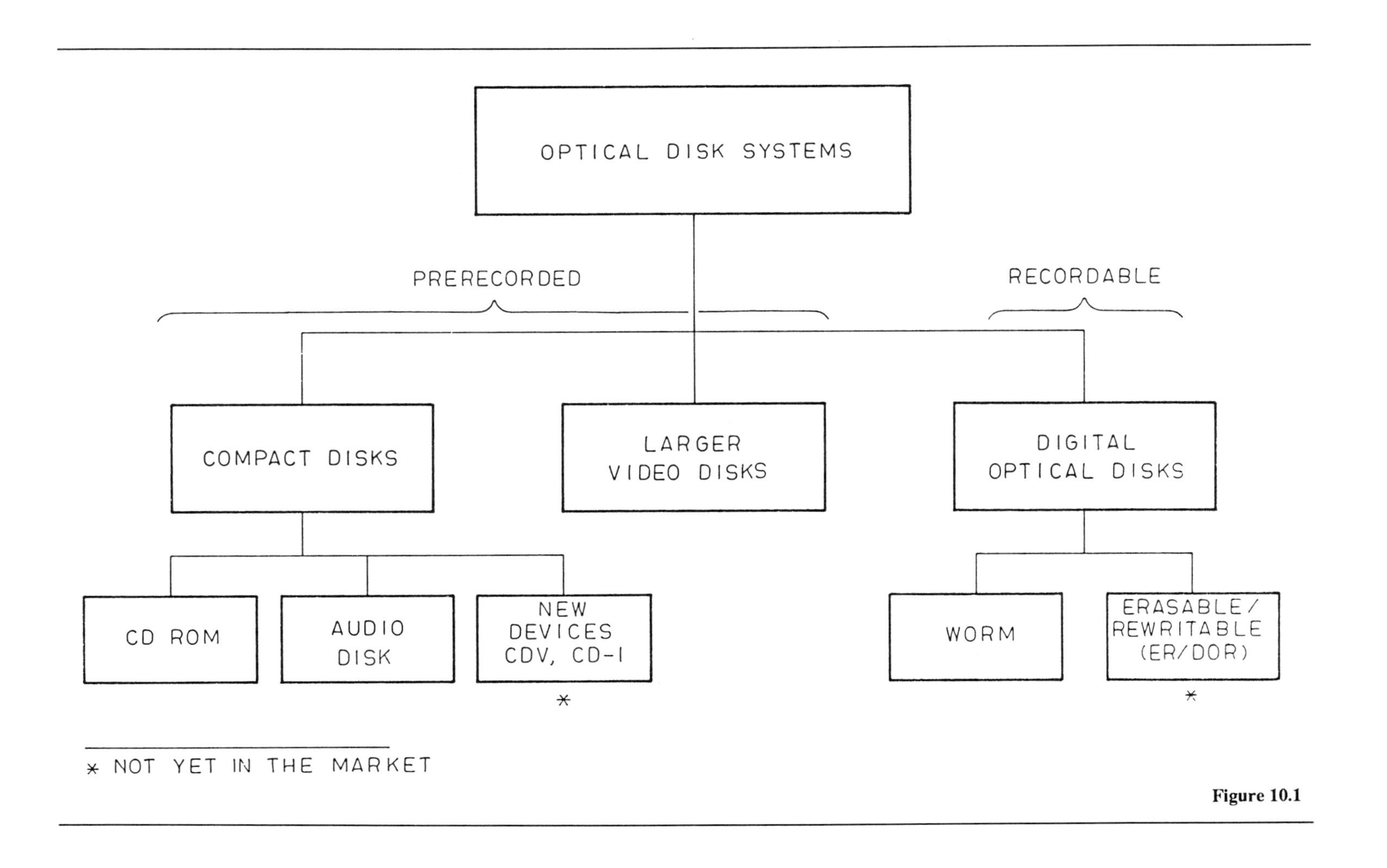
OPTICAL DISK SYSTEMS
PRERECORDED
RECORDABLE
COMPACT DISKS
LARGER VIDEO DISKS
DIGITAL OPTICAL DISKS
CD ROM
AUDIO DISK
NEW DEVICES CDV, CD-I
*
WORM
ERASABLE/ REWRITABLE (ER/DOR)
*
* NOT YET IN THE MARKET

Figure 10.1

Let us notice that all three are in full evolution. There are coming developments, particularly in classes No. 1 and No. 3, which promise to largely alter the way we are currently looking at optical disk technology and its implementation.

In a high technology environment, optical disks are destined to play a focal role. This statement is valid both at the central computer operations level and for workstations. There are, quite evidently, constraints.

The theoretical limit in optical recording seems to be constrained by the wavelengths of light emitted by the laser. Today, spots less than 3/4's of the wavelength cannot be detected. Media, laser and positioning technology combine to yield the performance ratio feature by current state-of-the-art.

But the business opportunity is significant. In Chapter 9 we have seen the reasons for and the importance of dedicated database engines for computer centres and, eventually, departmental computing. We have just as well spoken of terabyte databases. WS also have a need for large microfiles, and optical technology effectively responds to this requirement.

Table 1 further extends the classification which we have followed regarding optical disk products by identifying optical technology applicable to both the professional and the entertainment markets. For professional applications, which interest us in this book, two types are presently available:

1. *Read-only*, such as CD-ROM, videodisks, *and*

2. *Write once* read many times (WORM) memories.

These are the hardware solutions. Nevertheless, the fact should also be appreciated that proper information handling requires image management. This is the role of *image database management systems* (IDBMS). There is a growing interest in *Hypermedia,** the premise being that windows on the screen should be associated with *objects* in the database. Links must be provided between these objects:

- *graphically,* as labelled tokens; *and*

- *in the database,* as pointers.

Agile, user-friendly man-machine interfaces are, of course, very important. *Keyboardless* user contact employing touch-screen, tablet, mouse and voice-recogniser add to the user friendliness of the system. It also underlines the need for a modern IDBMS.

This is written on the premise that the convergence of new technologies is encouraging the rapid growth and application of *imagebases* - as contrasted to

* To be discussed in Chapter 11

classical databases. Not only is this the way in which technology develops, but also multimedia communication is increasingly important as graphics, image processing, video, text, data, voice and large memory systems transmission capacities are becoming integrated.

An image management system is analogous to a database management system (DBMS). It provides storage and retrieval of information, as well as query and security. It assures storage and retrieval of actual images of paper documents with 100 percent information recall and conveys the ability to store, retrieve and browse through groups of associated images in encapsulated form.

Optical Disk Technology

Optical disk technology has been looked at over the last several years as the mass storage repository of the future. It has the potential for storing information elements for less cost per bit than the magnetic disk, and exhibits the unique advantage of being able to store digital, audio and visual images concurrently.

One of the metrics of this technology is optical disk density. This is a function of:

- How well the laser spot can be focused;
- Positioning accuracy; *and*
- Media transition resolution.

Central to their being able to exhibit better bit densities than magnetic media is their signal-to-noise ratio characteristic. Also, the fact that laser beams can be sharply directed with the head at a comfortable distance from the recording media.

The recording process on a compact disk system is shown in Figure 10.2. In sequence:

- Digital PCM-coded information on a compact disk is scanned through a laser beam;
- The pickup gives digital PCM signals to be processed through a digital to analog converter;
- The output of this converter (step waveform) is handled through a low-pass filter which removes non-audio components (steps) from waveform;
- After filtering, the analog sound signal is transmitted to an audio amplifier.

Originally videodisks were developed by consumer electronic companies to compete with video cassette recorders (VCR) for a share of the lucrative consumer market for video recording and playback systems. They were first shown in 1973, and became commercially available in the USA in 1978.

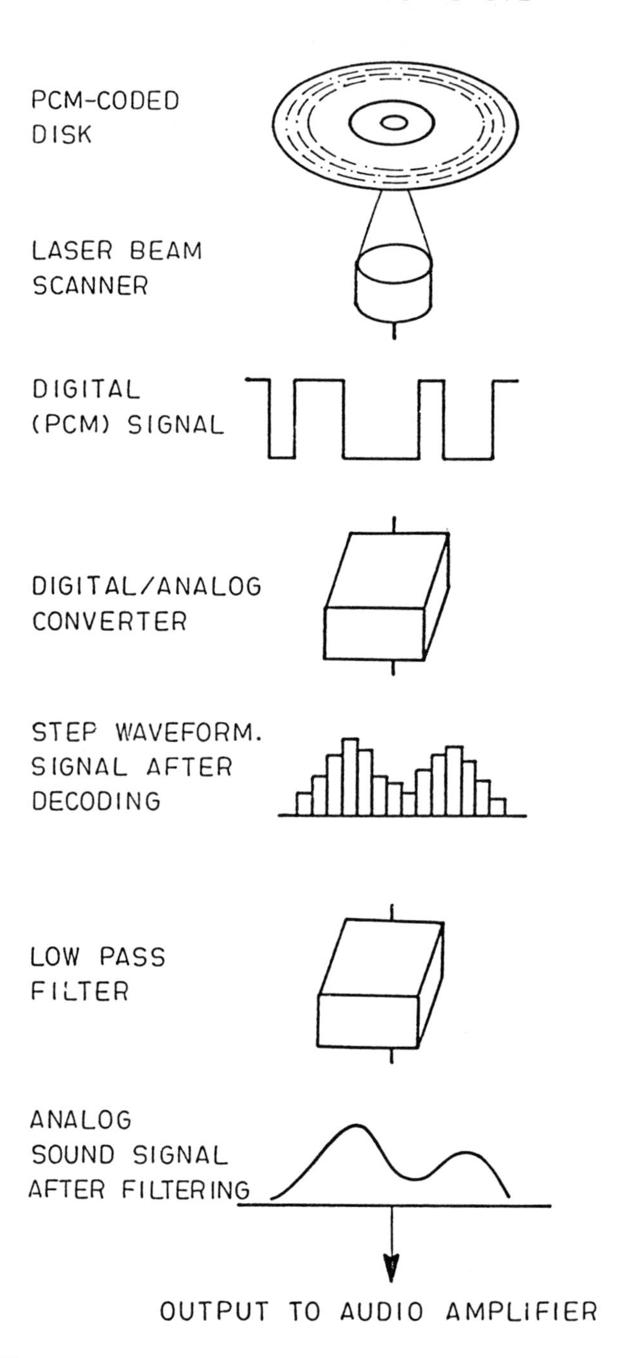
COMPACT DISK SYSTEM
PCM-CODED DISK
LASER BEAM SCANNER
DIGITAL (PCM) SIGNAL
DIGITAL/ANALOG CONVERTER
STEP WAVEFORM. SIGNAL AFTER DECODING
LOW PASS FILTER
ANALOG SOUND SIGNAL AFTER FILTERING
OUTPUT TO AUDIO AMPLIFIER

Figure 10.2

The advantages claimed for videodisks were that disk replication costs were less, and the players would be cheaper, than for VCR, and the image quality offered would be superior. Also there would be less wear during playback. It was also suggested at the time that the videodisk systems would be far more flexible, offering facilities such as freeze frame, random access and slow speed forward and reverse.

The fact that users cannot alter CD-ROM disks that have been prerecorded is both a limitation in a recording sense and an advantage with reference to safety precautions. Thus CD-ROM usage is proper for information that has already been published (statistics, public databases, dictionaries). Such usage typically addresses itself to information generated externally to the organisation, the CD-ROM constituting an efficient text and data carrier.

For evident reasons, CD-ROM finds a good market in the entertainment business. Here, however, lack of a recording facility prevented videodisks from taking off as a consumer item, although some features available with videodisks made them attractive to the institutional market. Today, they are being widely used for interactive training and point of sale systems -even though *mastering* remains a rather complex, expensive system.

Two main processes exist today for the mass production of replicate videodisks:

- Philips and 3M use cold photopolymerisation, referred to as 2P;

- Pioneer, Sony and others make use of compression injection moulding.

Both methods involve the pressing or stamping of disks. Also, with both of them, the more copies that are produced the cheaper they are, as the mastering and replication equipment is sophisticated and costly.

All videodisk systems currently on the market make use of a playback technique involving a low power laser and a photo-detector to collect light reflected by the disk. The laser pick-up device can be moved at random across the disk and will operate at a range of different speeds, or in still-frame mode.

Positioning, or seek, times for optical disks are slower than those for magnetic: 50 - 100 milliseconds (ms) vs. 16 - 20 ms. One of the major contributors to this is the mass of the laser and its optics. Optical disks also have to rotate more slowly due to write power considerations and high linear densities.

High linear density limits the rotation speed because of current I/O bus bandwidths. This should be rectified in the near future. Table 2 compares evolutionary developments in rotation speed and transfer rate, as well as long and short seek time, in the 1984 to 1990 period (the latter being an estimate).

Compact disks (CD) came later on the market than optical videodisks. Compact disks were first shown in 1980, and players were only launched commercially in 1983. However, their sales to the consumer market have outstripped the sales of videodisk players, with several million players sold in the first two years after launching.

Table 2 Trends in Optical Storage Media during the 1980s

	1984	*1987*	*1990*
Rotation speed in rpm	900-1800	1800	3600
Transfer rate in MBPS	1-2	2	4-8
Long seek time in ms	70-150	50-100	25-30
Short seek time in ms	2	1	1
Rotation time in ms	33.4	33.4	16.7

Compact disks are standardised in format and digital recording techniques (as designed by Philips) and in error detection and correction systems (jointly developed by Philips and Sony). It is this compatibility which led to the success of CDs in the consumer market.

The physical process involved in producing a compact disk are virtually identical to those for videodisks except that the material recorded on the disk is in a different format, and the master and replicate disks are single sided with a 4.75-inch (12 cm) diameter rather than a 12-inch diameter.

Plants designed to master CD can be easily adapted to master CD-ROM. The disks have the same physical dimensions, diameter and thickness, as well as chemical composition. Production of CDs is a three stage-by-stage process comprising:

- premastering;
- mastering; *and*
- replication.

At premastering, audio information is recorded onto digital tape to provide a standard input to the mastering process.

At the mastering stage video, audio or digital information is recorded onto the master, together with formatting and indexing information and (in some cases) retrieval software.

The replication stage involves the creation of a number of submasters from the master CD, subsequently leading to larger volume production.

Roughly two years after the launch of the first commercial CD audio player, Philips showed a CD-ROM (Read Only Memory) engine. This was a direct adaptation of the CD system for publishing and data processing applications. The worldwide standard which had initially been established for CD audio was extended in the CD-ROM area to cover the requirements of the data processing environment.

CD-ROM drives rotate at the same speed as CD players, in Constant Linear Velocity (CLV) mode. But there are also differences.

CD-ROM are primarily designed to store digital information. CD were designed to store digitally encoded audio signals. While CD and CD-ROM use the same basic control and display systems for locating/addressing information, there are differences in the number of tracks into which the program is divided, and also in error detection and correction (EDAC) techniques. CD-ROM drives require a further level of error detection and correction than CD players.

CD-ROM can be a valid publishing medium because of its high storage capacity and robustness. A single disk can hold the equivalent of 60,000 to 220,000 sheets of A4 information or between 500 and 1,000 medium-sized books.

Grolier, an American publisher, has put its Academic American Encyclopedia (30,000 articles, 10,000 pages) on one-tenth of one disk. The disk's 550 MB of data are equivalent to about 6,000 graphical images, or roughly 300,000 pages of text.

Five hundred and fifty megabytes on a single optical disk is equivalent to more than 1,000 editions of Gutenberg's first book. Volumes of printed text, reels of magnetic tape and drawers of microfiche can be reduced to a durable, small-size disk and the compact optical medium stores not only words, but pictures, graphics and images.

Like Gutenberg's books, CD solutions are affordable. Compact optical disk readers are available as microcomputer peripherals.

Interactive videodisk technology is essentially interesting to businesses using mountains of information that does not require constant updating. Companies like GM, IBM and McDonnell Douglas employ it. McDonnell Douglas, for example, estimated that it can put its 20,000-page DC-10 maintenance manual on a single disk.

Some of the drawbacks, however, are procedural, including the setting of standards. While there is one physical standard for CD and CD-ROM (4.7 inch media) and one de facto standard for optical laser videodisks, there are currently no standards relating to recordable digital optical disks.* They are available in a range of diameters including 5.25 inch; 8 inch; 12 inch and 14 inches and they make use of:

- A wide range of substrates (base materials) including glass, aluminium and plastics;
- Different recording mechanisms: pit forming, bubble forming, phase change;
- A variety of protective measures: single, dual, trilayer or quadrilayer recording surfaces;
- Grooved or non-grooved tracking;

---

* Write Once Read Many (WORM) times, discussed in Section 3.

● Preformatted, non-formatted or post-formatted disks.

The difference in normalisation is important, as the CD standard sees to it that all CDs can be played on all CD units. Theoretically, this is also true with CD-ROM, but because CD-ROM drives are computer peripherals, and users need a computer to access the data stored on the CD-ROM, a number of logical factors inhibit such compatibility.

1. The volume and file structures used on the CD-ROM disks. Precisely how the data is organized logically on the disk.

2. The fact that drives have different controllers. The commands that these controllers accept, and the format in which they must be supplied, are not standardised across manufacturers.

Differences also exist in terms of retrieval software. Incompatibility problems are not new when it comes to computer peripherals. But, it is a pity that a new, promising area of online storage starts with reduced portability among different systems.

Interactive Compact Disks

There are other variations in the compact disk range. An example is the Compact Disk Interactive (CD-I). In early 1986 Philips and Sony announced plans for a CD Interactive Media specification. But it is not certain that this specification will be generally adopted. Remember Video 2000 and Betamax?

CD-I represents a logical development of CD-ROM. Both CD-ROM and CD-I have applications in data processing. CD-I is a special use of CD-ROM whose data records can contain data in specific formats which feature:

● Sixteen kinds of audio;

● Two video resolutions;

● One microprocessor object code; *and*

● One file system.

As defined by Philips and Sony, CD-I is for the consumer market place. The High Sierra Group proposed another standard: CD commercial (CD-C) aiming to meet commercial data processing requirements. For the time being, all this is leading to confusion about the respective roles.

One of the problems is the fact that CD-I contains information like video and audio, as well as text and still pictures. They all need different error correction levels, the means by which the accuracy of digital signals is assured.

The issue is more complicated by the aim to make certain that a CD-I system can play both CD audio disks and CD-ROM digital disks, thus widening the market. But, it is not sure that the market appreciates such variation. The world has yet to digest CD-ROM. Still, if successful, CD-I and CD-C may change the nature of books, publishing and libraries, becoming the electronic printing press of the future.

Recordable disks are called Write Once Read Many (WORM) or Write Once Optical disks (WOOD). Both terms are somewhat misleading, as they tend to imply that all the data has to be written at the same time. In fact, users can continue to record data onto free sections of the disks over a period of time until the entire disk is full.

The basic technical characteristic is that a WORM optical disk can only be written once in the sense that:

- It is not possible to overwrite data onto a sector that has already been used; or
- To erase and rerecord data on the same sector; so that
- Updates are made by recording revised information to an unused portion of the disk.

Each sector of the disk is write once read many times. This makes WORM an ideal archival medium likely to replace magnetic tape in a number of applications as a data archiving medium. WORM will also replace microfilm and computer output to microform (COM), the input coming directly through computers or facsimile scanned images of documents.

The key to optical storage media replacing magnetic disks - as opposed to complementing them - is the development of erasable optical media. Re-writeability has been slower than expected in coming, although costs for read only have not come down as fast as expected. It is projected that re-writeable optical media will have a density somewhere in the range of between 1/2 and nearly the same as ROM.

Erasable Digital Optical Disks can be viewed as a third generation of optical disks which will give users the ability to not just record data on the disks, but also to erase and re-record many thousands of times on the same track sector, just as they do with magnetic storage.

Erasing and re-recording is essential for active computer applications where files need to be retrieved, modified and rewritten on a regular basis. The majority of operations and applications software packages assume the use of erasable media. The main advantages erasable optical disks offer over magnetic media are:

- Low costs drives;
- High storage capacity (in excess of 300 MB per 5.25 inch disk);
- Jukebox type devices offering automatic disk handling facilities with terabyte storage; *and*

● Durability in use as non-contact laser record/read is employed.

The processes involved in mastering and replicating erasable disks are almost identical to those for recordable units. The mastering stages are the same, but at the coating stage the chemical structure of the recording layer is different, so that the recording/erasure process can be repeated.

Although working prototypes of erasable digital optical disk systems have been shown, no erasable systems are commercially available at present, and hence exact specifications are hard to give. 3M and Kodak are among the leading firms in this area, together with KDD, Sharp, Hitachi, Ricoh and other Japanese companies. Two basic techniques are employed:

1. Phase change *and*

2. Magneto-optics

The first takes advantage of the phase change that occurs in a tellurium-based alloy film when heated. It switches from crystalline to amorphous form. These forms have different optical properties (reflectivity). But there is a finite number of times that the media can be cycled through this phase change.

The magneto-optic effect shifts the direction of magnetisation. Unlike the phase change approach, there seems to be no limit on the number of writes. The recording of data in magneto-optic systems involves the use of a laser to provide heat and magnetic coil to serve the magnetic field.

Prior to recording, the entire magnetic film of the disk is perpendicularly magnetised. Then, at the recording stage, a micron sized region on the disk is heated up by the write laser above the Curie point. As the material cools in the presence of an external magnetic field oriented antiparallel to the initial direction of magnetisation, a small region (one micron) is formed that is reverse polarised. This is the pit.

How sharply the spot can be focused is a function of:

1. The focusing lens, *and*

2. The wavelength of the laser.

Orange or red lasers are currently used. A spot of about 1 micron can be achieved using a lens which is reasonably low mass and not too expensive. The use of a blue laser, whose wavelength is about 1/3 that of the red, may give a 2x density improvement.

The positioning accuracy for optical disks is currently better than for magnetic storage media. Optical disks have a better servo technique due to a lot of development work that was done for the commercial videodisk market. They also currently have more tracks per inch than magnetic media. Optical disks offer the interesting possibility of spiral recording where a disk surface is just one big track.

The key functions of optical disk systems currently available on the market are:

1. *Image filing*

Reading documents, and then compressing and storing information as image data on the optical disk. Also retrieving document information from optical disk by index or number, then displaying it.

2. *Multiple retrieval*

This will be increasingly important in coming years as applications move toward *multimedia,* and compound electronic documents will dominate the automation of most office functions.

3. *Image processing*

The use of optical media facilitates enlargement, reduction, reversing, clearing, moving, and composition of and image, as well as the evolution of machines working as intelligent copiers and network switches connected by high-speed communication lines.

4. *Image transfer*

The range of applications spreads from the transmission of a retrieved image to remote facsimile, to the full multimedia implementation necessary in the modern office.

5. *Boolean search capability*

This can be implemented using AND, OR and NOT operators. It includes the ability to restrict a search to specific sections of a record, and permits proximity searching for words that are adjacent or near one another.

6. *Browsing and display*

For instance, browsing the dictionary of indexed words in the glossary. Browsing is the ability to use a workstation to view document after document or page after page within a document and to examine the images.

7. *Manipulation and editing*

Examples include right hand word truncation, enabling searches for all words with a common base prefix.

8. *Other functions*

These include copier functionality as well as volume record change between optical disks executing, for instance, the reformation of files. Also, the provision of complete HELP information, allowing even first time users to effectively conduct their own searches.

Whichever is the specific functionality of the application we are planning, it should be characterised by consistency, dependability and uniformity. These are important characteristics in every information system design. Dependability is vital to compound electronic document implementation, a task to which optical disks will most likely be put.

Value Differentiation Through Optical Disk Solutions

With both the CD-ROM family and WORM access time is an issue. Compact disk approaches are faster than floppies But, as stated, the current average access time at the range of 50-100 milliseconds is slower than hard disk access rates of 14 to 20 ms. Yet, for the majority of PC users, and for many applications, some 50 ms difference in speed is not an issue. Also, once the laser beam has been positioned, we are talking of two to four million characters per second transfer rate. This is 16 to 32 MBPS, which can easily saturate currently available local area networks (LAN).

Access time problems are not a new worry. As computer applications expanded and diversified, the amount of data involved has also increased. This brought up such issues as optimum data arrangement, and associated operational problems.

But, the speed and type of storage does not necessarily correlate with the capacity, price and intended use of the device. Today there exists no storage system which meets all the user's demands for speed, capacity and price. The gap between the access time of main storage in:

- CPU *and*

- magnetic disk

is especially wide, about *4 orders of magnitude,* and prevents the improvement of whole system throughput (Figure 10.3). Therefore, to maintain high performance and an economical memory it is necessary to create a hierarchical storage system composed of various technologies. One of these is optical disks. Notice, however, that while:

- Photonics storage partly complements and partly substitutes magnetic disk media

- But it can *fully substitute* magnetic tapes providing much better performance and, most importantly, eliminating the highly manual tape operations.

Another constraint in photonics implementation is thought to be the impossibility of erasing written information. Some potential users are waiting for erasable CD technology to become commercially available. But non-erasable videodisks present significant advantages for some applications.

In a banking environment, for instance, non-erasable optical memory has advantages on which we should capitalise. It constitutes an *unalterable document* which could be presented as evidence - provided that the lawyers are keen to convince the courts on this matter.

The non-erasable feature is vital for *logs* and *journals*. Also for *signatures* and a *payment system* documentation. Besides, even after it becomes available, the use of erasable technology may be unwise until we get a little further in terms of standards.

Given its compact size, CD-ROM is a good option for data distribution within our organisation - particularly where there is a need to distribute a large amount of static information. To take an industrial example, automobile manufacturers are putting parts lists in CD-ROM. This technology seems ideal for:

- management directives and bylaws,
- catalogs,
- parts lists,
- directories,
- financial or demographic information.

The cost of an optical disk is minimal and this is nearly true of the reader. The major cost item is the construction of the masters. Although such costs have come down dramatically, it is still fairly expensive to put on CD-ROM narrowcasted applications. Volume production must justify the implementation.

There is also the cost of *organisation* and *training*. A user organisation must itself format its data for CD-ROM mastering, or use a service bureau. The latter is unacceptable to some companies as control of the product or customer files is passed on to a third party. Where security is foremost, only feasibility tests could be conducted on a service bureau basis.

All these references point to the fact that organisa-tional requirements must be addressed in a serious manner. A lesson to be learned and applied in structuring optical disk projects is that automatic documentation systems should:

- Track procedures governing each file and file type;
- Manage approval and release of files from one step to the next;
- Archive files for future search, enriched with proper indexing;
- Secure files against unauthorized use; *and*

- Alert managers and professionals to changes in the file.

These are some of the goals that an organisation should set if it wishes to become a leader in cost/effectiveness regarding implementation of integrated multimedia information systems, of which optical disk technology is part and parcel.

A successful implementation requires clear goals and a global application perspective. As an example, the optical disk system instituted in a leading industrial firm consists of seven basic components:

1. Document Entry Station (scanner);

2. Editor Station;

3. Indexing Station;

4. Optical disk Storage;

5. Local Area Network;

6. Workstations; *and*

7. Laser Print Station.

Systems software has been necessary to drive the system, along with image management procedures, communications routines, text and forms generation software, as well as computer output to laser disk software.

The functionality of a document-image processor starts with its ability to capture an image from paper and, thereafter, process the electronic image rather than paper itself. Systems software imprints a unique serial number on the paper document so that the original paper can be bulk filed for storage, and obtained later if it is required. Software is also required to permit the extraction of data from the image for entry to mainframe and subsequent processing.

In this sense, the *optical scanner* has been approached as a crucial computer peripheral. In many applications, the amount of paper records to integrate into the photonics system see to it that document handling operations should focus on hardcopy input, its clearance and indexing.

At the same time, users of electronic publishing systems want the ability to reproduce photographs with complete fidelity to the original. In at least one specific project, the situation arose where a host computer had to be used for image processing.

Research conducted among computer vendors identified that, among the better units offered on the market, were reliable gray-scale scanners with a resolution of 400 dots per inch, combined with image processing hardware and software. Other equipment provided a resolution of 300 by 600 dots per inch at 64 levels of gray, featuring capabilities to create half-tones and control contrast.

After a dependable input solution was found, the document image processor converted paper documents into digitised images to be permanently stored on optical disks. The result has been high capacity storage that is cost/efficient coupled with retrieval that is fast and easy.

The systems solution also required software which automatically integrates, manages and routes the flow of stored documents from workstation to workstation throughout the organisation. As the project referred to was able to demonstrate, work gets processed faster on a local area network where users can share data and information, and can even work on the same document at the same time.

In the general case, electronic manipulation of digitally-represented document images includes: storage of images; scanning of images; retrieval of individual images, along with logically associated information; routing of images in a predetermined sequence; as well as creation of images through annotation or word processing tools. Their presentation on video and/or laser printing is an integral part of the systems solution.

When the organisational study is properly done, optical disks provide impressive advantages in handling images. This is one of the reasons why, in engineering, optical storage systems replace microfiche and aperture cards for archival storage. And, optical disk drives constitute an important tool in making large databases and sophisticated CAD systems possible on a desktop.

Within this perspective, graphics packages are essential. They must, however, be well documented, easy to use, flexible and professional. Another area on which the organisational study should concentrate is information distribution and sharing. At the present time, distribution options are still limited, being confined to low bandwidth transmission, except where islands of broadband networks or satellite connections do exist.

Since technology is in full evolution in this domain, a properly structured optical disk project should keep all options open. It should take a fundamental approach, studying all alternatives prior to coming to a conclusion - and to a firm proposal for implementation. As Chapter 11 will demonstrate, there are several prerequisites to be answered in an able manner. Haste with great care is the best policy.

Laser Disk Solutions and CAD

As the experience of the case study in section 4 helps to document, optical disks are a multimedia solution which integrates different forms of information on the same support - but this integration should be done with great care. The range of implementation areas where laser disk technology may provide effective approaches to information systems is in itself rather impressive. We have seen in this chapter references ranging from internal documentation, as a substitute for COM, to the enrichment of the WS database, and provision of the infrastructure required for multimedia integration.

Another field in which photonics can be profitably implemented is computer aided design (CAD). Applications ranging from mechanical and electrical engineering, to architectural design, office layout and networking can benefit through optical disk implementation.

The introduction of photonics in CAD is just as valid in manufacturing engineering environments as it is in banking. At the Swedish PK Banken, for instance, the use of CAD led to an "Intelligent Office Automation" (IOA) project. Here are the fields which were identified:

- For the last three years PK Banken has used CAD for the design and production of office forms.

As Dr. Gian Medri, the director of research, was to suggest, the stream of new banking products to face an increased competition makes it wise both to automate and to coordinate form production - including softcopy and hardcopy. A new form, however, takes 1 to 2 weeks to design through classical means. With CAD, it only takes 1 hour, and integrates nicely with the other forms used in the bank.

Formatting in the 1990s will not be the inflexible and monolithic approach we know from the past, but a polyvalent, dynamic approach adapted to the specific requirements of the transaction. It is precisely in this connection that EDI can be of interest to business and industry, as we will see in Section 6.

Increasingly, form management will require multimedia supports and can be effectively handled through CAD, enriched with optical disks and expert systems. Such a solution should be studied organisation-wide from production to the distribution and usage of documents stored on photonics devices and transmitted through photonics highways.

- CAD usage in DP operations for management of tapes and disks. Also for scheduling.

While indexing should remain on easily updatable (hence read/write) media, the contents of tapes and disks in archives can be profitably converted to photonics. There is a golden horde of Japanese examples relative to this implementation. Subsequently, the optical disk system can be endowed with CAD software as well as with expert systems constructs.

This transition has already started. One of the leading financial institutions which, in 1972, moved from paper to COM, in 1988 changed from COM to photonics. However, the preliminary study which led to this changeover demonstrated that, without appropriately rethinking the system, the results would be much less satisfactory than those expected. The major benefits to be derived from technology are usually embedded in system changes.

System changes have a deep impact on the way we are doing business - from research to the production and distribution of services. Among the important steps the latter study identified as preconditions to successful implementation of optical disk technology has been I/O interfacing.

The purpose of an I/O interface is to allow existing operating systems and application programs that employ magnetic disks to use optical disks instead, with minimal change. This is fundamental during the transition period from magnetics to photonics, which may last a whole decade. We will see a good example with Banque Privée Edmond de Rothschild, of Geneva, in Chapter 11.

In the case of PK Banken, this financial institution uses CAD quality scanners and computer aided engineering software to overcome the document input difficulty. Able solutions have also to be found for interfacing between magnetic, electronic and photonic online media.

The whole area of office automation (OA) is open to the implementation of photonics systems. In an engineering environment, CAD/CAM type applications use optical disks for the archiving of all drawings. Marketing activities of all types, too, can benefit through photonics. Integrating graphics and images into DP/WP serves the objectives of handling patents, court actions, transfers from supplier to consumer, or vice versa. Hence, for full *product documentation.*

Thus, from networking and databasing to DP and OA applications, office layout, form design and other office chores, CAD and optical disk technology can be put to advantage in a banking environment. Whenever multimedia approaches are necessary, or simply advisable, photonics supports should be added to CAD. The same is true of AI.

As an optical disk project gets focused, this enlarged implementation horizon should be kept under perspective. When it comes to high technology, a broad, properly chosen conceptual approach can make the difference between profit and loss.

Why Photonics Approaches are so Important. The EDI Story

A recent study indicated that about a tenth of the cost of goods and services traded in the European Community is wasted every year in useless or unnecessary paperwork. Shuffling even the necessary purchase orders, invoices, confirmations and receipts back and forth between companies costs an enormous amount of money.

It is scarcely surprising, therefore, that organisations are turning to more effective means to transfer documents.

- Three years ago, in 1985 only specialised industrial sectors applied Electronic Document Interchange (EDI).

- Now more than 1,000 of the larger European companies employ Electronic Document Interchange.

In the United States, where the retail trade applied EDI at an earlier stage, there are presently an estimated 7.000 companies fully using Electronic Document Interchange. The prognosis is that, by 1995, close to 180,000 companies will move to EDI in Western Europe, not far behind the projected 250,000 in the USA.

Hong Kong, too, is taking steps to establish an Electronic Document Interchange system. Early in 1988 a blue chip consortium - comprising leading banks; container and air freight terminal operators; a major British trading company; Hong Kong Telecommunications Ltd.; and mainland Chinese interests - put together a company called Tradelink Document Services Ltd., with the express purpose of setting the EDI ball rolling.

But is there really an Electronic Document Interchange *standard* by which sellers and buyers can make business transactions online computer-to-computer? If the question is asked in a flat way, the answer has to be: ''No!'' - and this because, in reality, there is no *EDI* but there are *EDIs*. A recent EDI study compiled by an Input, Mountain View, CA, based company diskovered through the sampling of 200 MIS managers (among Fortune 1,000 firms) that most corpora- tions are interested in installing EDI. Only 22 percent of those interviewed said they have no plans to implement the ''standard''. Which ''standard''?

There are many colloquial implementations usually clustered around a network which interlinks vendors to suppliers. An example is the Organisation for Data Exchange by Teletransmission in Europe (Odette), a London-based coordinating body made up of major manufacturers and suppliers. Odette starts its life by bringing together Renault, Fiat, BMW, and BL/Rover.

A competitive network to Odette has been worked out by Electronic Data Systems, which has hooked up with several value-added network operators to implement an open, pan-European online system linking General Motors to some 2,000 suppliers of parts and equipment. This network is expected to be the world's largest Electronic Document Interchange implementation to date. When fully operative, it will use direct links and existing networks to connect GM with its suppliers.

The plan for EDS is to integrate a number of network operators that currently link suppliers to GM. These include the United States' GE Information Services (GEIS), the United Kingdom's Istel Ltd. and France Telecom's Transpac. Still, this EDS network is now at the ramping-up stage and it will not be a universal one - but rather an EDI VAN.

According to what is so far known, the GE Information Services' global network will link suppliers in six European countries to EDS. Users in West Germany, Spain, Italy, Belgium, the Netherlands and Luxembourg will connect through the GEIS node to exchange documents such as invoices and delivery schedules. In the United Kingdom, Istel will be used to link suppliers to EDS, while French suppliers will be connected to EDS via France Telecom.

The need for open networks is clear, and evident to everybody. Just as clear is the need for a *standard document protocol.* The current state of the art, and its spontaneous approaches, does not suggest that the necessary *universal standard* is being worked out. In the end we are going to have many EDIs which are incompatible with each other and which need expensive gateways to communicate amongst themselves.

This is a pity. From online message and transaction handling, to the use of photonics storage media, we need a universal protocol - not fragments. Not only should such a protocol be generally accepted, but it should also be:

● Very *thoroughly studied,* and

● Unshakeable in both a *legal* and a *user-oriented* sense.

Such is not the case with the current fuzzy state of EDI. Volvo, for example, studied it and found several problems with the:

1. Timestamp;

2. Protocol requirements to handle transactions; *and*

3. Legal certification

Today there are no real purely technological problems that need to be ironed out. But, there are protocol problems, timestamp challenges and legal issues. In other terms, the real obstacles may be in the actual business case for implementing EDI or the legal implications of such a system - as well as organisational and procedural hurdles which must be sorted out.

Rationalising documentary procedures and settling political infighting could take longer than the actual technical implementation. This is the more true as the primary interest around EDI suggests a strong trading focus, and most observers expect trade to be the basis of any EDI system. Photonics approaches are so important not only because they represent the peak of technology. A keypoint, if not the primary point, is that they are new and, therefore, free from vested interests. As *compound electronic document interchange* will ultimately take place:

● Optical disk-to-optical disk

● Over optical fibre highways

now is the time to study the unification routines and the resulting implications.

This cannot be done by everybody taking their own unique path as has happened with Unix - where we ended up with a dozen incompatible versions (including the Xenix, Venix, IX, Ultrix and other dialects) plus two consortia:

- AT&T/Sun *and*

- Open Software Foundation (OSF)

at war with each other. Not only physical sizes (for optical disks) and logical issues (for communications and storage protocols) should be standardised, but also legal issues should be sorted out in advance, with acceptable answers provided to them.

A system which is insecure is a born failure, and I do not care how cost/effective the solution appears to be. Furthermore, the very high speed and great capacity of photonics magnifies the need for security - and makes cross-checkup mandatory. In early 1988, in England, there has been an ''inside job'' in the Electronic Payments System. A manual check found it out. So far as I know, no fundamental study has been carried out in connection to EDI to avoid exposure and to improve security.

Promoting a Universal System for Document Exchange

Optical storage may be the ideal media for document handling, but the universality of a system for document exchange starts with its acceptance. Take Hong Kong as an example. There are somewhere in the region of 5,000 to 8,000 companies that generate half the city-state's trade, with another 40,000 or so that account for the remainder.

Any solution which addresses only the 50 percent of the document exchange problems, by focusing only on the top 8,000 companies of the small city-state, will be half-baked - and consequently counterproductive. By the same token, within one and the same company, bringing EDI to the attention of DP/MIS is just the start. To be successful, EDI must be supported by all levels because it involves the entire firm and all of its offices.

It takes more than great pronouncements like: ''With 1992 in mind, no single VANS will meet all the requirements of one company or one industry sector on a pan-European basis'', to make EDI work. A document protocol structure has to be both adaptable and flexible to cope with the changes in commercial relationships which not only happen, but will continue happening, with greater frequency.

To achieve commendable results we must necessarily make sure that we speak the same language with our customers and suppliers. In basic terms, this means that the message must have the same:

- *Vocabulary* of accepted words and business definitions; *and*

- *Syntax,* determining the possible message segments and punctuation.

Correspondents also need to be sure of a correctly addressed electronic envelope with full details of the sender, to be handled in a reliable manner. These notions are part and parcel of an approach to universal EDI.

To complete the transaction, both parties need to be in agreement on packaging methods as well as sorting, and delivery instructions. Implemented the correct way, Electronic Document Interchange should:

1. Be available 24 hours a day, 7 days a week, regardless of time zones;

2. Carry appropriate identification, authentication and timestamps;

3. Fulfill all needs of legal certification;

4. Address both transactional and message exchange requirements;

5. Reflect technological prerequisites connected to the new storage and transmission media -photonics;

6. Feature a universal protocol adaptive by trade; *and*

7. Provide for automatic language translation.

Such prerequisites are far more important than those usually advertised in connection to EDI, that:

8. The information the system carries is up to date;

9. The input requirements are greatly simplified; *and*

10. The user organisation experiences substantial savings to the costs of paperwork.

True enough, this is in itself a substantial factor. However, we should take the global perspective not the parochial one. In the manufacturing industry, for instance, the cost of paperwork is put between 4 percent and 15 percent of the value of the goods sold. By how much will photonics and EDI reduce this figure?

We know by now, from past experience, that even when we take the paper out, the cost stays in if the long, hard look is missing. Hence the crucial questions to be asked in connection to photonics and EDI are:

- Do they help rationalise relations between companies, with favourable aftermaths on the cost of internal resources?

- Do they improve customer service? Can orders be processed faster, with less order chasing, while information for customers is of a higher quality?

- Do they accelerate the invoicing and payment cycle, which substantially improves cash flow and helps in improving financial results?

As it cannot be repeated too often, one of the deterrents to the wider use of EDI is standardisation. Some effort has been made quite recently in this direction, but it is too little and it is not properly focused.

EDIFACT (EDI for administration, commerce and trade) is a ''recommended standard''. There is also TDI (Trade Data Interchange) proposed by the United Nations Economic Commission for Europe (UN/ECE), and an American standard known as Ansi X.12. It is likely that EDIFACT will also be supported by the Japanese. But this is not certain. Neither is its protocol sufficiently normalized.

In other terms, not only physical media such as those connected to photonics storage, but also successive layers of logical support are necessary to *fine-tune* an effective compound electronic document system. The overriding need for the observance of logical requirements and standards is based on the fact that:

- In a knowledge society information volumes tend to rapidly increase. This emphasises the need for efficient supports.

- Both document storage and exchange should be considered, as production, storage and distribution of information are an indivisible process - both for practical and for efficiency reasons.

As we will see in Chapter 11, these considerations have led organisations with a headstart in photonics storage towards the retructuring of their optical disk projects. Such organisations also found that they have a vested interest in the adoption of a standard. Will Electronic Document Interchange be the one? We shall see.

As an example of what is necessary, Figure 10.4 presents a layered approach to enduser-to-enduser communication integrating datacomm and databasing resources, from electronic mail to applications programs. In the last analysis, what we are talking about is efficient, integrated information management for the 1990s.

Ample evidence documents that the costs of poor data management are high. In a recent survey in America, respondents listed *excessive or unnecessary paperwork as the biggest cost,* followed by too many late communications and wasted personnel time. The same survey also showed that companies are spending millions of dollars per year trying to establish and control information files manually - in spite of the massive computer power that they have installed.

In another research, a pool of more than 100 American aerospace, defence electronics and manufacturing companies revealed that engineers:

- may be able to design, but

- they cannot efficiently manage the data that they generate.

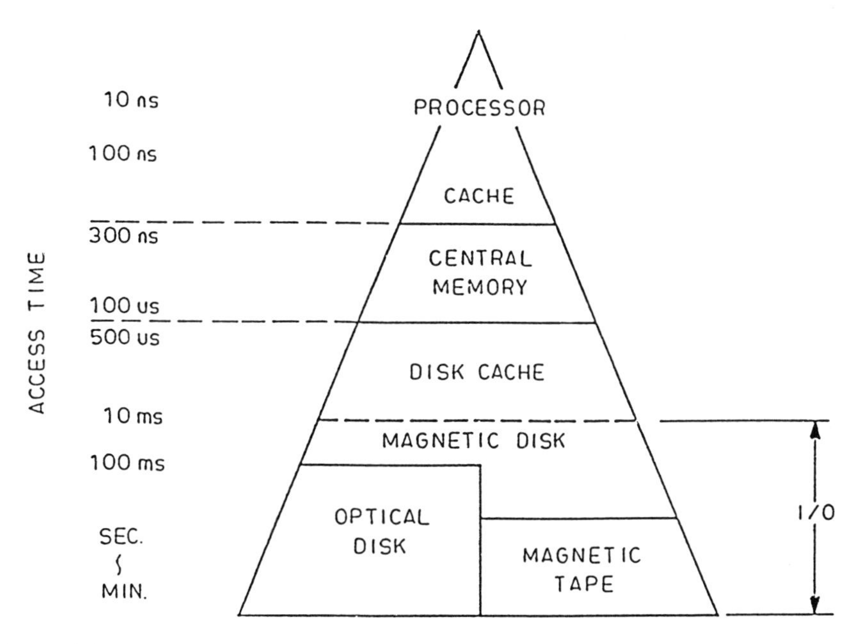

Figure 10.3

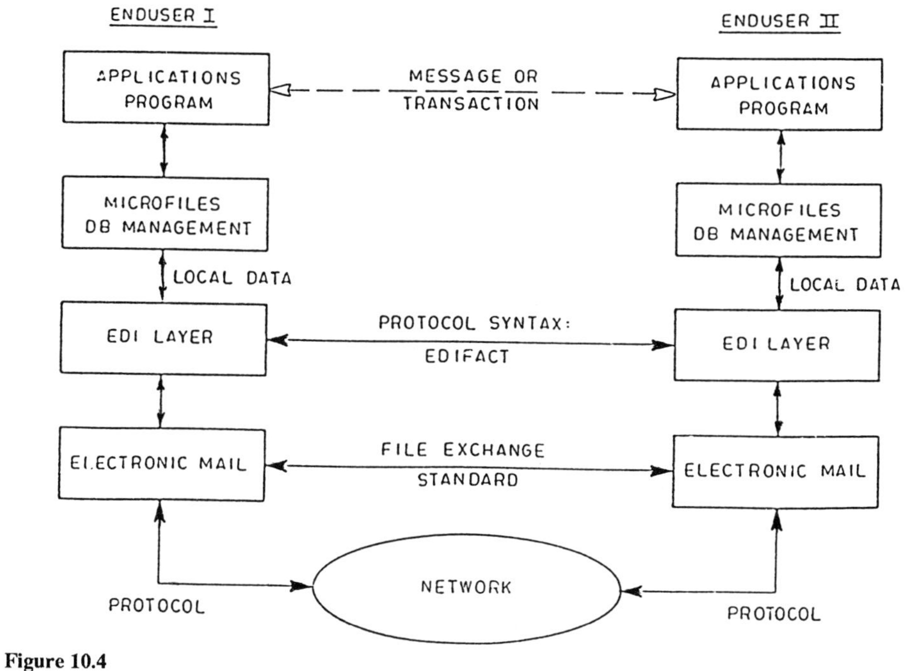

Figure 10.4

Of the 266 managers who responded to a survey of the Fortune 1000 companies, 258 cited ineffective engineering data management as their biggest problem. Not suprisingly, more than half of those polled have already set aside funds for the study and solution of their design data bottlenecks, and 69 invested dollars in implementing automated CAD/CAM/CAE data management. *

The use of photonics storage should be seen as an integral part of this experience. Truly automated systems would track and control computer files, microfiche storage and paper files, *by integrating all of them on digital optical supports* - if we are able to establish from the beginning the proper norms and procedures.

Correctly organised, and tuned to the observance of developing standards, photonics storage media can be used for:

1. Generic file systems, providing cross-file data compatibility;

2. Large information databases, including the integration of text, data, graphics, image and voice;

3. Online storage in substitution of currently-used magnetic tapes.

4. Image and document storage for archival purposes;

5. The conversion of *save* files (programs, data) typically done through magnetic tapes;

6. Journaling applications, using the non-erasable features as an asset;

7. Extensive microfile support at WS level;

8. Backup systems to CAD/CAM and other applications;

9. Training in engineering, manufacturing, sales and service;

10. Integration into two-way video and audio systems;

11. Determining the original source and final destination of transactions and messages; *and*

12. Electronic publishing, as well as authoring and editing tools.

Some organisations are ahead of others in this implementation range. In Japan, the Tokai Bank has used optical disks in connection with its mainframe operations since early 1986. Hitachi employs juke boxes to replace magnetic tape filing. Bloomingdale's Department Store, USA, put optical disks to the same application.

---

* Which, by the way, is also plagued by a notorious lack of normalisation. See D.N. Chorafas and S. Legg "The Engineering Database,", Butterworths, London and Boston, 1988

A local bank in Sendai, Japan, replaced with photonics magnetic disks for image filing purposes - and also used photonics in a tax filing application to replace paper records. This, too, is a promising field of Office Automation, 1990s-style.

Fujitsu Computers attached optical disk jukeboxes to its ''M'' Series. It also uses jukeboxes as a standalone for document handling purposes. Nippon Telegraph and Telephone (NTT) employs optical disks in an AI-supported online Telephone Book, easily accessible to client queries. Not only is optical disk technology well implanted in Japanese business and industry, but it also experiences polyvalent areas of implementation.

A practical example of *using optical disk for Computer-Based Training* (CBT) is offered by the Betriebswirtschaftliche Institut der Kreditgenossen-schaften, Germany; also by British Airways. The British Airways experience is interesting because of the personnel problems which may arise in the implementation of new technology.

Management commissioned a project on the use of an interactive videodisk for employee training, as well as for the training of travel agents. However, the British Labour Unions objected to the reproduction and repeated use of lessons given by expert aircraft engineers. At the same time, expert travel agents had to be emulated through actors, because the agents themselves did not wish to be recorded for replaying. What is near-sightedness called in labour union jargon?

11

Designing and Managing Optical Disk Systems

Introduction

Chapter 10 has shown that it is not enough simply to select the best photonics storage system currently available on the market, even if it is endowed with first class software. There is a whole methodology that we must institute for file handling, with a view to optical disk storage.

An impressive amount of preparatory work needs to be done, starting at the mainframe level if, for instance, we wish to substitute COM with optical disks - and, in general, to make optical files easily accessible by the enduser. The necessary preparatory steps for file searching at the user's workstation must definitely start at the level of central operations. So long as they are connected to mainframe-held files, they should be executed in batch as a background job.

Preparatory work at central operations is only one phase of what is necessary for the best possible application of a new technology. Alhough the required study steps start at the mainframe level they do not end there. There is plenty of scope in optical file cabinet organisation, as well as at the workstation level.

Let us start with some basic premises. The chances are that photonics storage will be first used in our organisation either to substitute for one of the older, established processes - for instance COM, or magnetic tape - or in a brand new business domain. Among the most promising fields in the latter are those processes which are being phased-in, themselves being a substitute of older processes which are being phased-out. Figure 11.1 helps to explain this reference.

If our organisation addresses itself to one of the new business areas, then we may as well decide from the start to use photonics storage. This will be a far better solution than converting to magnetic media as a substitute for paper-based records - such as

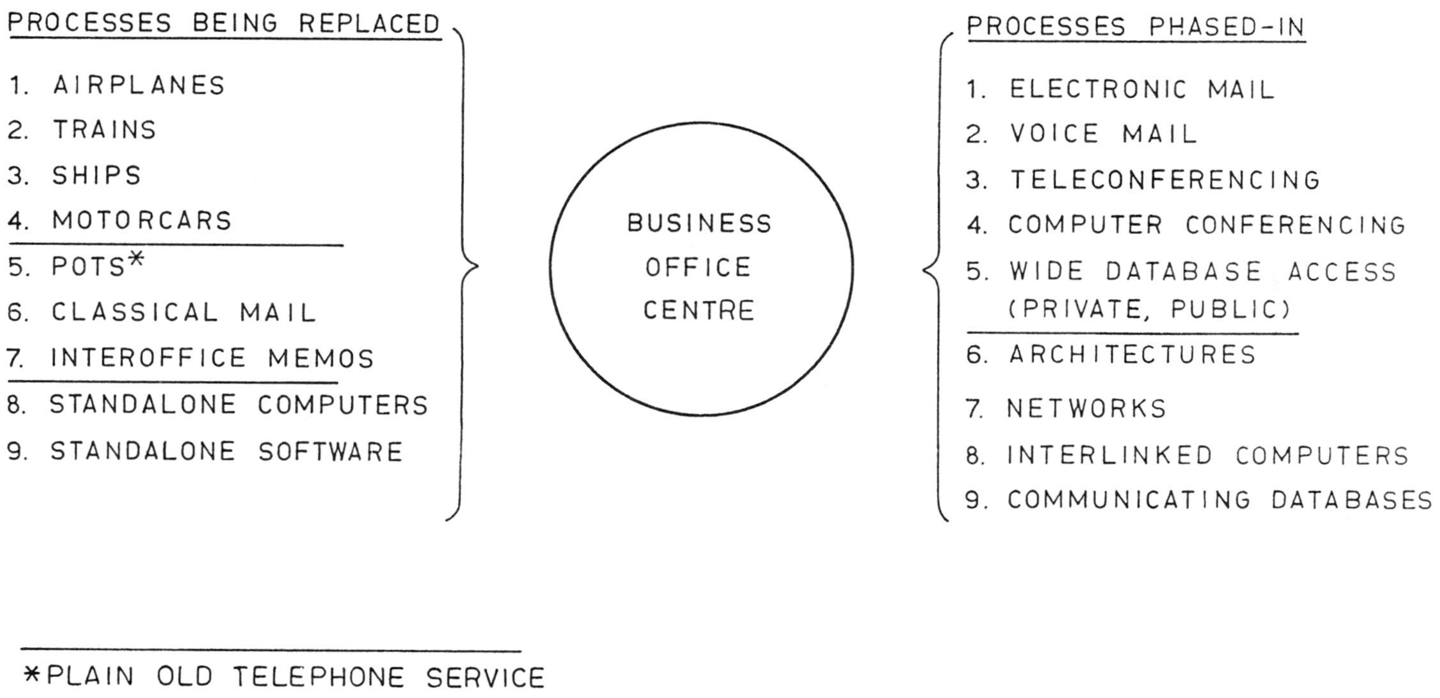

PROCESSES BEING REPLACED

1. AIRPLANES
2. TRAINS
3. SHIPS
4. MOTORCARS
5. POTS*
6. CLASSICAL MAIL
7. INTEROFFICE MEMOS
8. STANDALONE COMPUTERS
9. STANDALONE SOFTWARE

BUSINESS
OFFICE
CENTRE

PROCESSES PHASED-IN

1. ELECTRONIC MAIL
2. VOICE MAIL
3. TELECONFERENCING
4. COMPUTER CONFERENCING
5. WIDE DATABASE ACCESS (PRIVATE, PUBLIC)
6. ARCHITECTURES
7. NETWORKS
8. INTERLINKED COMPUTERS
9. COMMUNICATING DATABASES

*PLAIN OLD TELEPHONE SERVICE

Figure 11.1

memos kept on messages from plain old telephone service (POTS) and classical mail. In this sense, electronic mail, voice mail, teleconferencing and computer conferencing can be stored on optical disks, handled through jukeboxes and the recordings made available throughout our organisation by means of a LAN and workstations.

An optical file cabinet (or optical disk jukebox), is a random-access file system for write-once optical disks. This storage can be patterned after a conventional office file cabinet where all versions of an information element (IE) are retained, but:

● The most current version of each item * Is the easiest one to find.

The Optical File Cabinet uses only a small part of the space taken up by old versions of files. More importantly, it provides a uniform, fast, dependable and cost/effective file access means.

● Old versions of files may be required later for legal or other reasons.

● New versions will be more frequently accessed than older versions.

The streamlining of systems and procedures is a fundamental step if we want to replace ageing versions, or to supplement their information content. This calls for database management approaches, which constitute the kernel of the present chapter.

As long as CD-ROM and WORM technology is the one available, our organisational perspectives, and the methodology that we will apply, must capitalise on the CD-ROM and WORM advantages while also accounting for their limitations and constraints. As underlined in Chapter 10, write once optical disks have several benefits over magnetic disks:

1. They are easier to handle at enduser level;

2. They are not susceptible to media failure;

3. Data is never rewritten by overwriting;

4. High performance of proximal (nearby) seeks may exceed that of magnetic disks; *and*

5. Cache copy-back policy is applicable.

Through a number of examples we have seen that photonics storage is a modern, elegant approach. It is very compact, thus permitting an impressive memory size at WS level. It risks no demagnetising effects, whether unintentional or because of computer crime. It is removable, easily transportable and hard to damage, *provided that proper care is taken.*

The methodology we should develop for the competent handling of photonics storage must capitalise on the fact that optical disks are resilient to media failures, thus providing a stable memory. As such, they are an appropriate and convenient choice for file systems.

Given that no information is overwritten, the history of the file system is preserved on the medium while auxiliary data structures may be required to retrieve snapshots of memory contents. In principle, disk overhead is minor, as we will see later.

Also, contrary to some preconceived notions, the read performance of an optical file system is not necessarily worse than that of magnetic disks. On the contrary, photonics storage exhibits an advantage for proximal seeks:

- A well organised optical file system is spread out over the entire disk, while

- The active part is localised in a window of near contiguous blocks.

It is precisely within this window that seek performance is potentially superior to the seek performance of magnetic disks. Given that, under current technology, the transfer times for magnetic and optical disks are approximately the same, WORM can be used instead of magnetic disks in data-logging applications.

Photonics storage provides important advantages when used as a substitute for other archiving means, such as COM. These advantages derive from greater capacity and flexibility of photonics storage, as well as a much shorter access time than, say, COM - provided that the appropriate preparatory work has been undertaken.

The nature of such work, together with its prerequisites and obtained results, is shown in the following case study. It will be appreciated that the information on this implementation is significant but, at the same time, non-competitive in its nature. It reflects on problems and results rather than on managerial decisions. Even so, it can be of significant help in bringing the reader's attention to crucial items which can make an optical disk project successful.

An Optical Disk Project

The Digital Optical disk (DOD)* started at Banque Privée Edmond de Rothschild on the premise that classical batch, classical realtime and (by now) classical Infocentre implementation - as well as traditional approaches to office automation - cannot fulfill document handling requirements. Photonics were, therefore, seen as the way to integrate different information technologies into a multimedia system able to improve the cost/effectiveness of administrative activities in a modern bank.

It was precisely this awareness, and the quest for efficient solutions based on high technology, that led management to the decision that the study to take place should examine the whole process of document handling:

---

* In French: DON, for "Disque Optique Numerique"

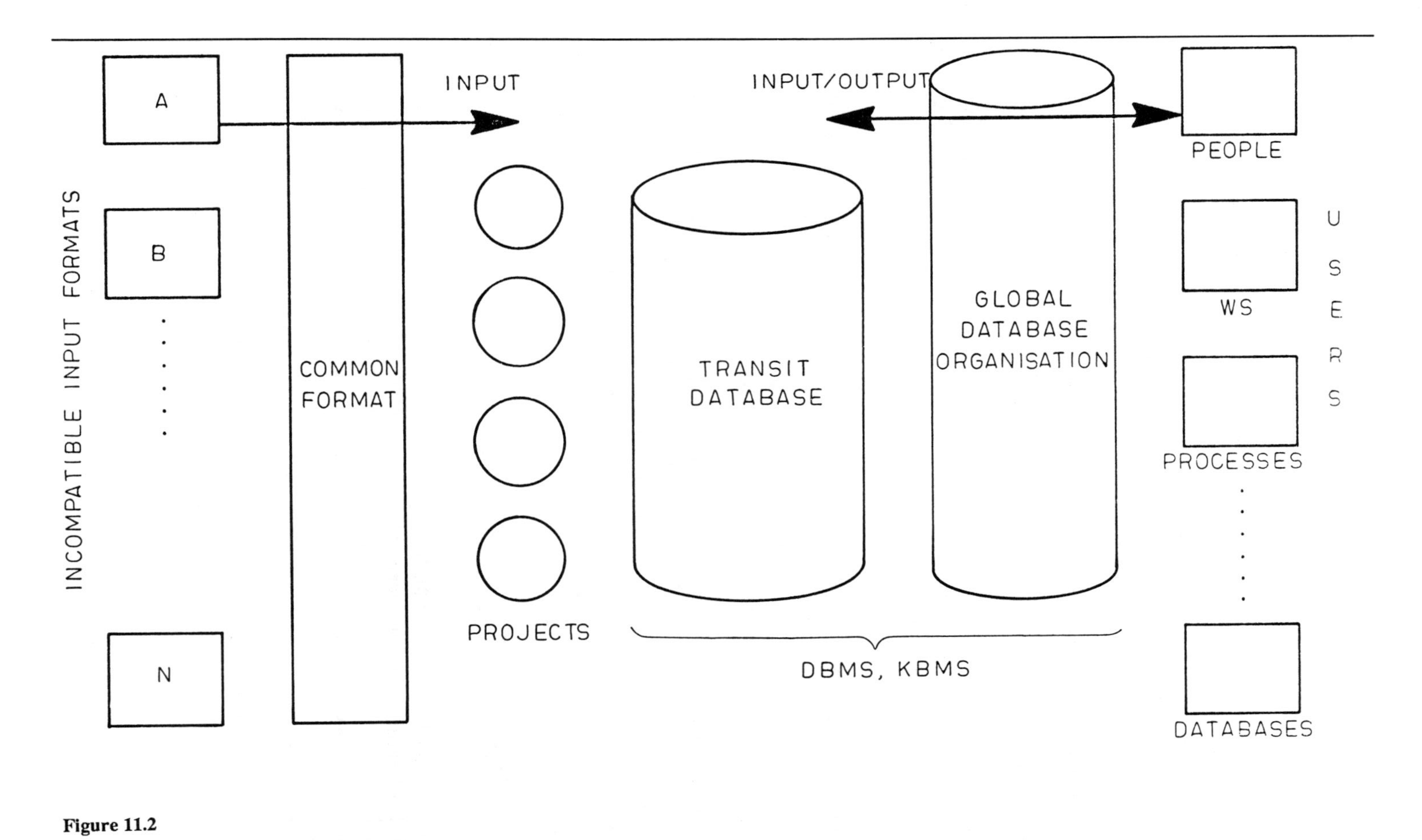
INCOMPATIBLE INPUT FORMATS
A
B
N
COMMON FORMAT
INPUT
PROJECTS
TRANSIT DATABASE
GLOBAL DATABASE ORGANISATION
DBMS, KBMS
INPUT/OUTPUT
PEOPLE
WS
PROCESSES
DATABASES
USERS

Figure 11.2

- From mainframe input (both realtime and batch)

- To electronic message systems (telex, Swift, etc.), *and*

- Paper documents coming from clients, or from different departments in the bank.

This range of document handling activities has been investigated with a view towards full integration. In other wrods, in a fundamentally serious manner as a photonics study should properly be done.

Figure 11.2 comes from a similar study and demonstrates the generic approach which should be taken in examining multiple document input with a view to creating homogeneous:

- storage, *and*

- output formats.

As such, it ports attention to necessary organisational and information systems perspectives: (1) conversion to a common format; (2) project emphasis and resource allocation; (3) the creation of a transit database; (4) global database organisation; (5) user interactivity, including people, processes and databases.

One lesson learned from this implementation is the attention that must be paid to *automatic indexing.* An internal feasibility study regarding manual indexing of documents - for subsequent research and retrieval -indicated that this will be neither advisable nor feasible in the long run. It had to be done automatically at mainframe level.

As the Banque Privée study pointed out, after a document is looked at, a minimum time of *15 seconds* per document will be required for manual indexing alone. This can add up to *tremendous delays and costs* when first thousands, then millions, of documents are handled. And millions of documents is the relevant statistic for many large organisations.

We have spoken of a multimedia input approach with incompatible input formats. This is typical of any industry in general, and banking in particular. A point where Banque Privée seems to be ahead involvesthe handling of telexes and Swift messages. These come directly into the computer and, therefore, integrate into the optical disk implementation.

Although this bank did not elaborate on expert system usage in this connection, there are at least two successful experiences in converting telex messages to Swift format by:

- The General Bank, in Brussels, *and*

- Citibank, in New York.

Both use expert systems for the conversion of telex messages into Swift format. Dr. Georges Wanet, of the General Bank, was to mention that the hit ratio is better than 80 percent, while only 5 percent has to be done manually. The balance (15 percent) is handled through an operator at the online WS, assisted by the expert system.

With the input problem addressed in an able manner, the next key objective of the optical disk project at Banque Privée Edmond de Rothschild has been *image automation requirements* of two classes of documents:

1. Copies of those sent outside the bank -principally directed to the clientele; *and*

2. Documents staying within the bank for administrative and other purposes (journals, memos, statistics).

The premise has been that both classes are growing in terms of volume. Most particularly the second one, where the need for information is rapidly increasing. This conclusion was also reached by Citibank and Bank of America. The difference is that both Citibank and Bank of America *saw the solution in the implementation of 5th Generation database computers.*

There are advantages in both approaches. The database computer integrates better with the existing mainframe programs and associated culture. However, while it relieves the classical data processing of:

- many of the database management chores that are necessary today, as well as

- spooling and the costly, error-prone manual input/output of magnetic tapes,

it does not handle *compound electronic documents.* Hence, it complements, but does not necessarily overlap, with the optical disk solution.

True enough, both approaches, optical disk and database computer, have similarities. The most notable is the elimination of the delays and high costs included in the way of handling computer resources in a manual fashion -which still prevails in most organisations.

Learning from the Rothschild Bank experience, an optical disk project should capitalise on the incentives for creating an *internal standards concept assuring compatibility* between systems to be installed in the organisation, whether at home or abroad. Attention should also be paid to controlling some of the more dubious claims for compatibility of products made by vendors - that do not respond to reality.

All told, the optical disk project at the Rothschild Bank seems sufficiently well advanced for us to learn some lessons from it. This helps avoid redundant effort. The following references identify where the system experts' attention should focus:

1. The *problems* which it has posed in terms of *computer storage* during automatic indexing procedures;

2. The need to have *computer resources* available nearby to assist the photonics conversion process; *and*

3. The *modifications* which have been necessary to the Cobol programs for data access reasons.

Furthermore, *legal problems* such as copyright, privacy, and Courts' acceptance should also be carefully studied. Some countries do not recognise the legality of documents that are not on paper.

Lessons from the Experience at Banque Privée Edmond de Rothschild

Lessons from the optical disk project at the Rothschild Bank can now be learnt, as the development work has been successfully completed. This includes the transfer mechanism which moves data from the mainframe to the optical disk via a local area network and a PC interface, supporting *document indexing* and the construction of menus. Both are accomplished through software running on the mainframe.

Since the optical disk equipment is attached to the LAN, any workstation which also operates on this LAN can look up a page on the optical disk, examine its contents on video, and subsequently print the page if necessary. Client account statements are created online. They are transferred:

- To the printer for the client, *and*

- To the optical disk for internal bank use.

Results available so far do, however, point to certain problems:

1. *Integrated Archiving*

The Central Information File (CIF), General Ledger, Securities and Forex are operations which have long been automated. Problems and opportunities derive from this fact. While in the *online* environment the integration of several operations has been a reality in regard to account handling, the challenge remains in respect to *batch.*

In a typical batch environment, each department of a financial institution has been responsible for its own files. This approach reflects longstanding organisational solutions. Hence, the problem posed in integration for optical disk processing. Another challenge with batch programs relates to:

2. *Relatively old concepts which still persist in a batch environment.*

Since its beginning, the optical disk project at the Rothschild Bank has aimed to eliminate dead-wood type information. That is, information which is:

- Either obsolete but still around, *or*

- Does not contribute to the efficiency of bank operations.

This, however, means weeding out a significant part of what is still processed and stored under batch. That this is resisted by the data processors should come as no surprise. Let me add to this reference, that the meeting with the Dai-Ichi Kangyo Bank in Tokyo pointed to the same problem but for a different reason.

The plan of Dai-Ichi was to convert all batch into realtime, thereby eliminating all batch operations on mainframe computers. It is quite interesting to note that the problems encountered in this conversion are most comparable to those found in the conversion to DOD. In that sense, they could be regarded as generic.

Another problem which relates to batch, and requires streamlining for optical disk implementation, is:

3. *Online indexing done directly by computer.*

The concept behind this operation is a negation of what has been known for 35 years with grandfather-type data processing. Batch, for instance, features:

- spooling on tape, *and*

- background jobs.

Now, background jobs must be sorted, catalogued and indexed on the mainframe for conversion to optical disk storage. For automatic indexing purposes, such work cannot be done on tape (at spooling time). It has to be done on disk, thus posing a number of disk requirements. Another lesson that can be learned from Banque Privée results from:

4. *The emphasis placed on the elimination of manual intervention in the archiving of documents,*

whether of external or internal origin (to the bank). Fifteen years ago I had a similar problem with a group of financial institutions during the conversion to COM and microfiche from paper-based documents.

Still another important lesson is the discharging of the central computer site from operations concerning the exploitation of these documents, which are now addressed directly by the enduser. As stated, the enduser's workstation is online to optical disk storage, interconnected through the local area network.

A fifth challenge has been:

5. *The changes necessary to batch programs.*

Indexing, for example, poses no problem for one-page documents, but:

- Some documents to be carried onto optical disks comprise 20 pages, *and*

- Not all of these pages come up at the same time, as they are the result of different procedures running on the mainframe.

The experience of Banque Privée underlines the wisdom of a first-class approach regarding system control. A sixth lesson concerns:

6. *The resolution of operational problems, given that some procedures take hours on the mainframe.*

At the Rothschild Bank, such requirements called for the development of an infrastructure which has gone well beyond the optical disk application itself. It is beyond doubt that a successful optical disk implementation calls for all types of operational and data processing difficulties to be overcome. Full integration should always be our goal:

- There is no profit if half the application is on paper and the other half on optical disk.

- Neither are there profits if the different departments of the bank start filing away (on paper) copies for retrieval "later on".

Everything should be handled through photonics, and it should be made accessible to every workstation through LAN interconnection. These are the major lessons learned by Banque Privée - to which it was added that *a careful study should go beyond the level of the account statement in itself.*

- While every transaction generates an account statement, it can also lead to statement modifications.

- For instance, annotations, reversal of a transaction, or mapping of the transaction into a subsequent operational area are examples of such modifications.

Hence, the life of every transaction class should be carefully studied, leading pointers should be established and the transactional operation carried on while observing the integrated nature of the system.

4. *Design Choices and Recording Solutions on Optical disks*

Design choices and recording solutions should consider the *half-life* and *full lifecycle* of the information elements to be recorded. These vary from one organisation to another, but the half-life over which most accesses are made tends to be close to 1 year. The lifecycle sometimes reaches 10 years, or whatever is implied by law for the retention of a document.

Lifecycle decisions should be present throughout the whole design process of photonics storage, independent of the origin of information. As shown in Figure 11.3, such origin may typically be:

1. *Mainframes* downloading into optical disk storage.

This is the case which we have considered in sections 2 and 3, along with advisable procedures in order to respond to requirements in an able manner.

2. *Scanner input* followed by edit station(s).

Until online information handling truly dominates text and data processing, scanner input will remain a key component of optical storage. There are design prerequisites to be observed in this connection - from the dependability of digitisation to document indexing.

3. *Gateways* to and from a distributed environment.

With online systems, the necessary editing activity should take place either at the point of origin or at the centre - the former being the better alternative. In this case, the communications gateway should be interfaced to photonics storage through computer- based editing routines. Editing standards, roles and procedures should be universal throughout the system.

A key element in these standards is the decision on active half-life and lifecycle of which we spoke in the opening paragraph. It is only evident that beyond the most active half-life, a great deal of stored records will be superceded. Hence our approach should aim to conserve disk space both by:

- updating on a block-by-block basis; *and*

- using copy-back strategies through cache.

We should also aim at a procedure for recovering optical disk data in case of controller failures. Both write operations and read operations must be considered in the "save" function.

Statistics can be helpful in devising a strategy. Experience accumulating to the present day tends to indicate that read exceeds write by a ratio between 200 percent and 1,000 percent. A recorded document will tend to be read, on average between 2 and 10 times.

Likewise, a valid system design will take full account of the *limitations of WORM media*. For instance, although read operations have a low residual error rate: typically 10-12, their raw error rate tends to be high. Error detection and correction (EDC) codes are, therefore, necessary to make optical storage viable.

Since information is stored on optical disks for subsequent retrieval reasons, a careful choice should be made when selecting a method from among the alternatives currently feasible for storing data on write-once media. These include:

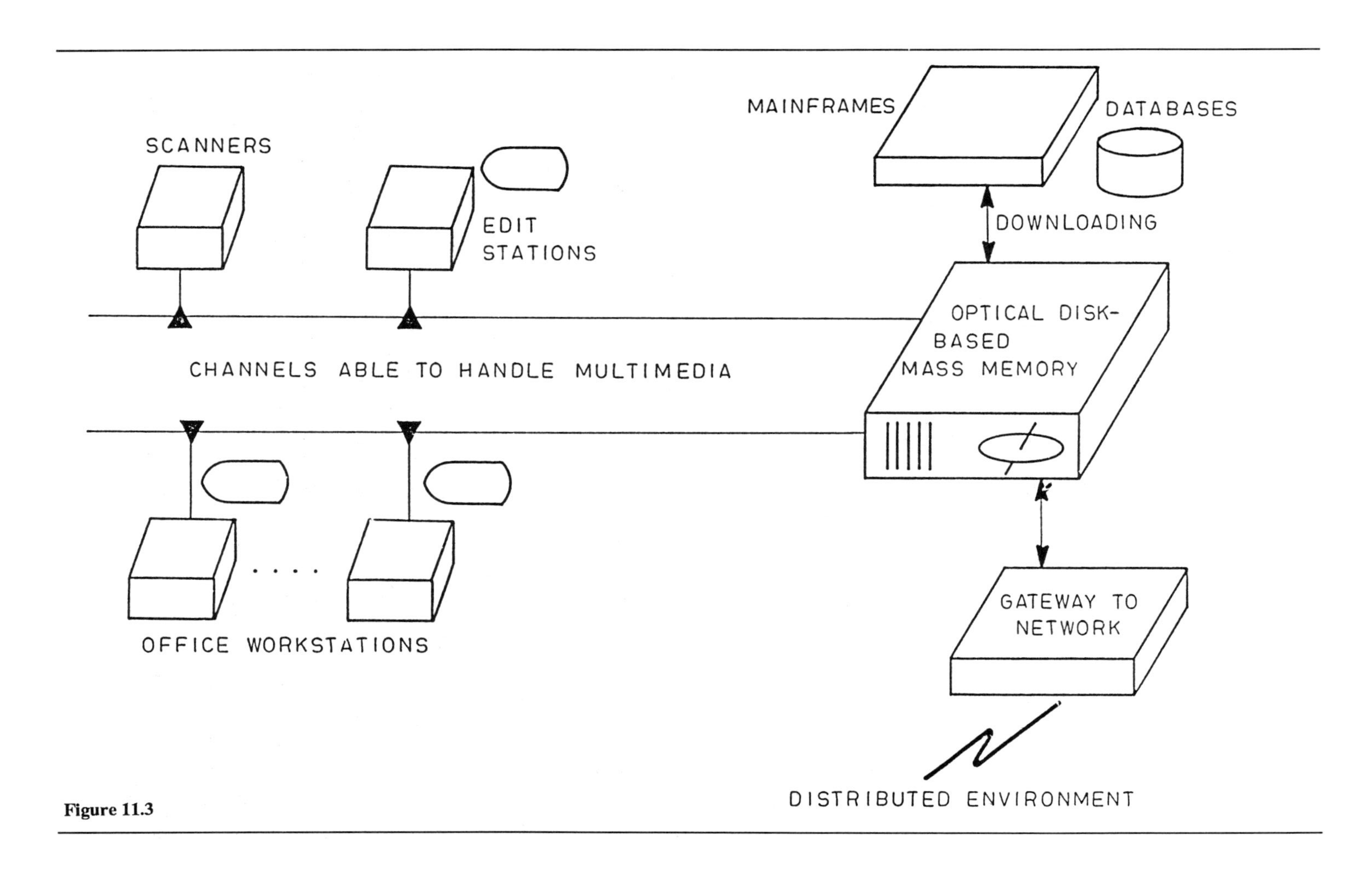
MAINFRAMES
DATABASES
SCANNERS
EDIT
STATIONS
DOWNLOADING
OPTICAL DISK-
BASED
MASS MEMORY
CHANNELS ABLE TO HANDLE MULTIMEDIA
OFFICE WORKSTATIONS
GATEWAY TO
NETWORK
DISTRIBUTED ENVIRONMENT
Figure 11.3

1. *Append-only*. New data is added to the end of the file;

2. *Copy-only*. The entire file is copied elsewhere on the media whenever it is updated;

3. *Hybrid approach,* storing data on optical disk and auxiliary information on magnetic disk;

4. *B-trees* for database applications, as used with magnetic disks.

Each approach has its domain. The *Append-only* approach is suitable for files in which new records are added to the end of already-written records, for instance, log files. *Copy-only* is more appropriate for files that will rarely be updated.

In the *hybrid optical/magnetic* approach, *as comes* data is stored on photonics. Information needed to make the optical data intelligible is stored on magnetic media. The risk, however, is that a failure in the magnetic media will cause the data stored on the optical disk to be lost also. Therefore, what is stored on magnetic disk should also be dumped on photonics, thus increasing dependability.

In any case, it is also wise to optimise performance through cache. As with a magnetic disk, a file cache is an in-memory collection of disk blocks (packets) belonging to files being accessed by the process:

- Writes by different processes are done to the file cache.

- Many of these blocks may, however, be cancelled before the file is closed.

Thus the optical policy is to delay writes from the file cache to disk until a block is about to be removed from the file cache.

Note, howver, that this approach is unsafe for magnetic disks, since some blocks may never be eliminated from the cache. Yet, it is valid for optical disks, where periodic *timestamps* are taken. In this sense, it conserves disk space by accumulating updates.

An update will coinvolve all structural information, which is also stored in WORM. We should use data structures that can be recorded on a write-once optical disk, yet dynamically enlarged and altered if files are updated. This brings under perspective the need for organisational solutions. Several have been tried in actual applications.

One of the better methodologies is that optical disk storage is organised into units (blocks or packets). These are contiguous, read and written indivisibly.

- When a unit is written to disk, the corresponding EDC is also written in a conventional location;

- Thereafter, the error-correction code of the information element (IE) cannot be changed;

● Thus, once written, individual disk units are never updated, even when there is room for more information contained in them.

This approach is familiar to us from *packet switching*. An X.25 protocol, for instance, has *header* (flag, ID, where from, where to), *body* (the message proper), and trailer. The *trailer* is a *cyclic redundancy check polynomial* (CRC) which provides error detection. The packet is, however, handled as one unit, like the IE in the database.

A principle of modern Information Technology (IT) is that *a message is a file in the database,* and every file should be designed as a message. With photonics we have the opportunity to apply this principle. So, let us immediately choose a packet structure for IE storage on optical disks, including the CRC protocol.

On this ground, from the present point onward reference will no longer be made to storage blocks but to *packets*. There are, however, two consequences of this approach.

1. Waiting to allocate list structures until it is time to store them on the media (a process known as lazy allocation) will not work.

Its implementation demands the establishment of a *preallocation* strategy.

2. Altering IE requires the new data to be written to new disk locations.

Available controllers for WORM mostly treat unused, unwritten packets differently from packets that have been written. An unused packet has no error correction code associated with it, so attempting to read it returns an error. This feature has important practical applications.

● *If* the write-once optical media is written to sequentially,

● *Then* the unwritten area of the disk will be delimited by the first unwritten block.

However, some controllers do not protect written blocks (except for EDC), from being overwritten by software error. Therefore, a read-before-write policy is necessary to protect written IE from accidental corruption. In such an approach, new data is written only when a read test signals an unwritten block.

Organisational and design perspectives *should* also consider access requirements. Theoretically, the average seek time for optical disks is an order of magnitude larger than average seek time for magnetic disks, but the seek times to blocks near the position of the recording head are an order of magnitude smaller than with magnetic disks.

The reason is technical. An optical disk head seeks into nearby blocks through lens rotation. This is, in itself, a more efficient mechanical operation than the read/write head movements used by magnetic disk controllers.

● Optical seeks into nearby blocks are known as *proximal.*

● Seeks into distant blocks, which require head movement, as *distal.*

A proximal window is the average number of disk blocks that can be scanned by purely optical means. That is, without moving the disk head.

Seeks within the proximal window are *local seeks*. Local seeks are an important design criterion in photonics storage. Therefore, it is always important to know the proximal window size (in MB, or thousands of pages) supported by the optical disk controller.*

Proximal window size and software support should be key criteria in equipment choice. This controller-level facility is not only vital to single disk but also, if not primarily, to jukeboxes. With present technology, the latter can be configured with up to one terabyte, and they will result in a very inefficient application if we do not optimise their implementation at the design phase.

A System View of Optical Disk Applications. Hypertext and Hypermedia

The solution described in section 4 is the best methodology that I found to be developed and applied by optical disk users for storage and retrieval. However, though valid in its principles, it is incomplete because it fails to take a system view. It therefore needs to be complemented.

The complement is *Hypertext.* This is the best methodology currently available for a distributed multimedia environment with nodes and links - though the technology behind it is not yet settled. It is in full evolution. Hypertext is an approach to information management in which:

1. Information elements (IE) are stored in a network of nodes connected by links;

2. Nodes can contain text, data, graphics, video - also source code or other forms of IE; and

3. These nodes may be *workstations* (with or without optical disks) but may also be central and departmental computers as well as *Photonics storage servers.*

Hypertext solutions aim to provide the global administrative discipline and rules to be observed, although each node with optical disks:

---

* Depending on a number of factors, 1 MB may give from 600 to 1,000 pages. A proximal window of 4 megabytes is a reasonable minimum to be supported. Some equipment features 8 MB.

- Servers, *and*

- Workstations

should *also* observe the rules which have been outlined in section 4, and which are specific to the good management of optical disk storage. Hypertext concepts constitute a higher-up layer than those already examined, but let us start with the fundamentals.

Hypertext is a discipline to manage documents in which information is stored in nodes connected by links. Each node contains *packets*. The links join the nodes to one another to form a *network of packets*. They are usually shown for each node as a:

- *from link* pointing to the node just read, *and*

- a set of *to links* that indicate the multiple nodes which one may select to read next.

But, a hypertext document differs greatly from a conventional paper document. In most conventional paper documents:

- Physical structure, *and*

- Logical structure

are closely related. Physically, the document is a long linear sequence of words that has been divided into lines and pages for convenience. However, logically, the document is also linear. Words are combined to form sentences, sentences to form paragraphs, paragraphs to form sections, and so on.

If the document has a hierarchical logical structure, as do many expository documents, that hierarchy is presented linearily. The abstract (overview of the whole) comes first, followed by the introduction, and the first Chapter, which divides into sections, etc. Linearily organised documents lead readers into reading them linearily: from beginning to end, following a fixed sequence. Few paper documents separate logical structure from physical structure. Examples are dictionaries and encyclopaedias.

- Physically, these documents are linear sequences of independent units, such as articles on specific topics, or entries for individual words.

- Logically, they are more complex. They are not read from beginning to end, but rather by searching into their contents.

The reader is also likely to encounter various *cross references* to other entries, including a list of *see also* references. To follow those pointers, the reader must locate the appropriate volume, find the appropriate entry, and then find the relevant portion.

Designed for electronics (and photonics) documents, Hypertext works in a similar manner, providing most of the flexibility of reference works, with cross-indexing and see-also references, as well as adding a number of new features. Therefore, it is most appropriate for an enduser interface to optical disk-based materials, as well as:

- Electronic mail and bulletin boards;
- Electronic publishing;
- Online manuals;
- Online help for other software;
- Expert systems implementation;
- User interface to other programs;
- Software engineering; *and*
- Operating system shells.

A more complex, AI-enriched, form of Hypertext, known as Hypermedia, is applicable to such demanding implementations as:

1. Financial modelling;
2. Financial analyses;
3. Market-trend studies;
4. Cross-referencing;
5. Global general accounting;
6. Cost accounting;
7. Financial eavaluation of P+L;
8. Tracking of multidimensional references;
9. Backoffice work; *and*
10. Project management.

I heard of at least one case where Hypermedia has also been used in computer-assisted foreign language translation.

Most Hypermedia and Hypertext systems can be characterized by *chunking information* into small units. These are variously called notecards, frames, and so on. My suggestion is to retain the term: *packets*. In Hypermedia systems, such units may contain text, data, vector graphics, bitmapped images, animation and, eventually, sound. Typically, units of information are:

- Displayed one per window, *and*

- Interconnected by links which permit users to navigate in a hypermedia database by selecting links in order to travel from unit to unit.

Through the creating, editing, and linking of units, users build information structures for various purposes. Furthermore, in shared Hypermedia systems, multiple users may simultaneously access the Hypermedia database. This is particularly true of distributed systems.

During Hypermedia implementation, the enduser moves from node to node by selecting:

- The desired *to link*:

- An embedded *cross-reference link,* or

- The *from link to* return to the previous node.

For many documents, the *to links* collectively form a hierarchical structure analog to that of many conventional documents. But, the user has to have freedom to create the structure that he needs, rather than always being pinned down to preestablished ones.

This freedom of choice according to ad hoc requirements makes Hypertext documents much more flexible than conventional documents. At the same time, it provides a valid way to solving the problem of updating large databases - an issue for which major vendors have still to present valid solutions.

Since the unit orientation characterises Hypermedia/ Hypertext:

- Information elements resident in the individual nodes can be updated;

- New nodes can be linked into the overall Hypertext structure; *and*

- New links added to show new relationships.

In some systems, users can add their own links to form new organisational structures, thus creating new documents from old.

This significant possibility, and the change which it brings away from old linear approaches, identifies a basic difference between Hypertext and conventional documents. When considered together, the Hypermedia-supported facilities are producing a qualitative change in the way in which we conceptualise information resources.

It is this shift in perspective that is creating such a wealth of new possibilities. Therefore, to better appreciate what is involved in the new diskipline, I am including a brief reference to the historical development of Hypertext, and to the best solution now available: CMU's *Knowledge Management System* (KMS).

Capitalizing on the Facilities of Hypertext

The man who invented the basic notions behind Hypertext (back in 1968), and who coined the name, is Douglas Engelbart. At the time he was working at Stanford Research Institute. The second major contributor is Ted Nelson who extended the notion to: Branch, Link and the ability to make arbitrary associations among distributed IE, a decade before MIT's Media Laboratory performed *Hypergrams*.

Some people, however, trace the beginning back to the late 1940s when Dr. Vannavar Bush, of MIT, described *Memex* as ''A device in which an individual stores his books, records, and communications - which is mechanized so that it may be consulted with exceeding speed and flexibility. It is an enlarged, intimate supplement to his memory''.

The first practical application of this idea did not come right away. It rested for 20 years until Engelbart and Nelson framed the concept in an applications oriented sense. It also had to wait another 10 years before microprocessor technology permitted serious efforts in a distributed environment. (Interestingly enough, inthe late 1960s, Doug Engelbart worked on a timeshared SDS 940, the sort of computer which we consider to be today no more than an obsolete PC.)

ZOG is considered to be the first practical application of Hypertext on record. Work on ZOG began in 1972 at Carnegie-Mellon University (CMU), which today also has the best solution, KMS.

- While developing ZOG and KMS, CMU researchers logged 10,000 man-hours and created over 50,000 nodes.

- Throughout this period they applied what they learned to iterate the design of these systems, creating scores of intermediate versions.

What is now called ZOG-1, the project which started in 1972, was developed for a summer workshop for researchers in cognitive science. It allowed the participants to easily interact with one another's programs by providing a uniform menu-selection interface.

In 1981, at the request of Westinghouse, CMU faculty and researchers formed a team to develop a commercial version of ZOG. The first real-life implementation was that of a computer-assisted management system for the Navy's newest nuclear-assisted powered aircraft carrier, the USS Carl Vinson. The hypertext system developed out of this project is still in operation.

Throughout the 1980s, interest in Hypertext brought forward a number of other projects, the better known being:

1. *Intermedia,* developed by a research group at Brown University and tracing its ancestry to Ted Nelson;

2. *Document Examiner*, engineered by Symbolics, providing online access to user documentation;

3. *HyperCard,* by Apple Computers, which is relatively simple. However, Apple's aggressive promotion of it has been changing Hypertext from an esoteric concept known to a few hundred people to a household word in computing, to be used by millions;

4. *KMS,* by CMU. Its aim is to help organisations manage their knowledge. This knowledge-based Hypermedia system is concerned not only with the productivity of the individual, but also the productivity of groups - from small work groups to an entire organisation;

5. *Neptune,* a Hypertext system for computer-assisted software engineering, developed by Tektronix;

6. *NoteCards,* an ambitious system of the past decade, developed at Xerox PARC;

7. *WE*, a Hypertext authoring system worked out at the University of North Carolina that produces conventional paper as well as electronic documents modelling human cognitive processes.

Most of these systems have been designed to help people work with ideas. Their intended users are intellectual workers engaged in analysing information, constructing models, formulating arguments, designing artifacts and generally processing ideas. For this reason, they provide the user with a network of nodes interconnected by links.

The network serves as a medium in which the user can represent collections of related ideas. It also functions as a structure for organising, storing and retrieving information. The software includes facilities for displaying, modifying, manipulating, and *navigating through this network.* Also for *browsing.*

In NoteCard's case, for example, a browser is a NoteCard that contains a structural diagram of a network of NoteCards.

- The cards from this network are presented in the browser by their title, displayed in a box;
- The links in the network are represented by edges between the boxed titles;
- Different dashing styles distinguish different types of links; *and*
- The diagrams in Browser cards are computed for the user by the system.

By far the more sophisticated approach is provided by the AI-based Knowledge Management System (KMS). In essence, it is a distributed Hypermedia for workstations, based on previous research with ZOG.

KMS supports organisation-wide collaboration for a range of applications.

1. Its kernel consists of a conceptual data model;

2. Its database is organized into screen-sized, WYSIWYG workspaces, called frames and containing text, graphics and image items;

3. Individual items can be linked to other frames, or used to invoke programs.

4. The database can be distributed across a number of file servers and be as large as available disk space permits.

The KMS user interface employs a form of direct manipulation designed to exploit a three-button mouse. A combined browser/editor traverses the database and manipulates its contents. Running under Units 4.3 BSD, KMS accesses and displays frames in subsecond speed.

Thus the Hypermedia node in KMS is the frame. In this screensized workspace the user can place text, data, graphics and image items. Each of these items can be linked to another frame or used to invoke a program.

The most important characteristic of a KMS frame is its *spatial nature*. Like space in the real world, space in a frame exists whether or not any objects occupy it. A frame may be completely empty. This is different from the degenerate way in which space is represented by word processors and mail systems which, relative to KMS, are primitive.

Frames unify into a single construct what are distinct levels in traditional computing environments. This unification of formerly diskreet island functions:

- simplifies the users' conceptual model considerably; *and*

- eliminates the need for many commands.

Precisely because of its unified command set, KMS can be viewed as a reduced instruction set interface, seeking performance through simplicity. But, unification also serves another key subject which has been our goal for years. That of providing the enduser with one, homogeneous interface - instead of many incompatible presentation screens and commands.

Operating Systems Features and Photonics Storage

The role of an Operating System (OS) in computer technology is well known. It does not need explaining. Less known are OS functions with photonics storage, starting with the handling of the logical disk as a strict linear ordering of IE packets. There is an indirect relationship between:

- Logical IE, *and*

- Physical storage

when a file is physically placed on write-once media. In every case, the approach which we choose must preserve the expected logical disk interface to the operating system by recreating a strict linear order, for instance, in the form of the leaves of a tree (B-tree).

Such an approach will see to it that we access the logical disk through a tree-shaped arrangement of pointers, the file system tree, whose leaves contain the logical disk packets seen by the operating system. The root packet always contains pointers toward a second level of optical disk packets.

An alternative solution, which I would personally prefer, is the use of *flat files*, therefore of a *relational approach.** Its implementation is both:

- Simpler, in a design sense, *and*

- More flexible - from design to end usage.

The prerequisite is that a classification/identification (CC/ID) effort is done a priori. This will permit the construction of a relational database matrix which is subsequently very helpful in the organisation of the database structure, as well as during its exploitation in day-to-day operations.

Furthermore, it is quite advisable that, in the case of flat relational files for packets, small expert systems are used to assist in:

- Interfacing to the applications programs (AP), and

- Engineering the photonics storage access chores.**

Simple IF, THEN, ELSE constructs can do a good job in both connections. They are very easy to develop, consume very little of WS resources, and permit simple AI approaches which can make a commendable contribution to end results.

Attention should be paid to the process of logical to physical translation. Typically, during operation, the OS requests logical packets from the database.

---

* See also D.N. Chorafas: "Databases for Minicomputers and Networks," Petrocelli Books, Princeton, NJ paragraphs.

**Access proper will be done by the DBMS resident in the controller, as we will see in the following paragraphs

- In a magnetic disk file system, no further computation would be needed, because the logical and physical block locations coincide.

- With optical disks, an additional computation is necessary to locate the requested logical packet.

The database management functionality should be contributed by a DBMS. For instance, one of the photonics file cabinets (jukeboxes) on the market features the Oracle relational DBMS which is resident in its controller. The availability of a resident, relational DBMS should be one of the crucial characteristics in equipment choice.

Because available DBMS functionality does not necessarily include AI capabilities, in the preceding paragraphs I have suggested the writing of simple expert systems as interfaces. We should do everything current technology allows in order to be more efficient, and to provide a better service to the enduser.

There is no unique way in which the OS and associated utilities convert the requested logical packet number to a physical packet. Much depends on supported functionality and on the size of the system. Typically, with large files, it is necessary to partition a larger physical disk into logical disks, each of easily manageable volume.

Another challenge is the handling of growable information elements. IE written to a write-once medium cannot be subsequently altered in place. They must change place by growing. To do so, our system may use a continuation list allowing the file to evolve dynamically. This brings under perspective the need for *continuation lists.*

A continuation list is a sequence of optical disk packets whose entries consist of:

- Information elements, *or*

- Pointers to other disk blocks.

We discussed earlier the help that Hypertext solutions can provide in this connection.

Typically, the list is linked by the last pointer entry in each packet. Whenever a block belonging to a continuation list is written to the disk, a continuation packet is preallocated. Essentially, it is designated for future use. A pointer to the preallocated packet is included in the packet being written.

Let us once more underline that lazy allocation of the continuation packet does not work with write-once media, since the packet in the written list cannot be updated with a pointer to the next packet.

It becomes slowly evident from this diskussion that old information system experience on the part of some of the specialists who would work on photonics storage may be a negative asset. This is so because some of the things they know from good, old EDP can be disastrous with photonics. A change in culture is no idle recommendation. It is a ''must''.

Another important concept in photonics storage is the timestamp. In a system sense, it is wise to use a timestamp to record a checkpoint of the logical disk.

- *Timestamps* form a continuation list;

- The last written continuation block in the list is the *current timestamp*.

In the event of controller failure, the logical disk can be recovered in a consistent state by following the pointer chain in the continuation list to the last timestamp. Timestamp information can also be helpful in retrieval.

Thus, an expert systems utility to be attached to the controller's OS should timestamp packets, each defining a snapshot of the file system in a continuation list. The timestamp can contain pointers to the root packet (as well as to accounting information). It should also contain a pointer to the continuation packet.

Further, for housekeeping purposes, it is wise to have a simple expert system able to periodically perform checkpoint procedures flushing:

- all caches to disk, *and*

- timestamp sequences.

The checkpoints to be included in the system define the granularity with which snapshotted versions of files can be recovered. The last checkpoint also identifies the current logical disk.

At boot-up time, another expert system should trace through the timestamp pointer chain to the last timestamp, locating the currently active version of the logical disk. The last timestamp is easily identified because the continuation packet has not yet been written.

These are lessons learned by organisations which start having experience in the usage of photonics storage. Since it cannot be taken for granted that optical disk vendors will provide all of the necessary functionality through pages, we should be prepared to develop for ourselves some expert system constructs. But, we should also see to it that the vendor supplies all OS routines ready for use, and most of the utilities.

Updating the Photonics Files on WORM and Simulating the Outcome

Updating is one of the most intriguing features of a system solution to the organisation and management of photonics files. This is particularly true with write once, read many times equipment.

Updating the optical disk calls for a thorough systems approach. When a packet in the logical disk is replaced, several interior nodes of the system must also be replaced. A new timestamp must be generated in the process.

As we saw in the preceding sections, software should see to it that, when interior nodes are read, they are stored in a copy-back cache in memory. This:

- Improves the performance of the read operation;

- Limits the frequency of writing interior nodes to the disk; *and*

- Keeps the interior node information needed to update the disk readily available.

Solutions along this line of reasoning are relatively simple with small files. They are much more complex with large files. In this connection, locking file system procedures in RAM has proved to be helpful. File handling operations such as copy-reregister have proven advantageous, with cache providing the pivotal point.

Strategies regarding the able use of cache memories can be instrumental both in updating procedures and in minimising distal seeks. The operating system, for instance, may use copy-back cache diskiplines, updating cached IE. However, a copy-back procedure does not update the disk copy until some later time.

Precisely for this reason, the OS and its associated expert systems should periodically flush copy-back caches. To do so, the operating system should operate a copy-back buffer cache, while an expert system uses a cache of auxiliary information to organise storage on the optical media.

Caches may be flushed during checkpoints in intervals. The OS should provide the timing. For example, the time between the periodic recording of timestamps on the optical media. It should also ensure the necessary coordination of system components.

Besides being a great help with updates, implementation of the cache concept in photonics storage can improve reliability. Without a cache approach, when the system fails between timestamps, any data in memory and any data written to disk after the last timestamp is lost. The amount of IE that might be lost can be controlled by:

- adjusting the timestamp frequency; *and*

- storing on cache between timestamps.

Furthermore, the consistency of a logical disk can be improved by choosing an appropriate diskipline for page writes. The timestamp ensures that the on the optical media, which define the relationship between the logical disk and the physical disk, are always maintained in a consistent state.

- The logical disk is always consistent, because the system is updated before the new timestamp is created.

- To further enhance reliability, an expert system could complete a checkpoint by writing pages to the disk in the following order: logical pages, interior nodes, root node, and timestamp.

In the event of software failure, the logical disk is thus recovered intact up to the last checkpoint, always in a consistent state. After a hardware failure, the logical disk is recovered by following the pointer chain in the timestamp list to its last element.

In this sense, the next to the last element is the current timestamp. The following checkpoint uses the first available and readable continuation packet that was preallocated in the current timestamp. The first free disk block is easy to identify if disk packets are allocated in strict sequence.

We have emphasised the use of checkpoints. Some user organisations have found that very short checkpoint intervals, about every 20 seconds, eliminate many of the advantages of the caches. Very long checkpoint intervals on the other hand, about once every 20 minutes, risk losing large amounts of data if a failure occurs between disk updates. We must balance cache performance against risk.

Another operation being practiced, besides copy-reregister, is copy-remove. This, too, involves distal seeks from optical media. A key element which should constantly be kept in mind is that the files we access are successively updated and tend to be found at increasingly more distal locations. The latter can be avoided by locking the most commonly-used file system procedures in memory - which, of course, is a cache strategy.

It is also advisable to recall, that in a dynamic operating environment, an update to a file may be superceded while the file is still open. Therefore, an efficient use of the disk would be to accumulate updates and write them to the optical disk at checkpoint time, thus avoiding wasted writes.

Recording accumulated updates to contiguous storage is consistent with the observed access locality patterns for files. This is known as *access skew*, and it means that a relatively small number of files account for a relatively high proportion of accesses. Statistics on the pattern of system usage can be instrumental in subsequently improving its reliability.

Next to reliability, attention should be paid to *performance monitoring*. Each time the logical disk is check-pointed, an accounting block of performance data reachable from the timestamp should be written to disk for future analysis. At least one organisation found it advisable to record the following values whenever a timestamp is written:

1. Average rate of disk usage in packets per day since the last timestamp.

This provides information on the cumulative rate of media consumption.

2. Amount of space remaining on the disk on blocks.

Such a reference indicates the imminence of media transition to photonics storage.

3. Time in days since the present media was loaded.

Thus providing an indication of cumulative media lifetime, which is very helpful for administrative purposes. Other important statistics include:

- Storage and retrieval overhead for the system;
- Remaining active media life in days;
- Imminence of media transition to the user;
- Total number of timestamps on this media;
- Overhead of storing the timestamps; *and*
- Hit rate to cache of interior nodes since the last timestamp.
- Ratio of proximal to distal seeks since the last timestamp

This ratio is helpful in establishing the efficiency of strategies to encourage proximal seeks.

While statistics are valuable for future system design, as well as for tuning current systems, a critical support can be provided through experimentation. For instance, the objective of simulating a photonics storage system should be to optimise random-access file storage for write-only optical media.

A simulator can model the kernel of projected operations helping to verify that the algorithms proposed for system management are complete and correct. It can also verify that the cache of interior nodes can have a sufficiently high hit rate to ensure that node interactivity does not affect response time.

Other issues worth simulating are the storage overhead of timestamps and interior node linkage, as well as the effective storage capacity of optical media. A simulator developed by a company using optical storage incorporated a model of:

- Write-once optical disk;
- Cache for data pages; *and*
- Cache for interior nodes.

Average rates of reading from existing files, writing to files, creating new files and deleting file pages were used to drive the simulation. The parameters of the simulation were the size of the file cache, the size of cache of interior nodes and the checkpoint interval.

In our work we may need to repeat precisely the same type of experimentation, or we may alter it - but it is always helpful to know what other organisations did and the results which they obtained. While several of the photonics storage implementation problems may be unique to a given organisation, several others are common to all. Therefore, we should always be ready to learn from other peoples' experience.

12

The Future in Telecommunications

Introduction

Telecommunications encompasses all transfer of information, both internal and external to an organisation. This includes voice telephony, data transmission, electronic mail, telegrams, telex, facsimile and picture transmission. Information can be transferred through wire (cable) or by wireless technology (radio). These two alternatives are partly competing with, and partly complement, one another.

As the earliest practical implementation of electricity, *telegraphy* made it possible for the first time to transmit information in non-physical form and in large quantities. Electrical signalling developed rapidly. Hardly had the telegraph network, based on the code invented by Samuel Morse in 1836, established itself, than the telephone appeared on the scene. It was invented in 1876 by Alexander Graham Bell.

Two decades later (in 1897), Guglielmo Marconi sent telegraphic signals through the air. The thousands of miles of telegraph wires suddenly no longer seemed important: wireless was born. Then radio broadcasting was invented. The very first broadcast in 1920 by the German radio station Königs-Wusterhausen brought a new age of communications.

At an electrical exhibition in Berlin, only nine years later, was shown something more sensational than the new radio: Television. Telecommunications evolved within the short frame of a century: 1836 to 1929 is less than 100 years. The path has been:

- From the dots and dashes of the Morse code, to

- Television, then already in colour in the laboratory.

Telecommunications became an important part of defence operations during World War II. But prior to that the radio had been turned into a propaganda vehicle.

However, it was not until after the war, with the change of focus taking place in business and industry, that the true potential of communications came to the spotlight.

The next big breakthrough has come with lasers. Few people realise that it is not even a quarter century since laser technology was first proposed and developed at STL Laboratories, in Harlow, England. Dr. Charles Kao was the first to suggest the idea of transmitting signals by light beam over hairfine fibre-optic cable. Then, together with George Hockam, he went on to produce the first working demonstration of the technology in 1966.

With support from British Telecom, Standard Telephones and Cables Ltd. pressed rapidly ahead to deliver the world's first fibre optic cable to carry live telecommunications traffic. The technology offers two big advantages over conventional cables:

1. Once trapped inside the fibre, light cannot easily get out and so it travels for long distances without amplification.

2. Light pulses travelling along the fibre keep their sharp on-off profile; this is important when signalling works at over 2 billion switches a second, as Bell Laboratories recently did.

Bell Laboratories recently sent the information over 170 kilometres of fibre-optic cable without a repeater. The rapid increase in the demand for communications services has brought forward the need for more efficient media and structures than those so far available. The great steps that technology has been making during the last ten years have substantiated many of the users' demands, including high quality voice and image services, but also peer-to-peer, computer-to-computer communications, and remote access to databases. More is still to come in the years ahead.

Beyond doubt, rapid advancements in technology see to it that voice communications, data communications, word processing and data processing converge. The common ground is distributed information systems, linked through networks. The impact of this integration will transform our lives as well as our professional and personal interests.

Wide-spanning, broadband, any-to-any networks are the best telecommunications support, and networks are replacing computers as the focal point of interest and investment. The commanding authority is the intelligent network. In a communications and computers (C+C) system, the network compares to the Christmas tree on which computers are hung as ornaments.

The term "network" is often used in a broad sense to designate not only the circuits (and lines) for voice and data transmission, but also terminals, such as telephones, teletypewriters, switching equipment (PBX, computers, etc.), and operating personnel. Networks become increasingly important as telecommunications and information processing continue to penetrate the mainstream of the business and scientific environments, and to demand the best communications solutions.

With long haul links carrying an increasingly heavy information load, four alternatives are open in terms of physical media:

- terrestrial radio relay;
- radio relay via satellite;
- coaxial cable; *and*
- fibre optic cable.

Which technology is best suited has to be assessed on a case-by-case basis. In a general sense, radio braodcasting may never be able to replace cable technology, but within the cable variant there are alternatives: twisted pair, flat wire, coaxial cable and optical fibres. The problem with radio is that, with the vastly increasing demand for frequencies, the spectrum available for transmission will never be nearly adequate to serve all the users through wireless communications.

- Up to a point, radio and cable are alternatives;
- Plain old telephony can use either radio or cable;
- Radio programs will always be broadcast by wireless systems;
- Wireless technology is the essential media in cellular telephones; *but*
- Television will increasingly tend towards cable transmission because of quality, programme selection and broadband capacity;
- Teleconferencing also requires broadband for live transmission rather than frame freeze.

Optical fibre technology will have the largest potential in the years ahead, followed by satellites. Laser transmission is ideally suited for wideband connecti-vity, and photonics in general present better communications means than electronics, a fact which seems to have attracted the attention of the inventor of the telephone.

Alexander Graham Bell had also designed the photophone. Instead of a microphone diaphragm this device employed a very fine mirror which vibrated to the human voice. Acoustic waves of speech were converted into a modulated beam of light. A receiver at the other end reconverted the light into acoustic waves.

Bell's idea was excellent, but also nearly a hundred years ahead of technology.* Sunlight, the medium he proposed, is a mixture of numerous different light rays.

---

* We will return to this subject in Chapter 13.

As such, it is unsuitable for the transmission of messages. A cable which could transmit light beams had not yet been invented, and open air, the only transmission medium then available, is not pure enough to accomplish this mission.

But the concept of using light for communication was not abandoned altogether. In 1790 Claude Chappe built an optical telegraphy system which could make signals with moving indicators. These were then read by the receiving station through binoculars. By being passed on from station to station, a simple message could thus travel 200 kilometres in 15 minutes.

Optical communications systems were used for a long time by primitive smoke and fire signals, and also by sailors with their semaphores flags and morse lamps. These methods are still used for certain purposes today -although this is not the sense of optical communications which interests us in this book.

Supporting a Nation's Infrastructure

The first transport revolution started when Watt invented the steam engine in the 1780s. The second when Alexander Graham Bell patented the telephone in 1876. The third came shortly thereafter when Benz and Daimler developed the motor car in the 1880s. The fourth has been characterised by the business applications of the digital computer, starting in 1953.

The fifth transport revolution is telecommunications enriched with computers and artificial intelligence constructs. It took off in the early 1980s and will have the biggest effects of them all. It will also take more than half a century to reach maturity, if past experience is any guide. The one-hundred-year-old motor revolution still has unfinished business.

Yet, even today, we can feel the impact of this new development. Like the trains in the 19th Century, the telecommunications network has become a crucial part of a nation's infrastructure. Retail and banking networks, for instance, are now based on large capital investments in telecommunications. Without them, it is not possible to put into effect electronic funds transfer (EFT), point of sale (POS) solutions, and automated teller machines (ATM) networks for 24-hour-per-day consumer banking.

The development of a telecommunications facility depends on the:

- Capacity which it can support;
- Reliability that it exhibits;
- Quality of service it provides; *and*
- Usage made of it.

High capacity today is supported through optical fibres and satellites at the links, and computers in the switching centres. The reliability for which we aim is the ''four nines''. That is: 99.99%. As we rarely obtain it, we settle for 99.9%, or slightly better.

High quality calls for steady investments and, in terms of metrics, it means better than a bit-error rate (BER) of 10^{-10} or 10^{-12}. Large scale usage is a function of tariffs, culture and added value. In all cases, public interest is served through competitive forces, not bureaucracy. It is no accident that the French translate PTT* as ''Petit Travail Tranquil''. The sense could be given in English as ''petrified tasks and trivialities''.

Not suprisingly, the USA, England and Japan have deregulated their telephone business to give it added vigour. As long as telecommunications stays regulated or, even worse, nationalised:

- Investments are lagging;
- Tariffs are uncompetitive;
- Bureaucracy dominates; *and*
- The quality of service leaves much to be desired.

Today, the highest per capita communications markets in the world are in America and in Japan. If the American investments count for 10, the Japanese count for 4, while the whole Common Market features the same factor 4 - standing, in the average, at one third the Japanese level of per capita investments in telephony.

Investments in telephony are particularly important as the emphasis in communications is shifting towards digital networks. These include, not only voice traffic and data transmission facilities, but also intelligent devices for providing services which were previously impossible. Some of the most imaginative developments in the years to come will address themselves to communications by way of human interfaces. They will incorporate natural language - from machine translation to direct voice input to computers, as this chapter helps document.

Considered in all its aspects, from facsimile transmission to teleconferencing, the communications function is becoming one of management's most powerful tools. The emergence of telecommunications as a key discipline has several causes:

1. Growing demonstration of benefits to functions such as marketing, production and finance;

2. Explosion of telecommunications technology, enriched by inventions in electronics and, more, recently in photonics;

---

* The public authority for Post, Telephone, Telegraph, which, in the general case, is a monopoly.

3. Increasing role of telecommunications at the corporate level, where it is easier to move information electronically than to move people and push paper.

New networks should be designed to support a wide variety of services from teleconferencing to compound electronic document handling, over and above data and voice requirements. Imaginative developments are called for, to optimise the solutions currently existing and also those projected for the near future. They are also required in order to capitalise on similarities and differences. Data traffic differs from voice traffic. It:

- May be queued for delivery;
- Calls for bit error rates three orders of magnitude lower;
- Is subject to both preprocessing and post-processing;
- Can be transmitted discontinuously in blocks or packets;
- Has different holding times; *and*
- In a number of cases requires multimedia interactivity for efficient service.

This means simultaneous parallel transmission and reception - something beyond the abilities of people, but not of computers.

The question is not simply how do we improve upon the telephone set of Bell and the exchange of Strawger. Telecommunications systems have outgrown, not only the classical telephone set, but also Bell's vision of a grand system. Not just men, but also machines require, systematic communication. But, networks cannot be simply turned on, complete, from the word ''go''.

Today, few people are immune to the rapid technical advances in telecommunications as the world turns from a manufacturing to a knwoledge-intense society. The telephone line has become a multifunctional tool, performing plenty of communications requirements. We are integrating voice capabilities into most of our office automation equipment, because most communication is in voice form and telephones are polyvalent terminals.

Electronic directories, consulted by telephone, are replacing paper telephone books, and consumers are teleshopping and telecommuting. A microprocessor-controlled security system telephones police in the event of a burglary. Consumers use home computers connected through the telephone line to distant databases, communicating with other users and their computers.

- Teleconferencing cuts travel;
- Banks are open twenty-four hours per day;

- Shops operate nationwide from only one outlet; *and*

- The world has become one integrated financial market.

Telecommunications advances are geared toward enhancing the range, speed, and processing of information. New methods have developed for collecting, storing, displaying, manipulating and moving voice, text, data and image.

Man-machine interfaces are changing as systems and their utilisation improve. Technology makes feasible cost reductions in microelectronics, software, digital devices and fibre optics. Conceptually, we have left behind the old world of hard metal contacts and moving mechanisms. We have entered the era of exceedingly small but intelligent structures. Yet, in telephony the mechanical devices are still around, and probably they will continue to be until the year 2000 - creating the problem of symbiosis between the new technologies and the old.

One of the keys to the developing capabilities lies in the use of electronic switching and transmission under software control. Another great issue is the *integration of traffic*. Voice, text, data and image must be carried in a general-purpose network able to handle *multimedia*. This network's speed, quality of service and mode of operation should reflect future, not past, requirements.

The decade of the 1990s will be crucial, not because of the number of networks to be developed and implemented, but because companies, both large and small, must make some major decisions about capital investments. And these will have to take into account the most probable shape of communications and information processing facilities during the next ten, fifteen, or twenty years.

Digital technology is fast, flexible, reliable - and getting less expensive. It is unusual to call coast-to-coast in the USA without passing through a digital trunk.

- Digital electronic switching replaces, not only electromechanical solutions, but also analog electronic approaches;

- High-speed facsimile replaces copiers;

- Optical disks replace file cabinets;

- Intelligent workstations replace dumb terminals.

The result is considerable savings in time, space, transport costs and energy - which amounts to greater competitiveness.

Change in the installed plant looks slow, not because technology is lagging, but because of the magnitude of the investment. There are about 650 million telephones (roughly 40% of them in the USA); an equal number of millions in television sets; 1.6 million Telex terminals; and thousands of data networks in the world today. Not surprisingly, 90% of them are serving 15 countries, documenting Pareto's law and creating a formidable telecommunications gap between the First and the Third Worlds.

Frustrated by bottlenecks and breakdowns, if telephones are available at all, the lesser developed (i.e. underdeveloped) part of the world is still attempting to install basic services. However, few people realise that, unless the majority of the world's population soon has access to a dependable communications network, its social and economic contribution will be severely limited.

These few words reflect the schizophrenia of the world in which we live. Instead of controlling their birth rate in order to actually reduce their population and thereby make available scarce resources and raise their standard of living, countries of the Third World drift in a sea of population explosion. No wonder the infrastructure gets weaker and more fragile all the time.

But the First World is not blameless. Instead of letting mother nature do its job by eliminating the population surplus through famine - then focusing their help on the Third World's infrastructure - the western countries do the opposite. They feed the Third World's weaknesses by feeding its hungry people and virtually strangle the Third World's strengths by leaving its natural resources underutilised.

As Paul Kennedy aptly suggests*, the world at large is caught in a triple challenge:

1. Provide military security, or some viable alternative to it.

2. Satisfy the socioeconomic needs of its citizenry, which are further aggravated by the demographic explosion, and

3. Ensure sustained growth through appropriate investments.

Challenge No. 3 is essential both for the present and for the future: affording guns and butter in the present (the former are utterly unnecessary, but keep the military happy so that they do not revolt); and avoiding economic, industrial and agricultural decline.

In fact a modern, efficient, high quality, any-to-any telecommunications network is instrumental in supporting both the positive goals (avoiding decline) and the negative ones (guns vs. butter). There is a pressing need to make telecommunications global, but it is not easy to adapt high technology equipment to the low technology conditions which characterise the majority of countries in the world today. Even the developed countries have to try hard to cope with their problems with the plain old telephone service (POTS). And this is, under no circumstances, an easy task.

Equally true, while politicians focus on *physical food* production and distribution, they forget the need for logical food. Yet, *logical food* is the upper layer of our

---

* "The Rise and Fall of the Great Powers", Random House, New York, 1987

civilisation's structure, the one that is absolutely necessary in order to sustain and further wealth.

Capitalising on Photonics to Stay Competitive

Because of transmission capacity, reliability, higher quality and lower costs, the 1990s will be a decade of transition from electronics to photonics. Leadership in communications will set the stage for the battle of the 21st Century to be fought along the lines of:

- Optical transmission;
- Optical storage; *and*
- Optical switching.

It is, therefore, very important to position our organisation in terms of culture, skill and knowhow, R&D effort, marketable products and financial investments to face the forces of the 1990s. Change will not come overnight. It will require a plan, a timetable, a concrete program, steady effort and lots of patience.

In a conference a few years ago in the United States, AT&T stated that photonics is one of the three *killer technologies*. (The other two are VLSI and software.) Due to this killer technology, the oncoming change will alter the concepts and the systems that we have cherished so far and with which we have become familiar. It will also make mandatory a great deal of research and prototyping - well beyond the mere understanding of the physical principles involved in the new science.

Just as we cannot conceivably produce large software systems without specifying just what the entire aggregate is supposed to do - and the role that each of its parts should accomplish - we cannot design photonics systems without clear implementation goals. The goals can be successfully set only if we understand the fundamentals. To handle transition in an able manner we must:

- Establish the point of departure;
- Comprehend the theoretical basis.;
- Focus on the work currently in process; *and*
- Project on the direction that future developments may take.

The advantages of fibre transmission on which we capitalise are small size and weight, low loss, immunity to electromagnetic interference, higher security than other media and greater bandwidth. It also offers more favourable economics than are provided by alternative solutions.

Fibre optic cables excell over copper cables, but they also have drawbacks. For this reason, not every copper cable can simply be replaced by an optical fibre. Among the attractive features of optical fibres are:

- Lightness;

- Low bulk; *and*

- Low transmission loss.

Comparisons with conventional coaxial cables of the same capacity reveal visible differences in the optical fibre's favour. Since it is light, and not electricity, that flows down the fibre, the signals are immune to inductive interference from electromagnetic fields. Optical cables are, therefore, suitable for use in power stations, railways and industrial plants. And, because of low transmission losses, they are favoured for long distance links, including transoceanic cables.

With its enormous information carrying capacity, fibre optics technology will play an increasingly important role in the developments to characterise the 21st Century.

- Fibre optic cables with laser light sources are already being used to carry high-capacity telecommunications traffic between large urban centres.

- As production grows and manufacturing costs fall, they will eventually replace the twisted pair cables that carry signals into the home and office, and open up the prospect of a host of new communications services, including the long-heralded video phone.

Optical fibres are also superior because they offer better protection against wiretapping. Although in principle it is not impossible to tap a fibre optic line, the process is technically much more laborious than with conventional wire cables. But, while a reduced multitapping capability promotes security, it also results in a limitation: difficulty in multitasking. This is crucial in replacing the current coaxial local area networks.

To understand the multitasking problem we should recall that the optical fibre cable does not carry electricity, but light. An optical fibre communications system consists of three parts:

- Light source;

- Light guide; *and*

- Light receiver.

The keyword is light. For a long time, scientists were in disagreement as to whether light was a stream of tiny particles, or wavelike energy. Then it was discovered that the stream of photons also had wavelike properties. Light is, therefore, both a stream of particles and a wave.

Electromagnetic waves apparently do not need a medium. Photons can travel through vacuum, but neither all light sources nor all wavelengths produce the effect required for communications reasons.

What physicists call *visible light* occupies only a small portion of the entire spectrum of electromagnetic radiation. Light is that range of wavelength to which the human eye is sensitive. For communications purposes, for instance, other wavelengths can be, and are, used which cannot be perceived by the human senses. Figure 12.1 shows the wavelength ( ) spectrum:

1. ultraviolet;

2. visible light;

3. infrared; *and beyond to*

4. microwaves.

It also locates the area where devices based on gallium arsenide may be of help: 1.300 to 1.550 nanometers (nm) wavelength. The lesser known area in Figure 12.1 is that of millimetre waves (mm) at around 100 GHz.

When we focus on the infrared area, we should distinguish between information communications and information storage:

1. For communications purposes, attenuation in decibel (dB) per km is very important, as it represents losses in transmission.

Here, scientists exploit a characteristic attenuation curve for fibre optics shown in Figure 12.2. At a wavelength of 800 nm, the loss is 2 to 5 dB per km.

Laser technology of the 1970s focused on the attenuation valley around 800 nm, but current developments make it feasible to work in other attenuation valleys located at wavelengths of 1.300 nm and 1.550 nm, where the losses per km are: less than 1 bB and less than 0.5 dB, respectively.

These references are particularly important for wide area networks. Optoelectronic elements are matched for these windows. There are, however, some other criteria to account for, such as *resolution* and *writing density.*

In different terms, apart of the fact that, as we move past the 1.000 nm wavelength technology becomes expensive, there also exist technical constraints to be taken into consideration. A 1.550 wavelength is not good for focusing. It gives poor geometric resolution. For this reason, for example, photolithography for semiconductors uses ultraviolet light where resolution is at its best.

2. For optical storage purposes, we are more interested in *resolution* than in the attenuation issue.

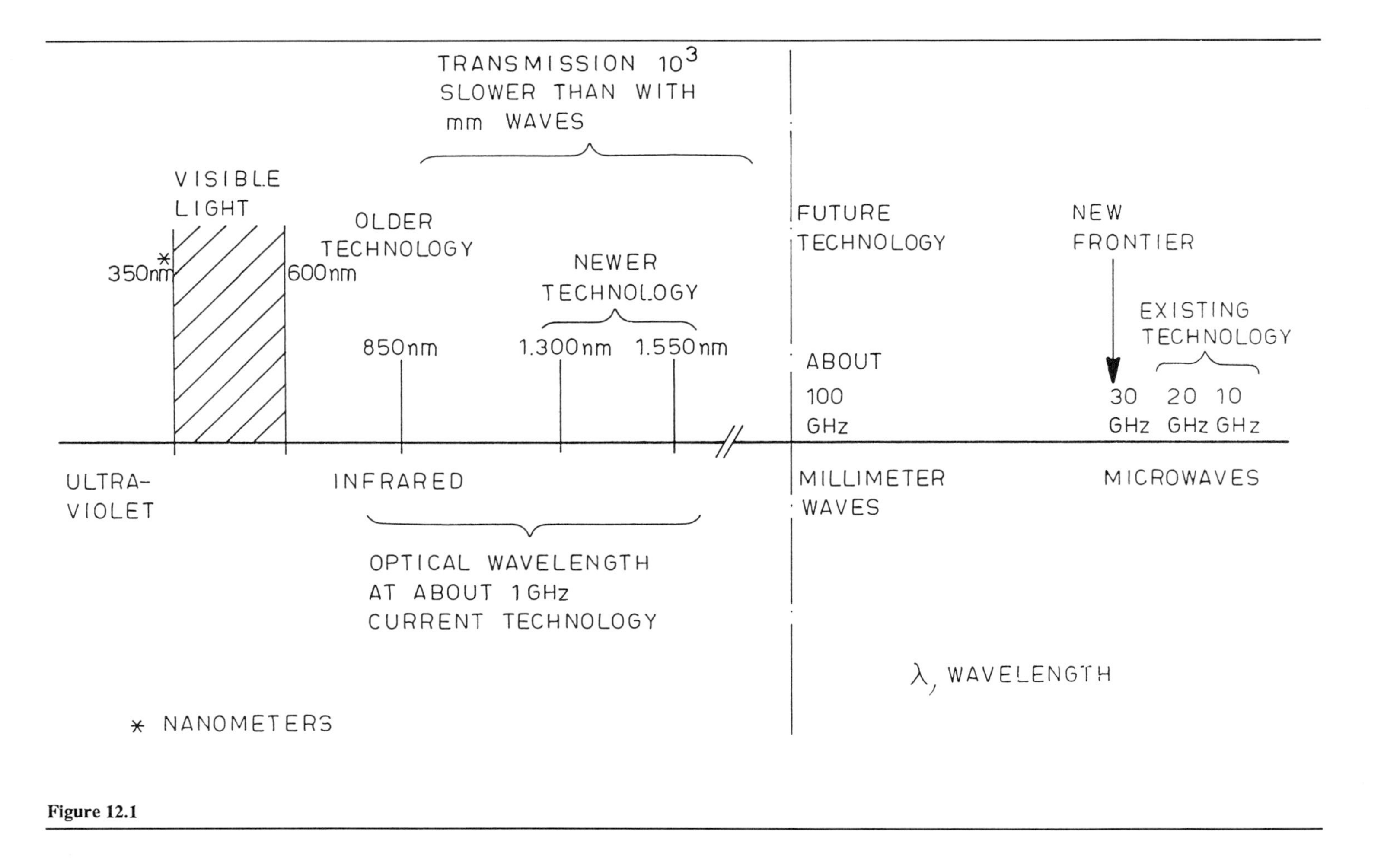
TRANSMISSION 10^3
SLOWER THAN WITH
mm WAVES
VISIBLE
LIGHT
OLDER
TECHNOLOGY
350nm*
600nm
NEWER
TECHNOLOGY
850nm
1.300nm
1.550nm
FUTURE
TECHNOLOGY
ABOUT
100
GHz
NEW
FRONTIER
EXISTING
TECHNOLOGY
30
GHz
20
GHz
10
GHz
ULTRA-
VIOLET
INFRARED
MILLIMETER
WAVES
MICROWAVES
OPTICAL WAVELENGTH
AT ABOUT 1GHz
CURRENT TECHNOLOGY
λ, WAVELENGTH
* NANOMETERS

Figure 12.1

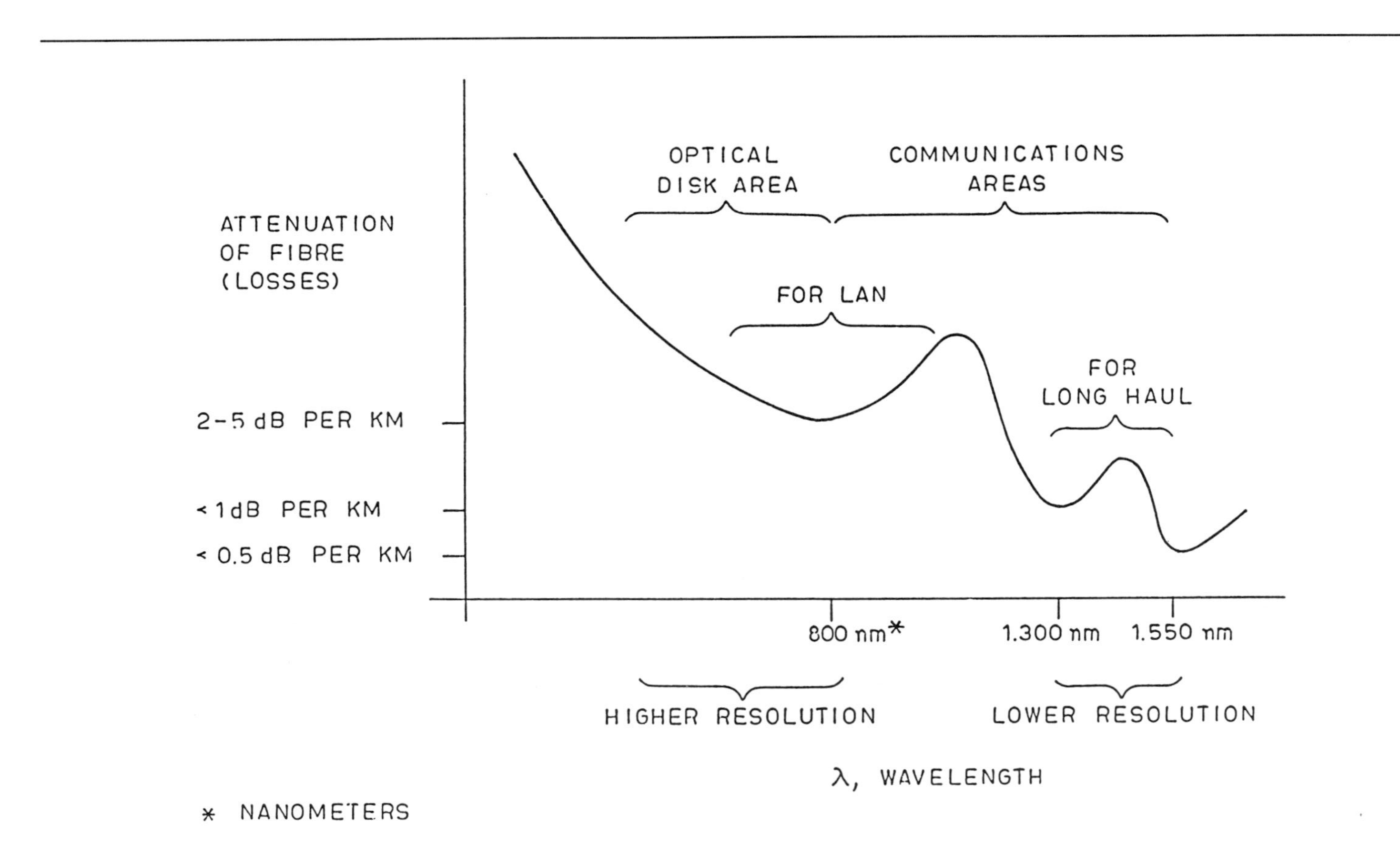

ATTENUATION
OF FIBRE
(LOSSES)
2-5 dB PER KM
< 1 dB PER KM
< 0.5 dB PER KM
OPTICAL
DISK AREA
COMMUNICATIONS
AREAS
FOR LAN
FOR
LONG HAUL
800 nm*
1.300 nm
1.550 nm
HIGHER RESOLUTION
LOWER RESOLUTION
λ, WAVELENGTH
* NANOMETERS

Figure 12.2

Carrier attenuation is no major criterion with optical disks. The key problem is *writing resolution*. For this reason, optical disk devices could get into the ultraviolet range in terms of implementation, but years ago the choice was made to work infrared at about 600 to 700 nm wavelength.

This choice fits perfectly the objective of using infrared for inhouse communications. While the band above 800 nm wavelength would have given a lower resolution, that below 800 nm provides a higher resolution.

There is a second major advantage to be noted with the 600 to 700 nm range. With the implementation of optical disks becoming increasingly popular, the cost of optical devices is steadily dropping. This provides relatively inexpensive laser media, operating at the 600 to 700 nm level, for usage in networking.

As this discussion points out, there is a significant difference in the choice which we should make depending on whether we wish to use optical technology for:

- storage, *or*
- transmission.

In the latter case there are further technical choices necessary in function of the distance that we wish to cover: wide area or local area. Our solutions should be cost/effective. Technology provides alternatives. We should be choosing from among them, always keeping our goals in mind. Another need in the any-to-any interconnection sense is to bring together currently-available distinct structures such as:

- The telephone network, *and*
- Computer networks.

This is the goal of computer-integrated telephony (CIT), which aims to produce an integrated desk. Two applications horizons are projected:

1. *Horizontal,* with global message centres, voice mail/text mail integration, database-driven automatic calls and fully distributed functionality.

2. *Vertical,* including telemarketing, worldwide financial services (banks, brokers), travel, and such other fields as maintenance service.

The key to CIT is voice technology management. This includes many distinct, but interrelated, areas: voice synthesis from text, delivery of store and forward voice, voice recognition and so on.

But voice technology has still some way to go, as underlined in another chapter based on the AT&T meetings. Therefore, even the most advanced current solutions have to make compromises, as shown in Figure 12.3. This specific approach links the PBX,

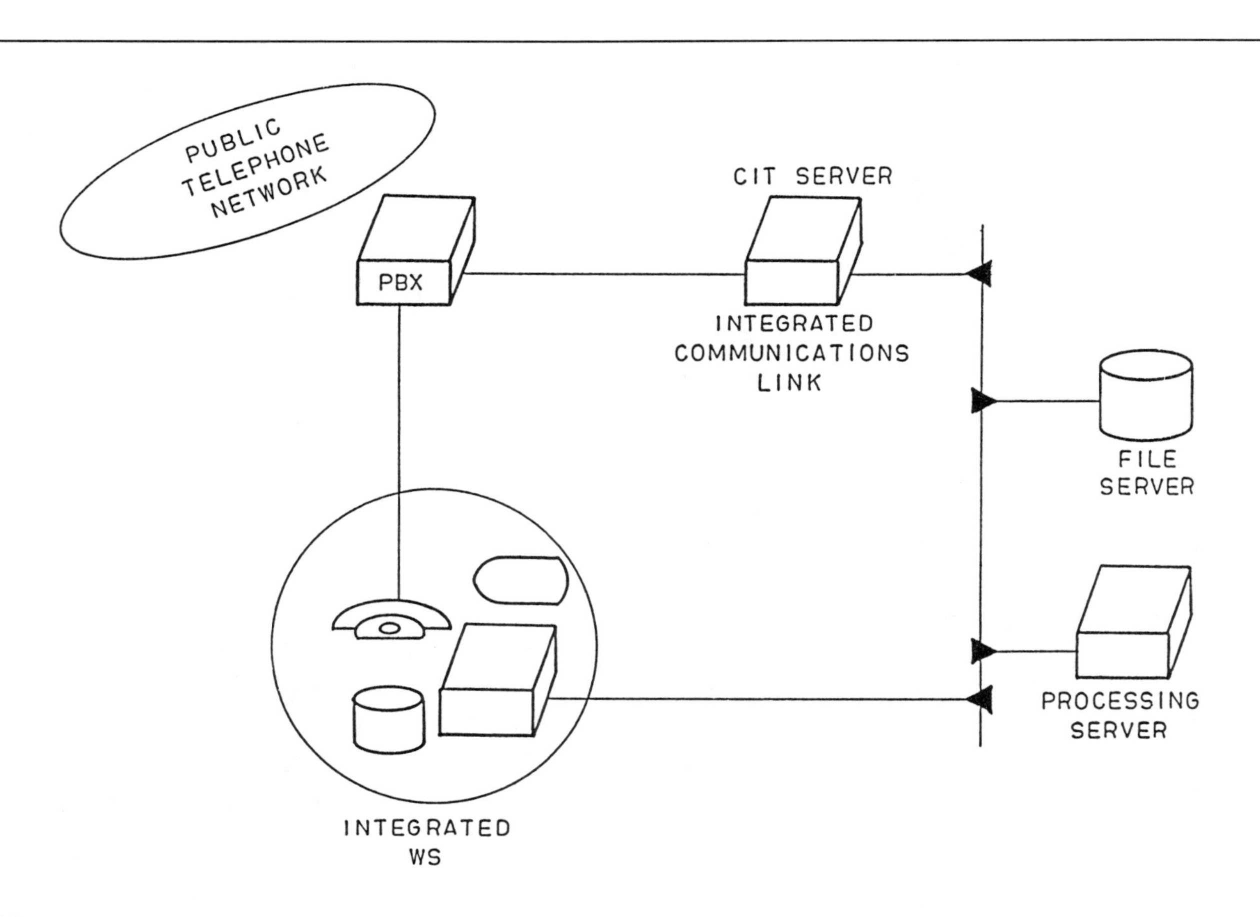
PUBLIC
TELEPHONE
NETWORK
PBX
CIT SERVER
INTEGRATED
COMMUNICATIONS
LINK
FILE
SERVER
PROCESSING
SERVER
INTEGRATED
WS

Figure 12.3

computer gateway and LAN, allowing for a relatively integrated desk, but the voice terminal and text/data terminal still remain distinct, even if served by one physical unit. Such a solution introduces the notion of a PBX server - a physical layer under CIT which is logical.

Research and Development on Advanced Communications

Today, most activity directed to fibre optics is at the level of 2 to 3 gigabytes per second (GBPS). A European RACE-financed project focuses both on this range and on a little more than 3 GBPS for local area networks (LAN). The aim is to obtain a common channel for *multimedia* able to handle moving pictures, voice, graphics, data and compound electronic document transfer. Multimedia communications need this capacity of ?3 GBPS.

Research and development on advanced communications (RACE) is central in the European research effort. RACE has so far brought together 230 different organisations to work on:

- telecommunications networks, *as well as*

- standards and services for telecommunications.

In the background of RACE is longer-term R&D involving pan-European telecom initiatives and including a significant variety of projects. Several of these projects work on defining an economic infrastructure that employs larger bandwidths to integrate different telecommunications services.

Multimedia is one of the main goals. To be precise, this involves the integrated transmission of image, voice, data, text and graphics on a high-capacity fibre-optic broadband network. The aim is that of enabling large amounts of digitised information to float around a barrier-free Europe.

Provided that the ongoing work is successful, the results obtained from RACE could lead to:

- A reasonable level of integration among separate telecommunication networks, *and*

- The application of fairly common standards in different European countries.

This statement is written on the premise that the outcome of RACE will be a set of common specifications for the European telecommunications infrastructure of the mid-1990s. In all likelihood, such an infrastructure will include the specifications of terminal facilities and advanced databases/data communications services.

The premise behind RACE projects is that multinational collaboration on a technological level can help:

● Reduce the cost of research and development;

● Facilitate pan-European standardisation; *and*

● Provide a stimulus for faster results to meet the challenge of competition.

It is too early yet to say whether or not these objectives will be met. Three main handicaps may reduce the effectiveness of the research effort - and have to be considered as such.

First, t*he lack of a common language.* Like Esprit, RACE is a multinational and multilingual project. Multinationality is good if it leads to cross-fertilisation, but research documentation written and kept in different languages makes knowledge transfer so much more difficult. Translation distorts some of the concepts in research papers - though many projects chose to use English as the lingua franca.

Second, *coordination and supervision* has not been allocated as yet the appropriate budget. Third, again taking Esprit as a reference point, European R&D projects are *pre-competitive.* They focus on basic principles, and do not get into product design. The problem is that void of marketing and implementation perspective, research risks becoming daydreaming.

This is particularly true when we keep in mind the very broad range of issues to which RACE addresses itself. Its R&D tie-ups include ventures by every major equipment manufacturer in areas ranging:

● *From* microchip and semiconductor technology; *to*

● Cellular telephony;

● The next generation of digital transmission systems; *and*

● Data communications protocols.

Therefore, it is advisable that each RACE research contractor sets himself - for his part of the project -an implementation objective. Rather than contradicting RACE goals, such a policy will strengthen them. It will also set a precedent as the future of R&D in high technology, in Europe, will undoubtedly include a proliferation of pan-European programmes and alliances between telecommunications and information systems companies - the recent GEC/Siemens effort to swallow Plessey being an example.

It is always important to capitalise on the installed scientific infrastructure, leading to the creation of advanced telecommunication poles throughout the European continent. To *succeed,* high technology projects should:

1. Devote part of the budget in basic and applied research; *but*

2. At the development side, use state-of-the-art components to produce products in short timespans;

3. Employ some of the best brains available in the R&D organisation, cross-fertilising them among different disciplines;

4. Follow a very precise implementation timetable, with the criterion being specific results; *and*

5. Focus on a niche market, carefully selected for its growth capability.

The biggest problem with niches, however, is that succeeding means being a trendsetter, a role which should be played well, or not at all. This requires, not only understanding the technical fundamentals, but also what makes the market tick.

Furthermore, many technological advances eventually get into existing products and are made standard equipment. Hence, product design for a niche market should also be supported by marketing expertise applied at the drafting board level.*

Both innovation and substitution should be examined. The same is definitely true of integration with other existing products in the marketplace, always with a clear view of the target implementation environment.

Lasers and Coherent Light Systems

The invention of the laser as a light amplifier, nearly three decades ago, radically transformed optical engineering. It made available a more powerful, but also purer, source of light, *monochromatic* instead of the mixture of wavelengths radiated by ordinary lamps.

We have seen in the introduction the problems that Alexander Graham Bell faced with the photophone when he tried to exploit sunlight. Laser light is *coherent*, whereas ordinary light is incoherent. It includes unwanted inputs - therefore *noise*. For this reason, ordinary light is useless for anything but the simplest communications.

Laser emits monochromatic light instead of the mixture of wavelengths radiated by ordinary lamps. And, since ordinary light is incoherent, whereas laser light is coherent, it is understandable that much of the scientific effort since the laser's invention has concentrated on raising the power of laser light sources. During the last decade, optical engineering has been experiencing significant breakthroughs which are expected to accelerate in the course of the coming years.

Coherence is associated *with lack of noise*. Coherent systems are characterised by a very intensive wavelength separation, leading to *Wavelength Division Multiplexing*

---

* See also Chapter 8 on this specific issue.

(WDM). In the fundamentals, there is no real difference between frequency division multiplexing (FDM) and WDM, except that:

- FDM addresses itself to electrical properties, *while*

- WDM is applied within an optical implementa-tion environment.

WDM represents wavelength modulation while FDM is frequency modulation - but WDM uses an FDM approach: phase-shift keying (PSK) modulation. In WDM implementation, PSK modulation is employed in connection with a receiver achieving demodulation in a Costas loop.* The advantage is that signal demodulation and phase tracking are feasible for semiconductor laser devices which are available today.

The optical cable spectrum is split, permitting the use of a coherent multiplexing approach. The bandwidth stands at terahertz. WDM provides the possibility to fully exploit fibre optic bandwidth capabilities.

There are two assets of coherence:

1. Higher receiver sensitivity.

Theoretically, 10 dB or more for the receiver for a bit error rate (BER) of 10^{-9} or better.

2. Very intensive WDM multiplexing.

The Heinrich Hertz Institute in West Berlin have developed a system with channel spacing modulated through phase shift keying, each channel at 70 MBPS. This is experimentally used for television distribution.

But even the most recent developments in WDM, as well as in single channel optical fibre, have not yet tackled in an able manner the fibre optics' classical weakness: multitapping. Fibre optic networks for local area implementation are typically organised along *star coupling* lines. This poses four problems:

- Reliability;

- Flexibility;

- Privacy/security; *and*

- Blocking.

Once the hub is damaged or, for any other reason, out of service, the whole multithread network depending on this hub is cut-off. There is no alternate pathing.

---

* The concept was originally developed in the 1956/57 timeframe. WDM is employing a Costas loop to demodulate (not modulate) the signal.

As a result, hundreds of WS may be isolated from central/ department resources and from one another.

This could be corrected through a horizontal structure, instead of using a single backbone on which local resources are attached. An example is given in Figure 12.4. Such an approach would help provide:

- Fail-soft capabilities;
- Higher system performance: *and*
- Directed traffic characteristics

What is more, this can be achieved through simple enough protocols, which we already know from packet switching. Such a solution will also enhance the flexibility of the installation, with the polyvalence of configurations further improved if infrared rather than cables (fibres, coaxial or twisted pair) is used to interconnect the node to workstations in a room.

One of the important issues in advanced communications is the choice of waveguides. Optical waveguides are typically made of two different materials. Optical fibres consist of two kinds of glass, each with a different refractive index:

- The core material is always the slower medium;
- The outer sheath, the faster.

It is also important that the core/sheath interface is highly polished, otherwise the mirror effect is reduced and light is lost. The effort is to contain the light in such a way that all of the light fed into the fibre is available at the other end.

Three types of optical fibres can be distinguished:

1. *Stepped-index*

These are of little significance in practice, but excellent for demonstrating the functional principle. A core is surrounded by a slightly thicker cladding. By the principle of total reflection, light is trapped in the core. What makes stepped-index fibres impractical is the fact that their transmission capacity in bits per second is too low.

2. *Graded-index*

These were developed to propagate photons in unison. Not only do they have one interface between core and cladding, but their cores are also graded. The center of the core is slower, but speed increases toward the perimeter. Thus, particles travelling the longer path propagate faster than those on shorter paths.

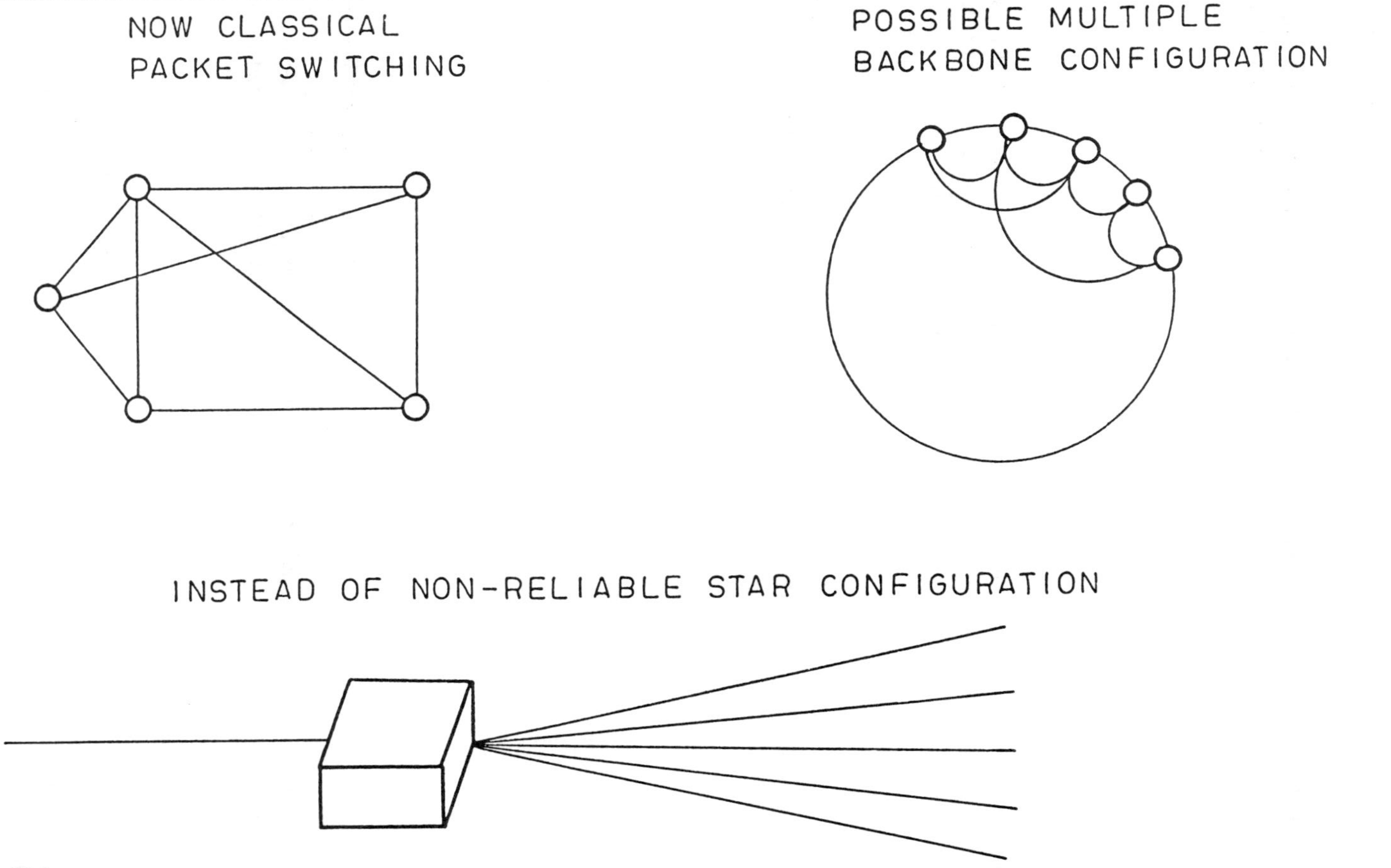

NOW CLASSICAL
PACKET SWITCHING

POSSIBLE MULTIPLE
BACKBONE CONFIGURATION

INSTEAD OF NON-RELIABLE STAR CONFIGURATION

Figure 12.4

3. *Single mode*

This is the best optical waveguide, as it represents the fibre with the highest transmission capacity. It is essentially a stepped-index fibre, but with a core so small that there is no room for the light signal to reflect backwards and forwards. It can only propagate in a straight line.

Another of the challenges that fibre optics technology faces is the need to exploit optical bandwidth with electronic components. As we said in the chapter on Optical Disks, most approaches today are hybrid: they involve both electronics and photonics. But we are able to obtain results.

Solutions are on the way. It is, for instance, projected that, in the future, gate delays will be reduced at picoseconds (ps) using gallium arsenide at lower temperatures. Today in laboratory work 1 ps is possible, thus making feasible speeds at the n x 10 gigabits per second level. At the same time, there is a shift in the wavelength on which we focus our attention for transmission purposes, as we saw in section 3.

Sometimes solutions which are chosen because they respond better to certain media, or in order to overcome a given limitation, bring to the foreground other weaknesses. An example is the risk regarding privacy and security with star networks as currently implemented through optical fibres technology.

In a star network, local or long haul, control of the hub can give the intruder access to all information passing through it. A distributed system improves upon this situation by sectionising information transfers. A communications aggregate, however, can only then be significantly better in a security sense if it uses efficient *encryption* algorithms.

Blocking is also of particular concern because hub saturation impairs the communication of WS attached to that hub - among themselves and with other resources. Here again, a star approach sees to it that there is no alternative pathing to relieve bottleneck situations.

In other words, efficient implementation of photonics rests as much on architectural solutions as on the development of more efficient light sources. Yet, a great deal of the scientific effort since the laser's invention has concentrated on raising the power of laser light sources - not on architectural perspectives.

Optical engineering has brought the laser invention to the stage where Bell Laboratories now claim a new record for the amount of information transmitted by optical fibres. We have already referred to the fact that this involves:

- 2 gigabits per second (GBPS) carried over 170 kilometres without a repeater;

- This being equivalent to about 30 pages of newspaper text per second.

Let us note that facsimile currently transmits at the rate of about one page every 4 to

6 seconds, depending on equipment type (Group IV). Facsimile transmission requires much more time with Group III, and reaches the 5 minutes level with the old analog approaches.

When we talk of laser solutions, a mid-layer should also be brought into perspective. This stands between the top architectural layer and the basic light source reference. This mid-level is important, as it conditions not only the structure of the light source, but also whether or not WDM technology can be used.

CURRENT SOLUTION

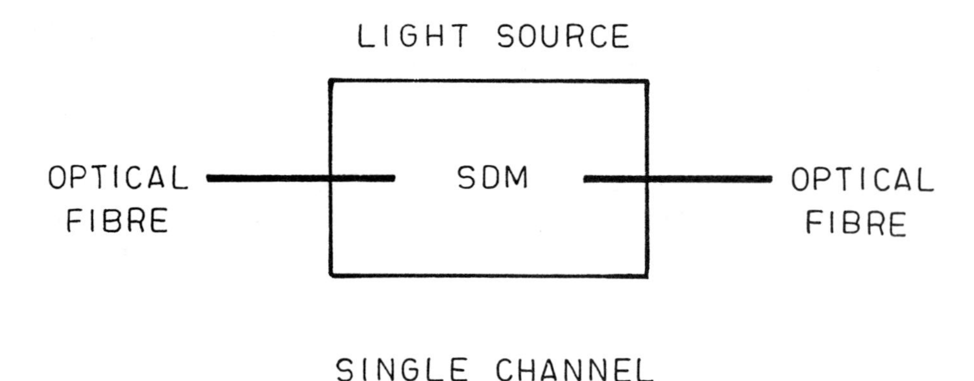

POSSIBLY BY 1995

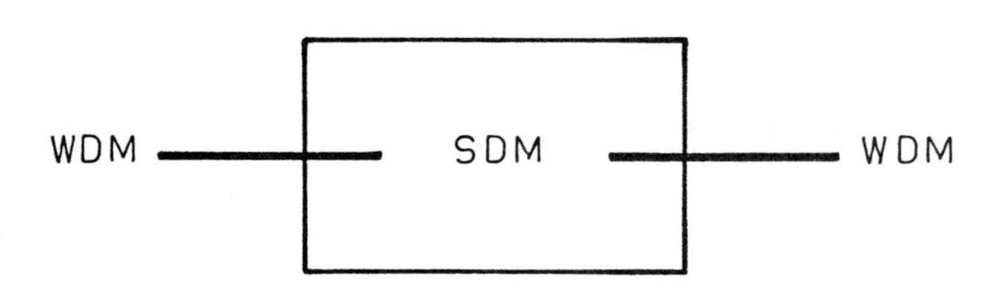

MULTIPLE CHANNEL

Figure 12.5

As Figure 12.5 shows, one channel connectivity per fibre is today treated through conversion to space division multiplexing (SDM). Up to 2 GBPS is obtained through such switching elements. By 1995 we may have a WDM-SDM-WDM solution.

One of the reasons why sophisticated WDM approaches are not just around the corner is that optical switching technology is still lagging somewhat behind. Another key reason is that the pristine light from the laser still has imperfections - for instance, the tendency to fluctuate in amplitude, which shows itself as noise.

Photonics Transmission at Bell Laboratories

The evolving photonics transmission system at Bell Laboratories for the first time used photonic amplifiers to boost lightwave signals and sent them over a record-breaking 223 miles (about 370 km) without converting them first to electrons and then back to light. AT&T also quadrupled the speed of its most advanced commercial lightwave transmision system to carry the equivalent of more than 24,000 simultaneous phone calls through one pair of fibres. This system has been placed into service between Chicago and Philadelphia.

A visit to the AT&T Laboratory for new laser technology provided evidence that Bell Laboratories are well ahead in their work to produce lasers for *7 gigabit per second* (GBPS) systems with *1.3 to 1.5 micron* wave- lengths. At the implementation level, so far 2 GBPS systems have been carried over the stated distance without a repeater, and the 7 GBPS application is coming.

In section 5 we have seen examples which exemplified what is meant by the capacity of a 2 GBPS channel. A 7 GBPS channel capacity is equivalent to over 100 pages of newspaper text per second. The difference between the coming technology and the current one is 1:350 - that is, between two and three orders of magnitude.

An interesting remark was made regarding the limits of optical fibre speed. At one time it was the laser which represented the constraint. Now the limiting factor is electronics. But, electronics switching will probably change into photonics switching before the end of this century.

If optical switching replaces electronics then there will be a big jump. Therefore, AT&T is actively working on optical switching, as we have seen. Until a breakthrough comes, gallium arsenide starts being used electronics can lead to jumps in speed of transmission (albeit smaller ones).

It is not the cable, but the repeater, which constitutes the bottleneck in optical fibre transmission. "Whenever you search for a problem, look at the repeater", advised the AT&T specialists. This is very true in electronics, in photonics and also in human communications.

As we also discussed in connection with optical disks, the currently available regenerators are hybrids of electronics and optical components. Fully optical solutions are some years away. Not only do the problems in physics have to be solved, but also dependable solutions found. With light emitters, reliability is very important.

Still another challenge is in system design: ''Put two devices acting differently side by side and you have a psychotic subsystem'', it was suggested during the discussions at Bell Laboratories. Yet, this is what happens worldwide, because designers fail to observe the need for simplicity:

- Simplicity of the process, *and*

- Complexity of each step

is a design guideline at Bell Laboratories. The visit provided evidence that this principle is observed. It will be wise also to use it in the development of computer-based applications.

We should always keep this architectural advice in mind when we talk about the structure and functionality of our communications and computers network. We have to remember what can be done with photonics, when we are planning for their implementation. These high capacity, gigastream pumps will make not only feasible, but also advisable, the use of optical disk storage devices. When this is done, both opportunities and limitations should be kept under perspective.

Optical Fibres as High Sensitivity Sensors

Throughout this chapter emphasis has been put on the role of optical fibres in communications networks, endowing them with gigastreams. But, while very important, wideband solutions are not the only ones to be served through photonics.

In contrast to *macrocommunications, microcommunications* focus on sensory duties. A recent article in Business Week,* emphasised a forthcoming revolution in *smart composites* wrapping, for instance, planes in a blanket of sensitivity. The point was made that a skin-deep grid of optical fibre sensors may (among other mass-made systems) be installed in future aircraft, making them ultrasensitive to the touch of enemy radar, or the weight of ice before takeoff.

According to this projection, signals from these sensors would be analysed by patches of special chips sprinkled over the *smart skin* of the airframe. In fact, the military brass has taken pains to convince the aerospace industry that smart-skin technology will be crucial to future aircraft, so the USA aerospace industry now invests millions of dollars of its own funds in this.

According to available information, improved safety is just one promise offered by smart skins. The new technology could make it possible to engineer a wide range of

---

* December 5, 1988

solutions in situations which have so far defied man-made approaches, such as the construction of buildings that counteract the destructive force of earthquakes.

In specific implementation fields now being anticipated, the optical fibres would function as sensors for collecting all types of data, including pressure, strain, and temperature. Optical fibre sensors are so delicate that almost anything that happens in their surroundings will cause some sort of distortion in the glass, which in turn affects the pulses of laser light passing through the fibre. Even the slightest variation can be detected by analysing the light coming out.

Military aircraft and other weapons systems are not the only ones to profit from the new technology. Civilian air travel can be a major beneficiary. For instance, during the last 10 years, windshear, microbursts and unexpected turbulence have been identified as the cause of 28 aircraft accidents, resulting in 206 injuries and the loss of 491 lives. Sensitivity sensors can provide a positive contribution to this problem, enriched with AI constructs.

Research groups are today investigating ways to *detect* and *predict* windshear in its various forms. Most are working to improve electronic detectors such as Doppler radar, but smart skin may be a better answer. The study of windshear is a natural outgrowth of development processes aimed to significantly improve air transport safety.

Expert systems and smart sensors can work in synergy, considering specific information regarding the flight, including:

- The type of plane;

- Its performance characteristics;

- Payload including fuel;

- The route; *and*

- Weather conditions on that route.

Flight and aeroplane information has to be combined with data on airports of departure and arrival, including weather history of the airport, current hourly weather from the weather service, topography of the airport, and dynamic information provided by the sensors of the aeroplane's smart skin.

While human ingenuity will still be at a premium, controllers can rely on the computer to guard against errors that might arise because of overwork or lack of attention. The AI-enriched computer can make a perfect backup and allow the dispatchers to improve efficiency -which cannot be done without dynamic information. How can a dispatcher decide:

- Whether a ''flight to be'' is under control?

- When to release an aeroplane at once or wait until the weather improves?

- When will weather conditions put the aeroplane at risk, given its, and the flight's, specifics?

- Or even predict a microburst or windshear problem?

A crucial amount of pertinent information is necessary, both on the aeroplane and on the process (flight, route, environment). Once all data is gathered, it must be overlaid on computer-based, AI-enriched cartographic mapping to check for terrain-induced waves.

In one application already in process (at the computer end), the artificial intelligence module sets up a 20 mile sphere of influence around forecasted areas of windshear. As flight plans are prepared, the route of a given flight is checked against the *no-fly area.**

If the route conflicts with the weather, the airfield delay simulation model is used to adjust flight time for a more appropriate arrival or departure. The index table is accessed to determine whether there is a low, moderate or high chance of shear.

If moderate or high is forecast, extrapolation is used to establish the amount of delay necessary for a flight to avoid the area. Additionally:

- The flight planning module recalculates flight plans using different routings, *and*

- Passes the information to the fuel management module for the determination of additional cost to operate other selected routes.

Such information is analysed and given back to the artificial intelligence module. Subsequently, based on company-set parameters, a new route or a delay is recommended. Smart skin implementation can go a long way in improving these AI-enriched air safety procedures which are already in process, by assuring realtime data entry.

Airframe producers, too, will profit handsomely from such implementation. McDonnell Douglas, for instance, believes that the new technology's initial payoffs will come in better manufacturing quality and productivity. By processing ultraviolet light through the fibres, the aerospace company expects to determine when the materials have reached their optimum hardness. Subsequently, the fibre network can be used to run nondestructive tests on panels, before and after they are installed on an airframe.

---

* Patricia K. Mortenson, *Airline Executive*, April 1988.

To the US Air Force, the concept of smart skins promises far-reaching possibilities. If successful in its implementation, this could resolve a problem now plaguing aircraft designers. In today's high technology air battles, electronics trigger alarms when a plane has been locked in by enemy radar, or a heat-seeking missile, and activate automatic countermeasures to confuse that radar or deflect the missile.

The trouble, however, is that there is not enough room inside a fighter for all these avionic systems. Therefore, some equipment has to be crammed into pods sticking out from the airframe, adversely affecting the plane's aerial performance.

The new technology of optical fibre sensors can be of help by permitting the chips today located in black-box equipment to be positioned around the fuselage and wings. They will be enriched with sensors, making the whole aircraft aware of everything going on around it. An entire weapons system could be turned into a radar antenna through the monolithic microwave integrated circuit (MMIC). These are the first chips that can handle the frequency and power requirements of high-tech radar equipment.

A still more advanced system employing artificial intelligence constructs might do in-flight damage assessment. The sensors would check the damage that has occurred and advise whether there is enough operability left to try making it back to base. The same could apply to evaluating the damage in orbiting spacecraft, or any other difficult-to-access system, such as deep ocean vehicles. In fact, next to aeroplanes, submarines may be the largest class of beneficiaries.

Smart skins may also enhance the near-invisibility of so-called stealth planes to radar. No wonder the Air Force has charged Hughes Aircraft and Rockwell International with assessing the technological problems and benefits of implementing the smart-skin approach. Composites suppliers, including Du Pont and Hercules, are assisting in the research, along with scientists at Virgina Polytechnic Institute.

A Competent Use of Satellites*

During the last 15 years, the transmission of television and other signals by satellite has become a technological commonplace. Communications satellites now circle the globe. They have made possible improved navigation and flight control, business teleconferencing, increased telephone service and worldwide high-speed data transmission.

The quality of satellite broadcasting significantly helped the rapid expansion of cable television, the proliferation of new programming, and a new industry enabling people to receive satellite signals in their homes. Consumers owning satellite antennae of varying shapes and sizes, use them to receive as many as a hundred different

* See also: D.N. Chorafas "Telephony. Today and Tomorrow.", Prentice-Hall, Englewood Cliffs, NJ, 1984

television channels. An estimated one million Americans take advantage of this facility.

Ten short years ago, no regular American television programming was transmitted by satellite. Today almost every viewer, whether or not he owns a satellite antenna, watches shows that have spent at least part of their transmission time passing through outer space. The significance of this change is not obvious, but it is vitally important. The satellite-television wave now getting under way has a potential impact on our knowledge society.

A great deal of credit for satellite-based communications facilities goes to Arthur C. Clarke. While most science-fiction writers spend their time predicting the discovery of planets ruled by human-like beasts, Clarke has largely confined himself to predicting things that actually come to pass. Writing in the British Wireless World, in 1945, he laid out the blueprint for the modern system of transmitting broadband signals by satellite.

The inability of television signals to pass through the ground, or bend around the earth, is well known. TV stations establish their transmitting atennae on mountains, towers or skyscrapers to increase their broadcast area. Even with such solutions, their range seldom exceeds fifty miles.

One way to overcome this limitation, Clarke wrote, would be to place TV antennae in space, on satellites circling the earth miles above the ground. Clarke pointed out that at an altitude of 22,370 miles, a satellite's orbital period would be twenty-four hours. A satellite orbiting at that height directly above the equator would seem to be motionless in the sky. Its signal could be received with a fixed antenna.

Satellites in this 22,370-mile orbit, now called the Clarke Belt, are geostationary. Clarke suggested launching three of them such that they would be evenly spaced above the equator. By using them to relay signals between stations on the ground, it is possible to send information almost instantaneously from any-to-any station.

Due to an early phase of a new technology, the first communications satellites ever launched were anything but geostationary. Built by Bell Telephone Laboratories, and launched by the National Aeronautics and Space Administration (NASA),

- *Echo* went into a low orbit in 1960. It was a big metallic balloon that served as a mirror off which radio signals could be bounced.

- *Telstar* was launched in 1962. It had an active transponder (transmitter/responder), but followed a low orbit.

Telstar is, however, a milestone because it carried the world's first intercontinental television broadcast. This broadcast was only a test, but it was intercepted in Europe.

- *Syncom II* was the first functioning geo-synchronous satellite.

By 1964, American television viewers watched part of the Tokyo Olympics courtesy of Syncom, but television was not a priority of the early communications satellites. Neither were communications satellites a priority in the early space programmes.

- *Westar I* was the first truly communications-oriented satellite. It was launched in 1974.

Despite all this, we were still some time away from consumer access to TV channels. The missing link was low cost satellite atennae. The technology was not yet ready. Another reason why satellite atennae were expensive in the mid 1970s is that the Federal Communications Commission (FCC) required them to be large, nearly thirty feet in diameter. This was intended to protect cable subscribers from receiving substandard pictures.

Earth station technology improved rapidly, and the FCC relaxed its rules. By using better receivers and amplifiers, cable operators were able to produce good pictures with smaller dishes. Costs fell. Now a cable operator using a small dish can set up a commercial-quality earthstation for less than $3,000.

Television is far from being the only profitable use of satellites. Weather prediction and navigation are two other important examples. So is agriculture. The point, however, with TV is that, with 600 million users around the globe, it has a basis for conquering a large consumer market.

Once a product achieves mass-market dimensions, costs drop dramatically and it becomes available at low cost to other fields. For instance, to the distributed landscape implied by agriculture. Low cost of the basic carrier also invites value-added solutions. Image-taking by satellite is an example. This is the second phase of commercial satellite exploitation, following communications usage.

The Americans were the first to launch a satellite of this type, the *Landsat* in 1972. The fifth, and most recent, launching was in 1984. The Soviets, Chinese and Indians have also launched satellites, but their pictures have never been shown.

Landsat created a new market as its images were distributed to all those interested, at very affordable prices. Gradually developers, town planners, cartographers, geologists, agronomists, oceanologists and ecologists have discovered the advantages of satellite observation of the earth. There is not a single field in natural resource management that is not concerned.

Image-taking satellites observe and reveal agricultural land usage; the spreading of forest diseases; parasite damage; ocean resources; river flooding; volcano activity; and desert expansion. They pinpoint air and water pollution; subsoil formation; potential mineral, oil and gas deposits.

Meteorological satellites, which perform remote-sensing because they observe the Earth from such a distance, can be placed in geostationary orbit. They are especially suitable for weather prediction through the observation of the movement of large cloud masses. Military observation satellites, which are regularly launched in the USA and Russia, make out details to a matter of feet, operating from very low orbits, of 62 miles or less. But, they remain in orbit for only a few weeks or months, and their photographs are extremely expensive.

As a general principle, the more sophisticated the tool the greater the need for its management. The orbit of a remote-sensing satellite has to be carefully chosen. If it is too far from Earth, the image focus will be insufficient, and details would not be crisp. If it is too close, the satellite would gradually be slowed down by the upper layers of the atmosphere, before falling back to Earth.

Let us also recall that satellites are communications engines. They are not the processors of information. While they are vital parts of a knowledge society's system, they must feed their information into computers. The images which they transmit must be interpreted, control signals derived from them, actions suggested and taken. All developments are important. Even more critical is that they are studied, not in a standalone manner, but as parts of an *integrated system.*

13

Distributed Communications and Networking

Introduction

Information is sent from one location and received at another through the process of communications. As we saw in Chapter 12, it may be transmitted over radio waves, optical fibres, coaxial cables, or other types of links. It will travel from origin to destination via switching centres, and these are increasingly becoming computer-based.

At the bottom line, communications between computers means the transfer of bit or byte patterns. Several methods are used for such purpose: asynchronous, synchronous, and bit-oriented being the most common. A serial data transmission line uses a single wire on which bits will be sent and received, provided that there is a physical interface, usually expressed in the recommended standard RS 232. Such use of serial lines can be thought of having two states: high or low; "1" or "O".

A serial data bit stream is a line condition that lasts for a specific value of time depending on channel capacity. This is valid whether the line works asynchronously, synchronously, or in packet mode. The transmission rate of standard RS 232 serial data communication corresponds to the number of bits per second that can be transmitted and received: 1,200 baud means that there are 1.2 thousand bit times per second (1.2 KBPS).

The actual amount of time that a single bit value takes is inversely proportional to the baud rate. At 1,200 baud, the bit time is 1/1.200 or 0.83 ms. For proper synchronisation within each unit of information, receiver and transmitter station must be set to the same rate.

An *asynchronous* communication is characterised by character-by-character transmission. This is preceded by a start bit and followed by a stop bit. These serve both as delimiters and for synchronisation purposes. Equally spaced clock signals control transmission of information in *synchronous* communications. The clock is used to keep the transmitted characters in step and running at the proper rate. There are no start and stop bits as in synchronous solutions, but there are flags, header and trailer sections.

Transmission between sender and receiver must obey a standard convention, or *protocol.* The protocol defines methods of data transfer and helps to ensure that communications between dissimilar equipment will work. In a layered protocol structure, such as the Open Systems Interconnection (OSI) of the International Standards Organisation (ISO), each layer has its associated protocol. ISO/OSI feature seven layers, top to bottom (Figure 13.1):

1. Applications;

2. Presentation Control;

3. Session Control;

4. Transport;

5. Networking;

6. Data Link; *and*

7. Physical.

There is only one physical level and six logical levels. In long haul communications, the physical level typically includes the modem. In a LAN it incorporates the bus and the bus interface unit. Synchronous and asynchronous protocols, of which we briefly spoke in preceding paragraphs, belong to the data link level -the lowest of the logical ones. Higher-up layers have more sophisticated protocols.

Modern networks support all seven levels of ISO/OSI in an:

- End-to-end

- Any-to-any

communication sense. That networks are important to modern business, whether manufacturing, merchandising or financial, has become a truism. Less appreciated is the depth of study necessary to establish and operate a cost/effective network.

The first essential step in a network study is concerned with understanding the current position, covering basic facts and figures, establishing the quality of communications links already in operation, the type of service being delivered to users (both internally and externally to the organisation), and evaluating where communications stands in *our* company relative to competitor corporations.

GOAL:

- INTERACTIVE SERVICES
- FILE TRANSFER
- BULK FILE EXCHANGE
- PRINT FILE EXCHANGE

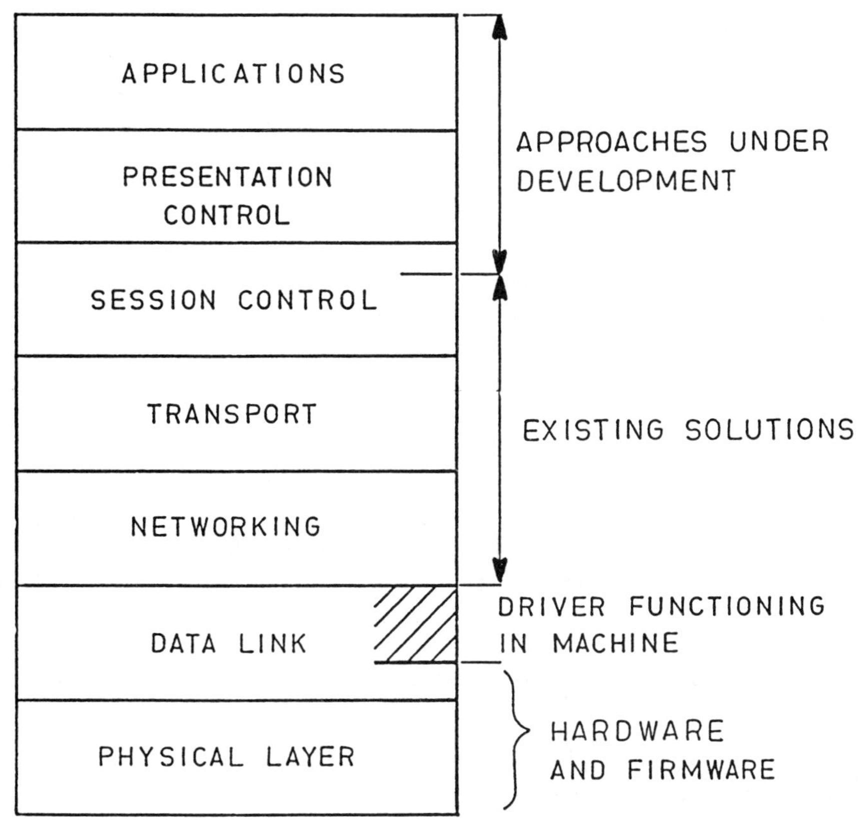

Figure 13.1

When we have obtained relevant information which is factual and documented, we can effectively talk of a second stage concentrating on the planning of the future network. In this case, our study should cover future needs for communications services, in the context of strategic plans; evaluate uncertainties and constraints impacting on communications developments; identify the state-of-the-art in the technological environment, including foreseable breakthroughs; and elaborate on the establishment of targets and associated implementation paths towards our goal.

Our goals should be properly set, defined in detail (not merely described) and include specifications beyond channel capacity. For instance:

- Any-to-any Configuration;
- Network Reliability;
- Service Quality;
- Security/Protection; *and*
- Cost/Effectiveness.

The role AI should play can best be exemplified if we keep in mind that both central and distributed signalling systems (as well as the network service logic as a whole) have to cope with port/socket-numbering plans and with addressing. Also with security algorithms.

The new communications system which we design must be able to support dependable terminal identifications, as well as qualify and quantify the oncoming multimedia message and transaction traffic. It should be flexible, to permit restructuring and reconfiguration of the increasingly intelligent resources to be attached to the network. And it should be cost/effective.

Global Networks

Since most of the voice, image, text and data messages eventually go over telephone company lines, networks are a big boost to communications carriers - both long distance and local. Hundreds of companies are trying to exploit this business opportunity, and not just the operators of switches and transmission systems - or their manufacturers. On the equipment side, a battle is raging over who will dominate the business of providing sophisticated computer systems that monitor and manage these networks.

The most intriguing competition, both for service and equipment, is shaping up between telephone companies (telcos) and computer makers who are starting to compete more directly than ever for the global telecommunications service. The strategy is to move the big companies:

- From older voice-grade telephone lines which are expensive and unreliable.
- To modern high-capacity digital networks which are cost/effective and quite reliable.

Goals run a range from increasing the traffic to selling equipment that manages value-added networks, such as error-free, packet-switching oriented information processing and transmission services.

At the same time, a substantial effort is being extended to make such networks increasingly intelligent. At the current state-of-the-art, network intelligence can be enhanced by incorporating:

1. *Expert systems at the nodes* and link interfaces able to provide supports ranging from diagnostics to better quality services;

2. The study and design of efficient network solutions through *AI-enriched CAD workstations;*

3. The organisation, maintenance and steady update of an *engineering database*;

4. *Rules* governing the *distribution* of, and the *internetworking* between, service logic modules supported at the nodes;

5. Other expert systems embedded at the nodes to help in *traffic management*, and in statistics concerning future *dimensioning studies;*

6. *Network Control Centre* (NCC) functionality, not only able to establish and maintain conditions for network operation, but also to support *quality histories*;

7. *Any-to-any*, user-to-user transport, enriched through artificial intelligence (AI) constructs in order to serve an increasing range of *office functions* in an able manner;

8. Rules and procedures for interlinking functional *switching and signalling* components to be dynamically managed through automated systems, themselves depending on AI; and

9. The *interfacing* of a number of physical and logical *subnetworks* through AI, so that the enduser sees only *one logical network* in his work.

As Ken Olsen, the Chairman of DEC, aptly remarks: ''We have to start thinking of the computers as peripherals. You start with the network, then you hang the computers on later''. Mainframes are just one more element of the system. The same is true of departmental and personal computers. Therefore, in the next decade, whoever controls the network will also control its user organisation.

Valid solutions account for the fact that converging *multimedia* technologies will dominate the picture in the 1990s. Taking 1995 as the 100-percent target, Figure 13.2 shows the relative weight of ''data only'' and ''voice only'' circuits today. Then, it extrapolates on the fast growing importance of multimedia in an integrated communications environment.

When it comes to networks, AT&T, British Telecom, Kokusai Denshin Denwa (KDD), IBM, DEC, Philips and Siemens - among others - have drawn up battle plans.

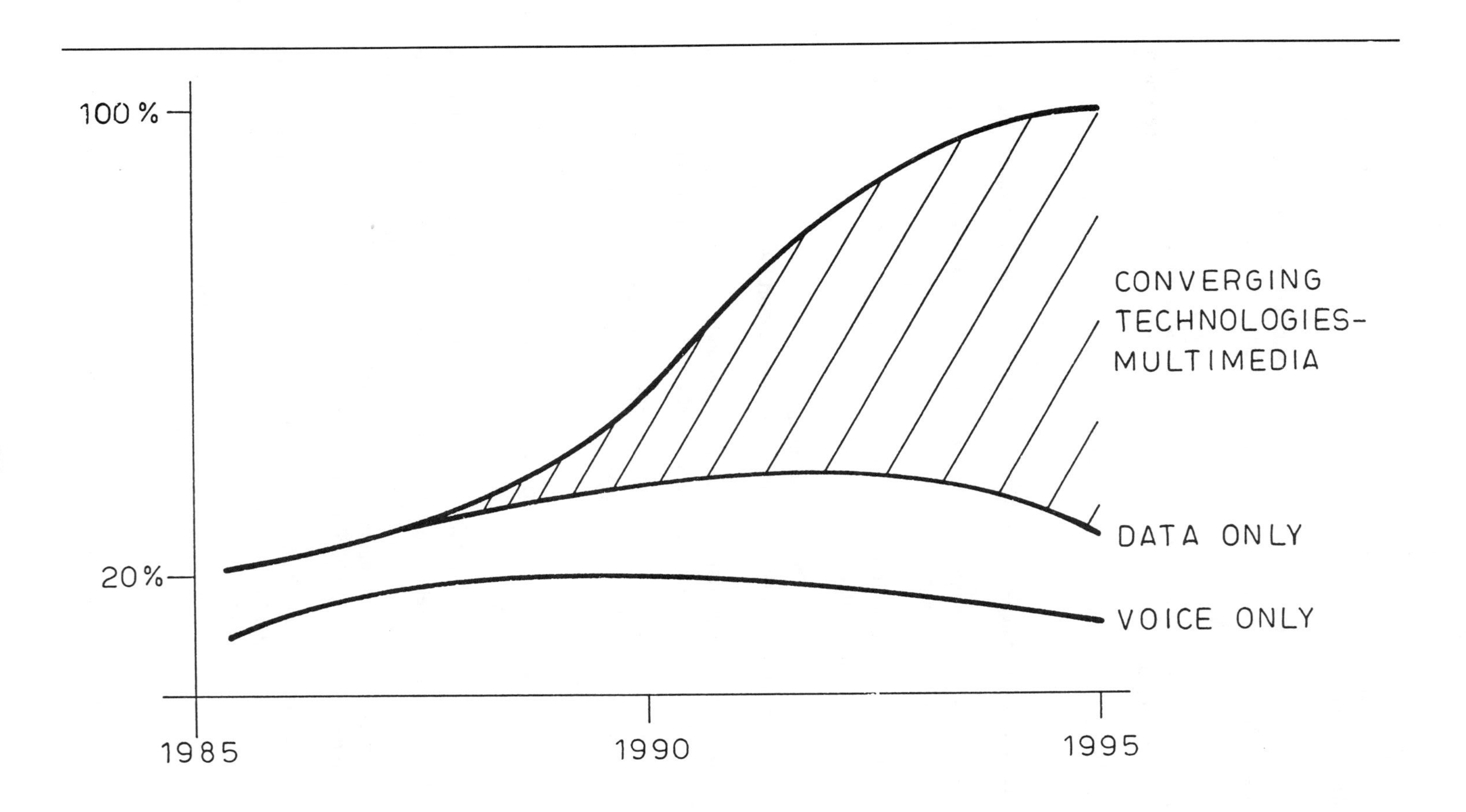
100 %
20 %
1985
1990
1995
CONVERGING
TECHNOLOGIES-
MULTIMEDIA
DATA ONLY
VOICE ONLY

Figure 13.2

In the 1960s IBM began building a global network that now extends to 145 countries, relying on satellites and private transmission lines leased from phone companies to interconnect an estimated *180,000 terminals worldwide*, and service in terms of messages and transactions some 400,000 IBM employees.

In just one year, in 1987, this network moved 3 trillion characters of internal information, replacing through electronic mail the voice telephone as the primary way in which IBM people communicate. And, if IBM also sells space on its network, so does Volvo and Toyota, Citibank, Chemical Bank and the Mellon Bank.

By effectively using its own network, Digital Equipment Corp. has bolstered its competitiveness in many ways, slashing its manufacturing staff and lifting productivity and profit margins. The size of the manufacturing staff, for example, has fallen by more than 5,000 to about 27,000, through reassignments and attrition, but also through greater productivity where communications play a key role.

DEC's private network has helped it to manage inventories better, cutting carrying costs in half during the last 5 years. General Motors' network links some 550,000 computer devices and telephones connecting 18,000 GM locations worldwide. It helped the company to save $175 million from 1985 to 1988. According to an estimate, GM spent 20 percent of the $500 million it took to build its network on the management systems that run it.

But in the general case getting people and companies adjusted to the idea of building polyvalent, multimedia, any-to-any, end-to-end networks is no trivial task. Even the concept of a communications system to which *everyone has access* is strange by old standards. This is one of the images which should change if we are to communicate effectively in an organisation-wide sense - and benefit from the resulting productivity.

For a computer manufacturer a value added network offers significant competitive advantages. It can be used to monitor and diagnose the problems customers are having with their computers. For banks, too, the network is a competitive weapon. It makes it feasible to reach customers online - increasing the sophistication of the service being offered, while at the same time reducing its manual component, hence the costs.

From moving payroll, inventory or sales records to providing instantaneous responses to client queries about investments - including sophisticated analysis leading to identification of undervalued and overvalued stocks - the emphasis is on online, interactive handling. Pricing is, of course, very important and we may be just prior to a global price war in the deregulated environments - while protected, heavily regulated markets will stagnate under high tariffs.

Cost/effectiveness is most important, not only in the day-to-day business, but also in the longer trend. From 1985 to 1987 international voice-and-data traffic grew by 38 percent in terms of minutes. By contrast, revenues grew by only 14 percent, indicating that price discounting has started. Like the computer market itself, and so many other market sectors, the international communications market is customer-driven. It will be even more so in the future.

In terms of using telecommunications facilities, the pace may well be accelerating, and there is historical evidence to support this statement. Private networks did not really gain momentum until the 1970s. The early motivation was cost. As big companies expanded around the world, they clamoured for better transmission lines, cheaper service, and faster installation than the telcos provided. In the USA, for instance, AT&T was slow to respond, and in the process it gave MCI Communications a good share in the USA long-distance market.

To save money through better management, companies spend millions to rebuild and upgrade their fragmented communications systems into a *cohesive global network.* An integrated communications system serves them well as they move heavily into complex markets and products with more value added. By and large, these are polyvalent networks of satellite and land-based communications links designed with the purpose of helping them keep on top of market movements throughout their operations domains. Success stories illustrate the range of possible gains that a global network can offer.

Compatibility in Network Design

Traditionally, network service logic has been directly mapped onto each physical switch. Each switch has been a node in the signalling network for the interchange of all control information. However, this approach is no longer sufficient unless enriched with network level intelligence.

Since any network will be composed of a number of subsystems which integrate into it, emphasis must be placed on compatibility. We can talk of different levels of *compatibility* observance: signalling and addressing, communications protocols, interfaces, the operating systems of attached devices and the languages used in programming.

Only recently have people and companies been beginning to care about having one reference system and networking solutions that usually work without problems. From the standards which we need to obtain better performance, we can distinguish:

- Programmer interfaces;
- Process-to-process exchanges;
- Device interfaces; *and*
- Generalised file interchange.

A programmer interface refers to the conceptual model and the syntax that the programmer uses when incorporating networking functions into application programs.

The calling sequence of procedure and library routines has to be standardised to permit interprocess communications among dissimilar processes.

A device interface focuses on the protocol used for communication between the device-independent and the device-dependent functions. It defines a device driver protocol consistent for all processes running on the systems and for information elements, permitting us to avoid the different acrobatics done today through conversion procedures between code sets, incompatible communications protocols, line disciplines, file structures and presentation modes.

Generalised, network-wide file transfer is necessary because of the plurality of types of devices that we use. These products are largely incompatible among themselves in design features, including file structure. Compatibility must be assured through interfaces, leading to an integrated online system.

Furthermore, a modern, globally integrated intelligent network should see to it that the signalling system carries the necessary call control information end-to-end in a systems sense. This means:

- Among different service logic clusters, *and*

- Between the service logic and the nodes/switches of the network's infrastructure.

A central resource may play a guidance role, but the communications network itself should be distributed. In the 1990s, the central resource will act as a big communications switch and network control centre - while intelligence residing at the nodes will support a distributed communications environment.

We are interested in a distributed control structure because it permits every node to be upgraded as requirements develop, without the stiffness and low reliability of a hierarchical or monolithic system. Resident in the unified logical network, AI components should see to it that new features can be introduced in the network, interacting with the nodes over AI-based interfaces. Such a network would:

- Make the service provider independent of the supplier of an individual network node;

- Offer lower total costs than classical hardware interfacing, hence greater cost-effectiveness;

- Speed up the introduction of new features, thus upkeeping the network's competitiveness; *and*

- Permit the network to grow rapidly with the new services as these are made directly available to the whole user base.

In the broadest possible sense, any switched network could be seen as a dynamic resource-management system, in which each node promptly reacts to user requests.

Because of resource sharing, signalling and switching are the pillars on which an intelligent network architecture should rest.

The AI constructs at the switching centres must ensure that two basic architectural characteristics are observed:

1. An *open vendor policy* permitting multi-sourcing, provided that the protocols chosen by the user organisation are observed by the different suppliers; and

2. The support of *global signalling* standards and of global name-space (ports, sockets) in a network-wide sense.

To produce a truly open network, new types of highest-level protocol standards have to be developed. This is what I call the 8th layer above ISO/OSI.* They should address all attached resources and functional components within the network.

CAN, the layer which Bank of America designed above the seven of ISO/OSI to help unify into one logical network its 67 diverse physical networks, has been projected on the following premise:

C: One *centre* concept for all systems people;

A: One *architecture* (8th layer), specifically taking SNA and making it open with a layer added above ISO/OSI; and

N: One logical utility *network* for all needs.

This logical superstructure over the 7 ISO/OSI layers permitted the Bank of America to integrate 67 physical networks into one logical network, thus making feasible a coherent, homogeneous and comprehensive system-wide view. Similar goals are sought after by the Manufacturing Automation Protocol (MAP) of General Motors, as well as the Technical and Operations Protocol (TOP) of Boeing, GOSIP of the US Federal Government, as well as EDI, as discussed in Chapter 10.

The principle always to keep in mind is that the need for integration is present in all firms. New networks add themselves to the old - which continues operations. But, only *one integrated logical network* should exist. It should be global, and supported through artificial intelligence.

Leading-edge organisations are now working along that principle. The new computer centre of Bank of America in Concord, CA, is itself an intelligent building, one of the new generation being developed across America, in Japan and in Europe. It is connected to the San Francisco headquarters of the bank through optical fibres, with an inhouse installation supporting protocols which cover all seven ISO/OSI levels.

---

* See also D.N. Chorafas: "System Architecture, and System Design," McGraw-Hill, New York, 1989.

Within the perspectives of an inhouse system, routing plans must be properly defined; the network service logic should be correctly programmed; both the physical and the logical network structure should be managed by means of advanced software routines - hence AI. Such developments would become a generic instrument to control not only the user-to-user channels, but also the backbone network for an extended pallet of services.

A carefully thought-out, AI-enriched architecture would represent a revolution in network development. Whether wide area, metropolitan, or local, current networks can be described as static and inflexible. We do not need to repeat the failures of the past. Flexibility will be at a premium in the 1990s.

With modern solutions, network service logic is already resident in software, and represents considerable intelligence. The next required step is to integrate expert systems with the aim of obtaining an agile, dependable and flexible structure. The provision of intelligent network services could speed up the process of change - and the network's competitiveness in the marketplace.

But, designing the proper network and using it as the integrator of computers and communications resources is a demanding task. It means first and foremost having an *architectural concept.* It also requires becoming familiar with hundreds of suppliers of network-switching, transmission and management gear, as well as with advanced computer technology. Keeping up with steady exponential improvements in products and services is a ''must''.

Even the best network installed today with state-of-the-art fibre-optic transmission lines might soon be out of date, since the means used to move voice, text and data over lines are constantly being improved. Layer technology is in full evolution. Computer information that only a couple of years ago travelled at 10 mega-bits per second (MBPS) soon will streak over fibre highways at 2 gigabits per second (GBPS) or more.

Also, telephone switches, data-switching devices, modems and multiplexers are getting smarter and smarter. With almost anything being installed today there is a faster, and usually cheaper, device that comes along. And the cost of just keeping up with this can be enormous.

Furthermore, one of the major headaches in setting up a global network is dealing with foreign telephone authorities. Prices vary, a leased phone circuit in the still-regulated European market costs, on average four times what it does in America. Within Europe there are tremendous tariff differences, such as that between Germany and the UK - the much lower tariff being to the latter's advantage and largely due to deregulation.

Inhouse and Metropolitan Area Solutions

Even the best lasers still have some noise, which sets a temporary limit to the capacity of any system using light to carry information. What the Bell Laboratories has recently done is to take this noise out of that part of the light which is actually conveying the information. The approach is known as *squeezed light.*

The possibility of handling light in this way has been known to science since 1965, when optical communications was merely a dream. By 1985 it was demonstrated how it may be possible to double the capacity of any optical communications system by using squeezed light which is the purer form of laser.

The idea is to squeeze the noise into the part of the lightwave where it causes least interference, thus giving a better signal-to-noise ratio. This helps to increase the stability of an optical switch operation at very low power levels.

Better stability means that such a switch will be more reliable as part of any future optical computing system. This is particularly important in wide area communications systems, and also for *optical switching.*

As we saw in Chapter 5, optical switches are key features in the concept of an optical computer more powerful by 3 orders of magnitude than the best electronic machines used today. Squeezed light also enhances the performance of a laser-based instrument: the optical gyroscope, a ring laser system currently used, for example, in jumbo-jets.

With ordinary laser light, this instrument is working at the limit of its sensitivity. This is acceptable for a commercial aircraft, but a more sensitive optical gyro would be needed to fulfill scientific and military needs, as well as to answer the requirements of the next generation of commercial aeroplanes.

If squeezing can be harnessed efficiently, optical engineers may be able to pack signals more tightly into an optical fibre. But, in order to do so, they will have to miniaturise the optical cavity. The next major scientific task is to discover where the fundamental limits to light devices lie. When we work in high technology we often hit fundamental limits sooner than originally expected.

However, contrary to what has been stated with regard to wide area transmission, some parts of inhouse networks can profitably employ lower technology. They thus gain the advantage of tested solutions with relatively low cost, while emphasizing:

- Connectivity to existing resources; *and*

- Cost/Effectiveness in implementation.

In an inhouse system, connectivity should go beyond the private branch exchange (PBX) and towards two other directions: central computer resources and metropolitan/ wide area networks.

With reference to metropolitan area networking let us recall that, quite recently, 16 of the 26 European countries grouped in the European association of postal and telecommunications agencies (CEPT) signed a memorandum committing them to installing digital cellular telecommunications networks by 1991.

European postal ministries would not spend the billions involved in setting up such a system if only a handful of energetic businessmen were involved. They are confident that the new mobile telephones will become a mass product - which is currently about to happen in America.

The mobile telephone network is an interesting case in point. Mechanisms have been established for automatic update, initiated from the mobile terminal of location registers. This permits a switch node to:

- Interrogate the register when a mobile called-party number has been received, *and*

- Get back relevant data for the direction of the call to the switch node from where he can be called.

Something similar should be applied with inhouse micro-cellular infrared systems. The location register could, for instance, feature the same type of number translation capability as is needed for the 800 service in America, the difference being that the called number should be translated to a network node (hub) number only - not necessarily to a new network port number.

The system architecture to be adopted should give the freedom to reconfigure and restructure, with economics determining the implementation in each case. Both normal load and peak load should be considered, as well as backup if load grows beyond capacity limits.

The market potential and benefits may be rewarding. With reference to the metropolitan cellular network, financial analysts expect the market for equipment such as radio base stations and exchanges to grow significantly, past the ECU 1.2 billion level in the mid-1990s. European postal agencies estimate that 15 million people will have a mobile phone unit on the continent by the year 2000 - which is an underestimate.

This is fairly well documented by what is happening in America today. Far from being the slow starter that some anticipated, the cellular business is one of the fastest-growing and potentially most lucrative new industries in the United States. It is projected that in 1989 USA sales of cellular phones should reach 1.6 million units, nearly four times the of total just two years ago.

Most significantly, there are presently about 3.5 million subscribers, up from 125,000 in 1984. These USA subscribers will spend $3.2 billion calling around the landscape - a figure expected to double by 1990 as cellular phone prices plunge.

Some financial analysts even think that cellular companies could one day challenge traditional telephone companies for regular calling volume. This is a concept which brings both greed and fear to the telcos, as they get increasingly unsure if their monopolies will hold for long.

Come what may, European governments have already started to divide offers for the basic equipment needed to run base stations and exchanges. According to official estimates the projected cellular network will have 5,000 of these base stations, mobile switching centres and 750,000 traffic channels before the end of the century. This is the metropolitan market perspective.

An inhouse market can stand at a multiple of what was just stated *if* cost/effectiveness is fully observed both from the user organisation's and the suppliers' viewpoint. For both cost and effectiveness reasons, four design rules pre-dominate all others:

1. The system should be any-to-any, as is so often underlined when the occasion arises;

2. Independence from physical media (including fibres) will be an asset;

3. Known technology should be preferred, provided that it satisfies communications requirements at affordable costs; *and*

4. Not only should the installation cost be flexible, but also subsequent operations should uphold the flexibility concept.

Organisational prerequisites must be served in an able manner. A fundamental reference is the elaboration of an Input-Output matrix leading to optimisation.

Assuming a programmable hub - even better, one enriched with artificial intelligence - it will be possible to combine in realtime "n" inputs splitting them into "m" outputs per specific destination. This subject is closely connected to the design of an *intelligent architecture,* including concentrating, switching and signalling.

Focusing Our Approach on Lightwave Communications

It was said in section 4 that optical fibres interconnect the new computer centre of Bank of America in Concord, CA, with its headquarters in San Francisco. Other financial institutions have followed a similar approach and have found the experience very rewarding.* The same is true of manufacturing companies.

The top criteria that successful network implementations have observed are:

---

* See also D.N. Chorafas and Heinrich Steinmann: "High Technology at UBS for Excellence in Customer Service" published by the Union Bank of Switzerland, Zurich, 1988

- Feasibility;
- System design; *and*
- Maintainability.

While design criteria have focused on:

- Capacity;
- Flexibility;
- Reliability;
- Bit Error Rate (BER); *and*
- Cost.

The task of the feasibility study is to tell whether or not the system with the technology being chosen - and, therefore, the solution being conceived - is *doable*. The necessary components may exist, but how are they faring in terms of *applications goals*?

In a practical, down-to-earth approach to networks, application goals have a great impact on system design. Networks perform best when they are designed in their fine detail with intended application and topology in mind. They are not user-independent. Neither should the design phase forget about cost/effectiveness and dependability.

Some of the design criteria we use may be contradictory. For instance, very low BER and contained cost. The fact that the bit error rate of a given approach may be lower than originally required does not mean that it does not have a place in our plans. Infrared transmission does present interruptions if obstacles are placed in the way or if people walk by. The latter:

- Could be partly compensated through choice of appropriate handshake protocols, for instance: ACK or NACK; *and*
- Could be used for less BER sensitive applications, such as inventory refilling and paging systems. Also for audio applications.

But, it could not be used for money handling. We should use single optical fibres for financial data transfer connecting the distributed file server or communications server to the workstations, and vice versa.

The protected fibre cable for a single WS is very thin, being only about *half* the diameter of protected twisted wire. It can easily be put under the carpet, thus presenting no major installation requirements. And the size of optical fibres cables becomes smaller and easier to handle. A hard optical fibre connection is very reliable.

It can also provide significant capacity:

- Both 50 MBPS and 200 MBPS rates are currently available in star connections;
- Multimode fibre is protected with plastic and can run over 2 km without a repeater; *and*
- The technology steadily moves towards higher rates and longer distances.

TAT-8, the transatlantic optical fibre cable, currently supports 420 MBPS and employs 1.3 micron lasers as sources over 60 km. Terrestrial installations run faster. What started at the 45 MBPS level of transmission, when practical implementation took place (in the Chicago area) in the mid-1970s, has now reached 1.7 MBPS. It will soon be 2.2 GBPS (the 2 GBPS mentioned in Chapter 12 is an average figure used as level of reference) and well beyond that level.

We cannot effectively reach these rates with coaxial cable. Fibres will dominate future network systems, and the earlier we come to master this technology, the better it will be. As it cannot be repeated too often, our continuing competitiveness depends on mastering photonics technology.

New products phase out some of the existing ones, even if they were recently installed. One of the major questions about TAT-8 is its potential impact on the trans-Atlantic satellite market, particularly on the services offered by the International Telecommunications Satellite Organisation (Intelsat).

The 115-member Intelsat consortium provides a variety of switched and dedicated communications services, including International Business Service (IBS), a private, digital international service offered at a minimum of 64 KBPS. It is provided - like Intelsat's other services - to endusers through the signatories' carriers.

The problem is that about 75 percent of IBS traffic -and by far the most lucrative part of that market - is in the Atlantic Ocean region. TAT-8 will siphon out some of this business, although IBS (correctly, in the author's opinion) bets that there will be a continuing demand for wideband digital services on Intelsat satellites worldwide.

The secret IBS worry, however, is that, under TAT-8 competition, communications services on the heavily travelled North Atlantic routes quite likely will not continue to expand at their previous very fast growth rates. Users will shift some of their traffic to TAT-8. For this reason, satellite carriers have not stood idly by as TAT-8 extended across the Atlantic.

On the USA side, Communications Satellite Corp. has been signing up customers to long-term leases at rates that are continually dropping. Comsat, the USA representative of Intelsat, has signed AT&T, MCI and US Sprint to separate leases. It is also applying to the US Federal Communications Commission for more reductions and creative tariffs. In short, the trans-Atlantic optical fibres highway has been *beneficial to the consumer even before it went into operation.*

At the other side of the implementation spectrum, in the inhouse installation, the fact, that for BER and reliability reasons, it is better to use fibres under the rag, attaching PC and other computer terminals, removes little from the flexibility and attractiveness of an optical solution. It:

- Provides a solid ground for data applications where quality of information is very important; *and*

- Makes feasible a flexible optical distribution system which can be reprogrammed under software control.

However, while there is no problem in the installation of optical fibres and lightwave sources in an inhouse system, there is a challenge with protocols. Today, point-to-point solutions dominate the scene. Networking has not been sorted out nearly so well as software programming of the network. Work on FDDI orients itself in the direction of establishing necessary protocols, but the 100 MBPS FDDI is not yet a generally accepted solution, although this might happen in the future with the newer protocol to support 140 MBPS, or better, capacity.

At the present time, the ANSI standard X3T9.5 for fibre distributed interface (FDDI) addresses itself to 100 MBPS token ring networks, including topologies and optical fibre interfaces - but it does not satisfy everybody. Although it is an American standard, most USA companies are quite reserved about it, suggesting that nothing less than 1 GBPS should be taken as a basis for investments. Any megastream will become obsolete too quickly.

Yet, as underlined in the preceding paragraphs, we do not need to be discouraged by the lack of multiple access, multidrop, or other protocols in optical fibre solutions. The star-type and WDM approaches that we discussed earlier provide good interim answers. Important from the user organisation viewpoint is that the solution to be chosen does not become obsolete in the near future. Neither should it become a bottleneck.

Furthermore, as with any system, a systems approach will be vital for successful implementation:

- First, we should be careful in identifying our goals and intended usage;

- Second, we should list all system components: those for which we have solutions, as well as the unknowns;

- Third, we should establish a hierarchy of the problems and solutions that we are looking for; and

- Fourth, we should choose from among various available alternatives to serve these goals.

As I will never tire of repeating, the system solution is important and it should focus on multimedia. This is both technically feasible and operationally wise. The systems that we design today should be projected to have a considerable lifecycle and breadth in market perspective. Multimedia solutions will be dominant in the years to come.

As a first step, multimedia integration should address itself to text, data, graphics and images. Not necessarily to voice. I have already mentioned that the discussion at Bell Laboratories leaned toward the concept that while, in principle, the technology for voice integration is here, the practice is still difficult.

In the laboratory environment we can do it. In fact, it has been done. But how it gets done as a widespread, dynamic system is another matter. The challenge is in the brainpower to be applied to the job and the cost/effectiveness of the end result - as there are many problems still to be overcome.

In other words, the technical drivers are there, but the implementation of voice integration is progressing slowy. As soon as we put the human into the equation, things change. We do not really know how people talk and hear, in spite of 113 years of experience in telephony.

Finally, it has often been underlined in this and the preceding chapter that the system approach should be enriched with artificial intelligence constructs. This is as valid of the constituent modules of the system as of the interfaces. Also, from the enduser's viewpoint, AI becomes very important. Smart WS are easier to use. And AI is also vital in the designer's task.

From Telephony to Integrated Consumer Systems

The 25 million people who live in the Tokyo metropolitan area have more telephones than the 500 million inhabitants of Africa. Sweden is the most densely populated country in telephony. While nine advanced industrial countries possess three quarters of the world's telephones, two thirds of the world's population has no access to telephones at all.

With the notable exception of Saudi Arabia, there is a correlation between gross domestic product and telephones installed per 1,000 population, as Table 1 documents. The figures prove that *telephones and development* are inseparable. Better telephones mean better trade, better health, and sometimes even better government.

This is the great dilemma faced by the current Perestroika rulers of the Soviet Union. The large scale use of communications and computers can greatly reduce waste and improve efficiency. But, their wide adoption not only implies heavy investments, it also challenges the intensely secretive, centralised, and bureaucratic system which controls the Soviet Union.

Table 1: Per Capita Gross Domestic Product and Installed Telephones per 1,000 Population

	GDP per head (in US $)	Telephones per 1,000 Population
Sweden	14,100	860
United States	13,200	790
Japan	12,080	580
Britain	9,660	520
Saudi Arabia	16,000	112
Mexico	2.270	81
Nigeria	860	7
China	310	5
India	260	4
Zaire	190	1

London's *Economist* once made the observation that 76 years ago, in 1913, Imperial Russia had a real product per man-hour 3 1/2 times higher than Japan's. However, Japan not only joined, but came to lead the communications and computers revolution. By contrast, Russia spent 72 socialist years stumbling and slipping backwards. As a result, the ratios have been inverted. Japan now has a real product per man-hour 3 1/2 to 4 times higher than Russia - and the gap increases as Japan expands its lead in telecommunications.

The lesson is clear.

- Second World countries should develop an open society structure and bet on communications.

- Third World countries should not only think about procrastination with its associated drop in standard of living.

To stand a fair chance of development, they should give priority to telecommunications. Able communications do not just mean telephony. The need for investments extends over the whole range of an integrated system with a tremendous amount of computer power - including personal computer power.

An example can be taken from the more advanced industrial countries of the world. The functioning of a knowledge society depends a on rich composition of information types: data, text, graphics and image, video and voice - as presented in Figure 13.3. Metrics for storage, information transfer and presentation are just as important as the provision of the facility itself. We have spoken many times about them.

The same is true of *integration*. A coherent introduction to the concept of modern communications demands that we bring order into the diversity and confusion that surround our perceptions of the new media. Such media typically use the telephone

INFORMATION TYPES

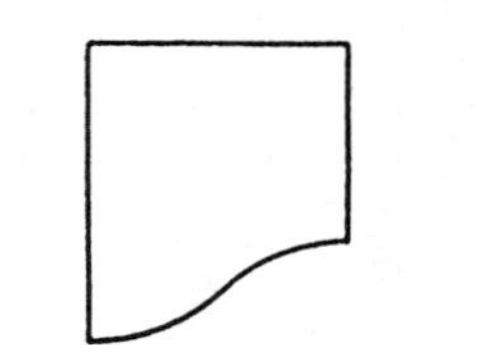

DATA

TEXT

2.000 TO 6.000 CHARACTERS PER PAGE

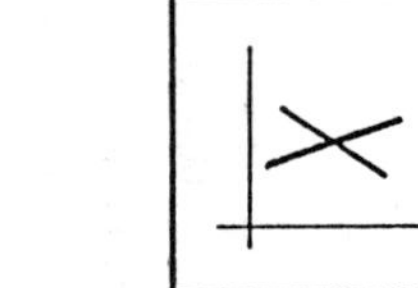

GRAPHICS AND IMAGE

200 PEL x 200 PEL = 4 MILLION PEL PER PAGE
= 50.000 CHAR. PER PAGE

400 PEL x 400 PEL = 16 MILLION PEL PER PAGE
= 200.000 CHAR. PER PAGE

COMPRESSED 10 : 1

VIDEO

30 IMAGES PER SECOND
80 MBPS

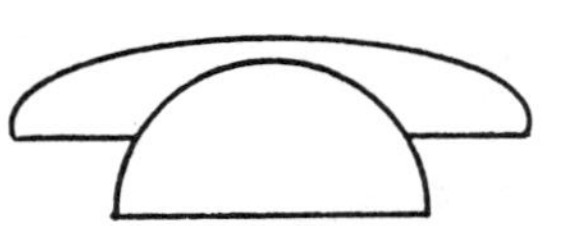

VOICE

64 KBPS WORLDWIDE STANDARD

Figure 13.3

network as well as the video screen. The user communicates with the computer of an information provider, having direct access to many different databases. That is the vital point of reference which, as stated, should be kept in perspective in every system design.

The communications service may use a television transmission network to send specially prepared, numerically coded information that a special decoder renders visible on the TV screen. This is known as Teletext. Broadcasting Videotex refers to information widely distributed by means similar to radio and television. Interactive Videotex designates the individual communication possible between the information provider and the user in the form of a dialogue between two partners.

The greed and irrationality of the PTT (European and other) has killed the potential of videotex and relegated it to be a non-entity. However, the process which was first launched in March 1979 in England, with Prestel, was significant and although it became a commercial failure (in all countries other than the U.K.) it is worth briefly reflecting upon it.

For text and data transmission purposes TV, the telephone, and two other elements are necessary: a modem and a decoder.

- The modem guarantees communications between the telephone network and the home TV or personal computer.

- The decoder decyphers the coded information transmitted through the modem.

Its task is to transform binary transmissions into letters, numbers and graphic images.

Since the contribution of information technology to household usage is not always appreciated, here is a snapshot of what it can offer. First, the system we just described is independent of time constraints. The information it contains and makes available can be tapped at all times and can be obtained in seconds. If I want to find out at midnight when the first flight leaves the next morning for New York, the system can tell me. Moreover, if the time is suitable, I can make the reservation directly via the computer.

Second, by being online the system has a high degree of timeliness. Its information can be brought up to date at any moment, and immediately. For example, if the last place on the flight to New York has just been booked, this will show up, and I will have to reserve on the next flight.

Third, the user can communicate with other users by sending and receiving electronic mail. This not only makes the message exchange process instantaneous (rather than the days the post takes) but it also significantly reduces the costs.

Fourth, the user can selectively call up the information that interests him from private and public databases. Lacking a 10-volume encyclopaedia at home, he can ask the system for the exact reference(s) which he needs in order to proceed with the project he is working on.

Fifth, in a society where financial assets have become more important than real assets, the user can bank 24 hours per day. He can interrogate his account, check on deposits and give orders of payments. Corporations do cash management online; consumers do home banking.

Sixth, the new media are available around the clock and, if properly approached, they demand a controllable investment in equipment to utilize the system. The costs involved are relatively modest, while they allow extensive dissemination of information.

Seventh, the new communications media are easy to use. No special training is necessary, as the solutions have been developed to serve a wide public. All these attributes add up to a technology that could be efficient, inexpensive and practical. But, it requires insight and foresight - and also an infrastructure. This takes time, money, knowhow and lots of forward thinking. This last quality is the one in really short supply.

Backwards into the Future: ISDN

The specialists of the ''Petit Travail Tranquil'' (PTT) tend to think that ISDN will be their saviour. ISDN means integrated services digital network - but some say that the anagram stands for ''I sense dollars now''. B-ISDN is a projected integrated broadband network. Neither ISDN nor B-ISDN is well studied, and wherever they exist the concepts do not really make sense.

ISDN includes all baseband services, as contrasted to television which uses broadband in the frequency spectrum. But it rests on old technologies and, therefore, it is destined to remain an interim non-entity. The 113-year-old twisted pair installed many decades ago by the PTT to connect the subscriber to the city centre is the wonder tool of ISDN.

If, after only 13 years of existence, videotex has become inefficient and is tending to be phased out of the market, think what a cheap approach (in a value sense) like ISDN has in store. The whole issue has been promoted by the unable who has been asked by the unwilling to do the unnecessary.

Contrary to what the PTTs may believe, ISDN is not new. The concept of an integrated services digital network was first discussed by CCITT in 1972. Yet, 17 years later there is still *widespread uncertainty* in the telecommunications industry, and particularly among users, as to what ISDN will mean. When it comes to real-life implementation, there are six deadly sins to account for:

1. The great difference between theoretical intentions and practical applications.

In terms of theoretical intentions, ISDN is supposed to extend the serviceability offered by today's global telephone network to all forms of information: text, data, graphics, image, voice and, ultimately, full-motion video. Through the (still to be seen) single access standards of ISDN, a wall plug will supply all present and future telecommunications services.

This rosy picture is simply make-believe. The reality is quite different. While equipment manufacturers and telephone operating companies have seized upon the ISDN slogan for dollars purposes (as a key marketing tool), a number of technical and commercial issues have yet to be resolved - starting with the international availability of an ISDN service which is still:

* Awaiting implementation of agreed upon internetwork signalling schemes,

* While commercial agreements on interconnection are also lacking; *and*

* No harmonised tariff policy appears to be in sight.

National divergences, technical incompatibilities and sheer incompetence on behalf of a bogus bureaucracy are in the background of such delays. The killing of videotex (which, after all, was a much more promising process) speaks volumes about what is in store with the ill-conceived ISDN venture. Besides, ISDN is by no means a dramatic implementation of new technology which leads us to its next deadly sin:

2. Industry requirements projected for the mid-1990s stand at 2 to 3 GBPS per connection - that is, at gigastreams.

This is more than 4 orders of magnitude the combined 144 kilostream capacity of ISDN. Table 2 presents a digital hierarchy based on the fibre optic DS standard (vs. the T for copper). One 2.25 GBPS channel would correspond to 33,600 64 KBPS channels of the ISDN variety.

3. Integrated services packet networks and metropolitan area networks (MAN), which the telcos will provide, call for a top-down grand design and a detailed bottom-up design.

By addressing itself only to the local loop and its obsolete equipment, ISDN is a patchwork. Patchworks are no answer to the communications *needs* of a knowledge society.

4. In the hurry to get the patchwork going, the PTTs forget that they will have to change switches and trunks.

Eventually, this will mean double investment rather than savings on current ageing plant. Another foremost challenge is the emergence of incompatible standards. While the Consultative Committee for International Telegraph and Telephone (CCITT) is due to agree on revised ISDN standards, the flexibility allowed by previous recommendations already has led to divergent national implementations.

In the United States, for instance, de facto ''standards'' are being set by Bell Communications Research and the development organisation's owners, the seven In

Table 2: Digital Hierarchy in Communications Channels United States Standard

Fibre Optics*	Bit Rate	DS3	DS1	64 KBPS Equivalent Voice Channels
DS1	1.54 MBPS	–	1	24
DS2	6.3 MBPS	–	4	96
DS3	45 MBPS	1	28	672
DS3C	90 MBPS	2	56	1.344
–	135 MBPS	3	84	2.016
–	180 MBPS	4	112	2.688
DS4	274 MBPS	6	168	4.032
–	405 MBPS	9	252	6.048
–	560 MBPS	12	336	8.064
–	1.2 GBPS	24	756	18.144
–	2.25 GBPS	–	1.400	33.600

* Vs. Coaxial T Series

regional Bell holding companies. In Germany, the Deutsche Bundespost's ISDN D channel protocol, 1TR6, differs from the CCITT's Q.931 because it was developed too early. France Telecom has derived its own standard in a similar way. British Telecom's signaling standards, DPNSS and DASS2, have also evolved from earlier CCITT proposals and are, therefore, incompatible with the newer ones.

5. Even the too little, too late of the ISDN recommendation is applicable only to a very limited topology: the local loop's connection to the city office.

Still, as reference No. 4 pointed out, nobody can guarantee that the recommended standard will be applied in uniformity. We have had that experience with X.25. Although these differences may not impede the growth of national ISDN networks, they have serious implications for the start-up of international services, a vital element in the creation of a mass market.

The crux of the matter is that the bureaucratic telecom authorities are not committed to standards or the user. They have not even solved the problems for themselves, much less for the manufacturers of telecommunications equipment and for the user population.

6. ISDN offers a 2B+D wire, but does not provide for multiple session capability.

Splitting the session is fundamental to competent WS utilisation. It must be done either through an X.25 node or a frontend. Current implementation experience indicates that X.25 is not performant in a network-wide sense of transactional communications.

an application sense, there are today some 16 divergent ISDN national specifications in Europe alone. All told, most progress has been made on standardising the so-called 2B+D basic rate interface, which incorporates two bearer, or ''B'', channels, operating at 64 KBPS, and one signalling, or ''D'', channel, operating at 16 KBPS. Still, significant variations exist in countries in implementing user-network signalling and procedures for supplementary services signalling.

Furthermore, not all of the PTT administrations are using a common physical connector. At the same time, practically no progress has been made in standardising the *30* B+D primary rate interface of 2 megabits per second, which CCITT says will come with B-ISDN.

7. A very important issue is *transmission quality* and dependability.

Transmission quality is expressed in bit error rate (BER). Voice quality lines range from the very poor 10^{-2} to the average 10^{-4} and 10^{-5} BER. The so-called data quality lines have a BER of 10^{-7} and $10^{-8.}$ All this is old plant. What is really needed is 10^{-11}, 10^{-12} or better, which the depreciated and degraded old plant of the different PTTs cannot provide.

These seven deadly sins tell why the way in which telco authorities (PTT and other), vendors and the trade press have disseminated information about the Integrated Service Digital Network ahas been absolutely irresponsible. A large number among them talk about ISDN indefinitely, but almost no one talks about it definitely. The pep talk emphasises the ''future'' wonderful applications of ISDN with its simultaneous data, voice and video, over the same pair of old, rusty twisted wires - the greener fields are in the future.

Despite the documented fact that the ''supported'' ISDN channels are substandard in terms of capacity - as well as of the other deadly sins we discussed - simultaneous voice, image and data over the same pair of wires is a feature, not an application. An application would be a method by which it can directly benefit business. Benefit to business has nothing to do with information about which carrier provided which ISDN version on whose switch. But it has a lot to do with how ISDN helps its users make business and profits:

- For a bank to sell its services online to its customers.

- For a manufacturer to coordinate its design, production, inventory and marketing activities.

The aim of multimedia communciations is precisely to respond to these *challenges* - and the ISDN fans have not yet provided any answers to this question.

Few attempts have been made to portray potential applications, and these few are unconvincing. By contrast, we are told that users are not pursuing ISDN as rapidly as expected - an argument which forgets that those who did find lots of problems are stuck with them. The telcos did not bother to weed them out.

Travel Related Services Co. was the first user of AT&T's ISDN Primary Rate Interface (PRI) service, which specifies how customers can use and control a T-1 link configured to CCITT standards to hook customer premises equipment, such as PBX, to a carrier's long-distance network. On paper this may look fine but, to take only one example, the University of Arizona in Tucson, AZ, which decided to pass from ISDN theory to application, found many unresolved problems - some of which simply do not make sense.

To start with, ISDN service initially will be used to handle voice and some *lower speed data* traffic, replacing two older data switches. That is hardly a breakthrough. ISDN will also be used for personal computer-to-personal computer communications, primarily for devices not attached to the Ethernet. This will mean a 64 KBPS transmission speed vs. Ethernet's 10 MBPS, and it is a first class example on how to stumble backwards into the future.

The university is investing in ISDN now, in anticipation of the day when Signaling System 7 (SS7) technology will permit the use of ISDN between local serving offices (SS7 is a signalling method used to establish paths through carrier networks). Even in this reduced level of implementation, the University is proceeding very carefully with lots of reserves: "We are betting that by the mid 1990s, we will have FDDI running on this campus at 100 MBPS", said a University spokesman. "It is horrible to say this, but in the days of 100 MBPS, the 64 KBPS of ISDN channels is slow stuff".* (It is also obscene.)

Another hilarious item from the same story is that, to fulfill the inefficient ISDN "standard", the local telco will rip up the existing underground telephone cable, install 36 miles of new twisted pair cable, and rewire all the buildings. Every room in each building will be equipped with dual phone jacks, so that telephones and terminals plugged into the jacks will use the twisted-pair wiring. It is tragic to say this, but it is true.

- The expense is great and unwarranted;
- The established channel is too narrow and substandard;
- In comparison to coaxial, system reliability is largely reduced.

About ten years ago, Brown University was wired with a coaxial local area network. It was installed in the same underground ducts which housed the telephone twisted wires. Then, a telco repairman accidently created a major underground fire.

The coaxial network was fully repaired within half a day. Made of twisted wires, the telephone network took between 2 days (as a minimum) and 7 days to repair. Part of Brown University was cut off from voice services for a whole week. Is this how the PTTs and other telcos see the future with the twisted wires (and twisted concepts) of ISDN?

* Network World, August 29, 1988

Serious executives and communications experts who examined how well ISDN fits *their* operations have come to the conclusions that a change-over is unfavorable. Therefore, they will not do it. Said Dr. Friedhelm Hesse:* ''I am not a fan of ISDN. It is:

- Not reliable,
- Too expensive,
- Too limited.

And it does not answer short haul problems in an economic way. There is no cost justification going to ISDN. In my operations alone, cost increase will be 30 million to 50 million Deutsche Mark (DM) per year if we change from X.25 to ISDN.''

As mentioned at the beginning of this section, it is frequently said in America that ISDN really means ''I Sense Dollars Now''. These fast bucks will be very bitter dollars, both for those who pay them, and for those who get them. Both will regret the day that they got into ISDN.

---

* From a personal discussion in Frankfurt, in the course of a research project, March 1, 1989

14

Increasingly Sophisticated Communications for a Knowledge Society

Introduction

A *knowledge society* requires very different communi-cations fromwhat was necessary in the industrial age. Future economic progress depends on the growth of the service sector, provided that we are able to *swamp costs* and *increase quality*. Both goals require an agile, powerful and sophisticated communications infrastructure.

Dr. Tom Stonier* aptly makes the distinction between *individual* brain intelligence and *communal* intelligence, which he associates with the evolution of culture. At their present status, and through their support for any-to-any communications, networks serve this communal intelligence. They permit the cultural evolution to transcend the biological evolution.

The biological evolution has been probabilistic and spans over millennia. The *cultural evolution* is possibilistic and moves at high pace carried through networking capabilities which both enlarge on and benefit from communal knowledgebases.

Intelligence, Prof. Stonier suggests, is not a property unique to the *human brain*. Rather, it represents a spectrum of independent yet interrelated phenomena:

1. Individual brain intelligence;
2. Communal intelligence;
3. Increasingly sophisticated learning processes;

---

* Postgraduate School of Studies in Science and Society, University of Bradford, UK

4. Cultural evolution transcending biological evolution;
5. Storage of culture as represented by books, libraries and museums;
6. Goal orientation rather than random selection;
7. Speed of cultural evolution;
8. Focused scientific effort through laboratories;
9. Acceleration of scientific developments; *and*
10. Research into brain functions.

The human brain might have developed through a process of natural evolution but, once it was in place, it has set its own mechanism in motion. Over the last two decades, this has been manifested through:

- Artificial intelligence (AI) devices able to map human knowledge, exhibiting neurological capabilities;

- The ability of such devices to comprehend context and culture - connected to the operations that they are carrying out;

- Advances in genetic engineering; *and*

- Developments in photonics and biochips.

Supported through communications networks, the merger of human individual and communal intelligence with machine intelligence will generate *new synergetic* forms. It will also make it feasible for new characteristics to *emerge* and be inherited in subsecond speed, providing *evolutionary discontinuities as profound as the origin of life, says Stonier.**

While these references to the impact of communications focus on the longer trend, and will *profoundly affect* our society, and human life at large, over the next half century, the short term focus is that of doing the current business much better. Competitiveness is the immediate reason why we design and install communications networks..

The steady effort to reduce costs in order to remain competitive finds, for instance, an excellent example in banking. In financial institutions, cost reduction has to be done in the face of an escalating volume of transactions and intense competition to bend the curve of rising staff numbers, thus absorbing technology costs and still showing benefits.

All banks today aim to derive a competitive edge by the optimal use of funds and very rapid response to market changes, placing stringent requirements for reliable networks offering:

---

* AI and Society, Vol. 2, Springer Verlag, London, 1988

- Wide connectivity between user communities, *and*

- Integration between the different players in the financial services supply chain.

Foreign Exchange (Forex) traders and securities' dealers need to exploit the world market for opportunities in financial instruments; equities, bonds, futures, fluctuating foreign exchange rates and so on. Innovative communications services are a critical success factor in all these activities.

In fact, in America, Western Europe and Japan, financial services represent the most demanding sector for communications suppliers. Diversity of needs, scope of networking and strategic challenges are in the background of this statement.

Banks, brokers and insurance companies compete with one another for corporate and individual investment services, but also for deposit taking, lending and money transmission on a national and/or international basis.

As each organisation has its personality, it requires very different communications supports according to the service mix: high volume retail or lower volume, but high value, commercial and investment business, national or international coverage, global 24-hour banking or primary retail trade with ATMs and POS, become focal points of the network.

Typically, the network which we design will have a physical kernel and outer logical layers as shown in Figure 14.1. It will be characterised by subnetworking, and will increasingly support the topmost layers of ISO/OSI. While specific requirements from one organisation to another always vary, all the aforementioned cases call for a *sophisticated communications infrastructure.*

Our ability to respond to the growing market and organisational demands for multimedia communications rests on three pillars:

1. The clear understanding of where we stand at this moment;

2. The goals which we wish to reach, and the financial, technological, as well as human resources, that we can apply to this work; and

3. The architecture which we adopt to provide us with an integration capability in the background, and broad implementation facilities at the foreground.*

Learning and comprehending our current position means taking stock of what we currently have in the organisation. This should cover: equipment and services, but also knowhow and costs, as well as management skill.

Positioning ourselves against business competitors calls for shooting at a moving target. The more our position rests on solid foundations, and the more it uses

* See also D.N. Chorafas "System Architecture, System Integration, and the Role of Artificial Intelligence" McGraw-Hill, New York, 1989

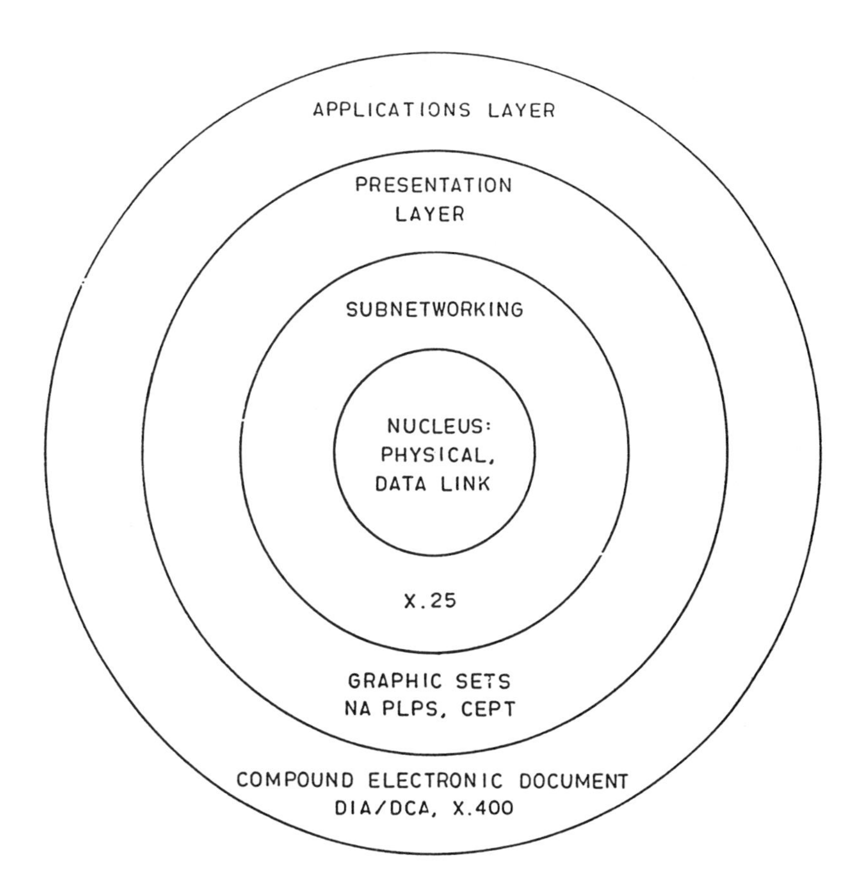

Figure 14.1

advanced technology, the better we will be able to face competitive requirements. Still, the environment is dynamic, our competitors also move, and we have to run fast just to remain at the same place.

The system architecture which we adopt, together with the network facilities that we have chosen, must ensure competent integration perspectives - from decision support to messaging and transactional implementation. It must also include the possibility that, at service level, supported communications media are measured and monitored. We will talk further about this issue in Chapter 15.

From the simplest type of interconnection: the PC-to-mainframe link, to the most complex, the any-to-any network, considerable preparatory work is a ''must''. To better appreciate the transition to higher levels of requirements we will start with the simplest possible link, then build up in terms of architecture, protocols, expert system and overall perspective.

Linking the PC to the Mainframe

One of the earliest preoccupations with personal computers as they integrated into the corporate culture was to link the WS to the mainframe(s). This has been typically done through a local area network, as well as point-to-point, multidrop or some other type of long haul connection. A WS cannot exist in a vacuum. Personal computing essentially means personal communications support.

The need for PC-to-mainframe connection is documented by xperience. At its employees own prodding, in 1983, a major French bank bought 300 PCs and installed them as standalones to an equal number of managers and professionals. Reaction in the first week was terrific. But, as the endusers learned to work with the new machines, their enthusiasm for standalone PCs shrank. Increasingly, they demanded online communications.

The requests were specifically oriented to the need for providing a database connection. Without this link being assured, after the fourth month of standalone PC implementation dark tubes* dominated the landscape. After a PC-to-mainframe connection was finally produced, WS utilisation again picked up momentum.

The engineering of PC-to-mainframe links has no secrets. For online connectivity, we must study:

- Electrical characteristics;
- Standards for new equipment;
- Support protocols;
- Interfacing software;
- System software for network operations;
- Provision of skills and knowhow;
- Finer programming interfaces towards the enduser; *and*
- Networking Control Centre (NCC) capabilities.

Solutions must be global and reflect a wide range of applications. Multimedia, data, text, voice and images be handled on the same network, on the same cabling. Attention must be paid to both physical and logical WS connection. The application

* "Dark tubes" is American jarbon, but converys well the message about personal computer monitors that are spint most of the time.

should observe security/ protection characteristics and be made in an integrated manner, including the incorporation of all existing programmable equipment.

The definition of gateways to all public networks (classical telephone service, X.25, and others), the choice of communications controllers, interconnection to other private and public computer resources - including public databases - all constitute part and parcel of PC-to-mainframe connectivity. The latter can extend to facilities relating to controls for smart buildings. For instance, controls or supervision of:

- Private Branch Exchanges (PBX);
- Power supply units;
- Lighting (offices and common areas);
- Fire control;
- Air conditioning;
- External and internal building security;
- Closed circuit TV; *and*
- Connection to a cable company, the PTT or the police service for global branch office control.

Every solution to a PC-to-mainframe connectivity problem must keep the user's requirement in sharp focus. The easy-to-use spreadsheets have given managers the ability to directly, without programmer intervention, manipulate budgets, analyse financial data, evaluate alternatives and present to themselves related information. A major challenge is, however, to provide proper interfaces to the transaction database(s) of the organisation, permitting us to realise the benefits of such software while preserving the integrity of the overall information system.

This is an example of the finer programmatic interfaces that we have been mentioning, and it has many aspects:

- Guarding against the accumulation of redundant, trivial or inappropriate data in the database;
- Utilising the built-in capabilities of the spreadsheet; *and*
- Avoiding the overproduction of reports, which has become a highly counterproductive practice.

Thus the PC-to-mainframe connectivity is not just electrical characteristics and a protocol. Its most challenging part is the institution of solutions, such as exception reporting, using graphics rather than overloading with tables, and helping users to tailor the facility to their purpose.

The value of spreadsheets, integrated software and graphics packages running on a PC is greatly increased when the user can access information already available on the mainframe database through his or her personal computer. Such access raises a number of issues relating to data link and file format:

- How much compatibility is needed between a workstation and the mainframe to allow downloading selected corporate data into the PC?

- Is an asynchronous terminal emulator enough, or is it preferable to provide a process-to-process solution? *and*

- Does the user need compatible PC-to-mainframe interfaces? Does he need the same file structure?

While the answers to be given must reflect the applications environment, as experience accumulates it becomes clear that (in the longer run) process-to-process solutions are needed for interactive communication.

Not only must the hardware and data link connections be made, but appropriate software must also be provided. Some products are accommodating access to Telex/TWX services. Certain vendors build entire packages based on this function; others include it as part of more comprehensive offerings. Possibilities range from simple teletypewriters (TTY) emulators to facilities that permit unattended sending and receiving of Telex/TWX messages.

Some interesting datacomm packages for personal computers support intelligent modems which permit automatic answer, automatic dialing, unattended operation and programs for directly controlling the modem. Facsimile reception/transmitting through a printed circuit board (PCB) add-on is increasingly interesting to a number of firms.

Data communications products linking PC to mainframe(s) vary widely. Some are full blown, integrated packages that make databases available for a number of WS applications. Others are only terminal emulators. Usually, what we wish to obtain in a PC-to-mainframe linkage is the ability to use our WS to login to the database of another system, accessing files and running programs on that other system the way that the enduser wants.

This implies that the software we select must employ our WS as a terminal (local system) to transfer (send or receive) files to and from any remote system, for one or more reasons closely linked to applications - both current and projected:

1. Queries of all sorts;

2. Spreadsheets;

3. Business graphics;

4. Word processing;

5. The handling of transactions;

6. Electronic mail; *and*

7. Communicating databases.

As with any computer and communications application, we encounter a series of tradeoffs, compromises and short-cut approaches to the PC and mainframe linkage. The technology exists, but the implementation is not straightforward - and this not for the lack of altern-atives, but because there are so many.

Without doubt, the simplest approach is to have the workstation *emulate a non-intelligent* computer *terminal.* This was the earliest method that computer users employed to interact with commercial databases. In such a timesharing mode the following are required:

- a terminal emulation software on the WS,
- a modem; *and*
- a communications line.

This may still be the most common form of connectivity, but not a very effective one. It allows the PC to look at the data, but not to store and manipulate it.

The time is long past for simple asynchronous terminal emulation. This generally takes place over ordinary public telephone lines, and let users have access to mainframe databases in a way which can satisfy searching or electronic mail - but not analytical-type applications.

One step up is the ability to *download data.* This applications perspective brought up direct bisynchronous terminal emulation, but it, too, is now past. The current solution sees WS becoming very *active local processors* and major users of inhouse data communications. They frequently call upon mainframe facilities to assist their own specialised applications programs.

The last reference is particularly true for central databases. Applications-wise, it ranges from controlled data entry at the point of origin, with immediate transfer to corporate files, to dynamic report generation with the authorised enduser able to extract information from files in any format and way of presentation that he chooses to define. Figure 14.2 presents some recent statistics on how online WS are used, and for which applications. Notice that:

- Online access to databases has now reached an equal percentage level with data entry, and will soon exceed it.
- Spreadsheet, statistics/analytical and graphics applications are becoming much more popular as word processing.

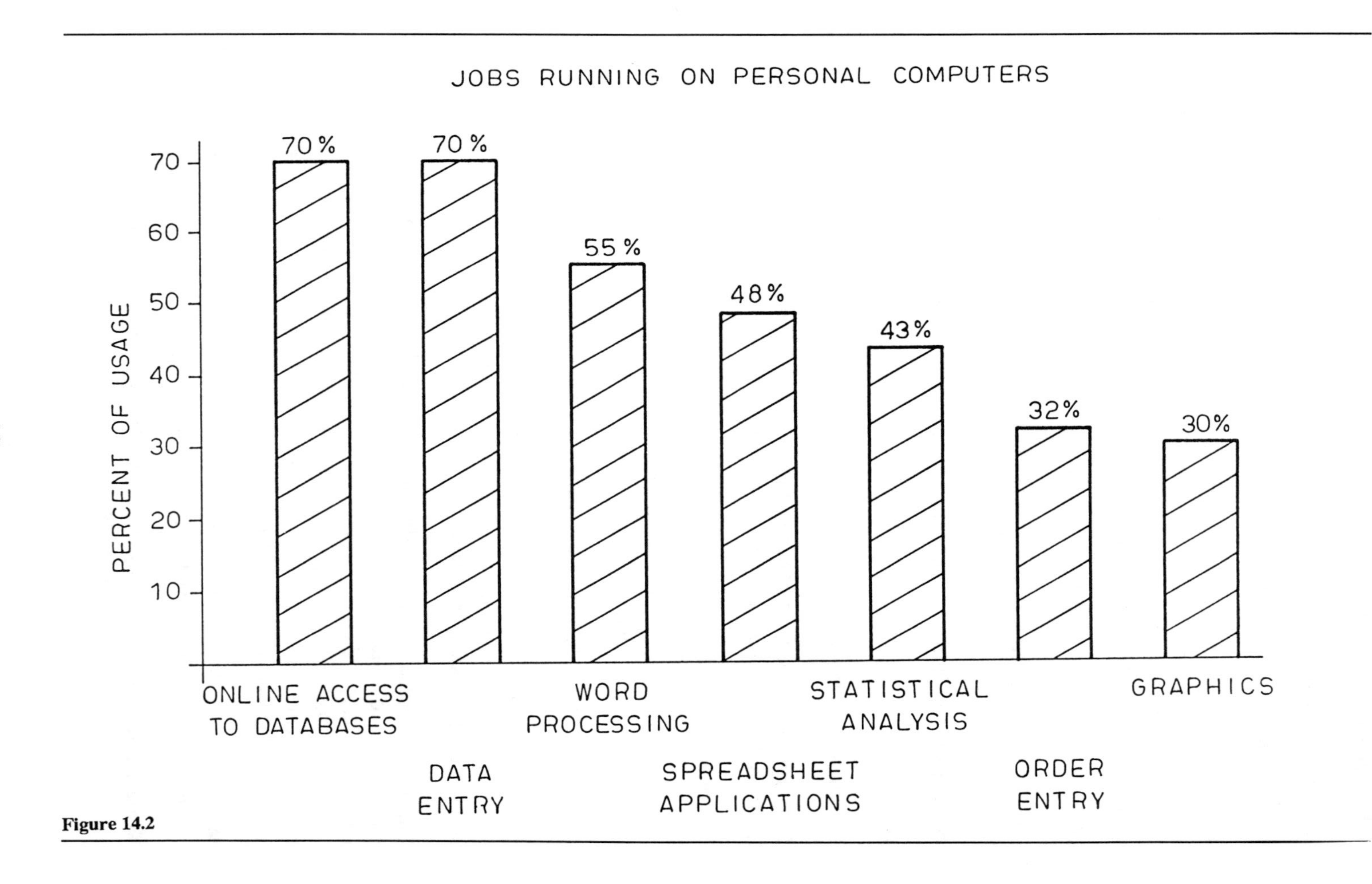
JOBS RUNNING ON PERSONAL COMPUTERS
PERCENT OF USAGE
70
60
50
40
30
20
10
70 %
70 %
55 %
48%
43%
32%
30%
ONLINE ACCESS TO DATABASES
DATA ENTRY
WORD PROCESSING
SPREADSHEET APPLICATIONS
STATISTICAL ANALYSIS
ORDER ENTRY
GRAPHICS

Figure 14.2

Such applications require much more than downloading, although the ability to download permits the enduser to call information from the mainframe, store it and work on it. With simple downloading, the data is in raw form and software is necessary to get it from storage into the spreadsheet.

Efficient File Handling

As applications get more complex, the enduser must be offered training on how downloaded data should be fed into various PC packages using different data formats. Also, he needs training with regard to what mainframe data is available and how to extract it. The user should also be trained so that he is able to determine when to use a financial planning package and when to use a simple spreadsheet, as well as the data required by each. The same is true of the features of a file transfer package.

An improvement over simple downloading is to add software to allow the *data to be formatted correctly*, so that data can go directly from the mainframe to the WS and into the proper slots in the desired program. Whether downloading or upline dumping, this type of communication requires more than handshaking protocols at both ends.

A valid approach is to have PC and mainframe run the same data structures, so that, at the personal computer, translation of data is not necessary. Process-to-process communications, like all other modes, can also take place between workstations.

The areas of activity we have just reviewed have in common data sharing. This should be a primary objective of the PC-to-mainframe linkage. Local networks are necessary for the provision of adequate data sharing facilities. The proper distributed database technology is finally coming of age.

Linked together through appropriate protocols, mainframes and minis will end up serving a database machine with a fully distributed functionality. WS will look into that facility. However, when choosing the solution which best fits an application environment, the greatest care is recommended in experimental verification. The operating systems, too, may present connection problems.

By carefully determining the primary requirements for mainframe to PC connections, professional analysts and designers put themselves in a position to offer a number of valuable services to endusers, and are also able to make system sense. System architects appreciate that the way should be paved towards system integration. Different reasons point to this approach.

1. The satisfaction of the growing ranks of corporate users. These users call their need to tie into a unified network: *critical;*

2. The reduction of processing loads on the mainframe, and the fact that the processing job is brought where it belongs: *At the workbench;*

3. The *tailoring of the application itself* to the individual needs of the user, together with an effort to considerably reduce *response time.*

At the WS level response time is reduced, because *computer power is now dedicated to one job,* and, therefore, has the ability to do it well. But, in the longer run, reduction in response time is associated with a competent approach to file handling on a network-wide basis - as well as with attention to the MIPS power of the dedicated WS in *relation to* the job to be done.

One of the best available mechanisms for data exchange is the use of file management software of a relational database management system (DBMS). By providing a uniform filing structure, a database system simplifies the exchange of data, while the flat file approach permits a better ad hoc access basis. Ideally, file transfer software should:

- Follow internationally recognised protocols, for which several projects have already been implemented on micros, minis and mainframes;

- Be independent of the systems by means of a transport and network architecture adaptable to individual DB/DC* requirements; *and*

- Ensure maximum use of network resources through continuous transmission flow.

Transfer tools should provide: simultaneous, multi-directional half- or full-duplex communications; automation of such exchanges; agile interface(s) with the application - making it feasible from every authorised site to request a file transfer in simple, comprehensible terms. There should be monitoring of the activity and the provision for operator intervention if necessary.

Recovery goals call for the availability of a log file. There should be automatic restarts in the case of incidents. A determination of priorities in transmission is a useful added value.

If the different computer systems concerned in exchanging files are all of the same type, or are owned by a single organisation, their software should be tailored in such a way that the files transferred between them are meaningful to each of them.

File conversion is necessary in diverse environments with a variety of equipment and file structures. The processing of a file on reception can become a multipass, multilevel operation, whose functions are defined either network-wide or at the local site only -relying thereafter on the network's data transfer mechanism.

Typically, a file transfer protocol structure consists of three phases:

- association;

- data transfer; *and*

- break, or dissociation.

---

* Database/Data Communications

The association assures the identification of the access rights to the remote device. It also provides for connection set up; establishment of file direction; conversions of file structure and data; and specification of additional operations on the transferred file.

The data transfer phase includes the local file operations, mapping into a virtual format, actual data transfer, and the administrative exchanges required to regulate it. The conversions of the file data and structure should be automatically done in this phase by the application process.

The break (dissociation) finishes the local file opera-tions and performs any additional functions on the transferred file (for instance, conversion of file structure, spool, job entry, etc.). On closing the connection, it indicates the final state of the transfer.

The format of the protocol should be simple. A protocol element would typically consist of a fixed length header part, a data part and a control part. The parameters in the control part depend on the type of protocol element.

Typical types of protocol elements are: initialisation of a process, with user access rights parameters; function and file descriptor control, for identifying the file by filename, properties and additional file operations; file data according to format; end of file and/or end of transmission. There should also be acknowledgement for synchronisation, error detection and recovery.

The networking solution is so much more useful when several types of application can simultaneously employ the services offered by the transport station, always assuring:

1. *Reliable* end-to-end transport;

2. *Transparency* of all network problems for the user; *and*

3. *Optimisation* of the costs as a function of the user's requirements and the conditions of the network service available.

Among the services to be supported, data conversion should be automatically done in the data transfer phase if the user does not want to transfer the file in a transparent mode. Among the types of conversion that could be required we can distinguish:

- *Text files,* with characters converted from or to ASCII/EBCDIC code;

- *Binary files.* Data formats are converted if the number representations of the involved hosts calls for it; *and*

- *Graphic files.* Standard graphic files should be automatically converted from and to each computer's representations.

The enduser should *never* be kept in the dark regarding the infrastructure put at his disposal. The best way to explain to him is to start from *a familiar ground*: the business oriented databases, then highlighting where they fail to address the graphics requirements - thus gently moving to define what makes the engineering database so special.*

Emphasis on the importance of standards can help give this issue the emphasis which it deserves. The definition of content is just as crucial. The engineering database, for example, should include not only graphics and text, but data in all its forms: results of analyses, tests, research activities.

Another peculiarity of the engineering database (continuing with the same example) is the need for storing and manipulating different versions of model data, depending on the application. The same is true of the ability to repair errors - a case which is particularly highlighted in some databases.

In an implementation-wide sense, engineering databases will be interconnected through networks. Networks will link, online, our firm to its customers and suppliers. They will transfer both text files and graphics files. They will transmit voice and data as well as images.

Networks have to be increasingly intelligent to deal with ever-growing user requirements, but also to handle ad hoc needs. If a complete file is to be sent to a remote place for storage and subsequent retrieval, then use by its originator, there seems to be no need to transform its structure of stored data. It can be simply accepted as a package by the store and returned, on demand, in the same state.

However, if the store is operated as a common service, providing backup and archiving facilities for several computers and their users, a uniform protocol is needed to govern the transfer of files. Such a protocol can treat files as packages whose contents are absolutely independent of the transfer mechanism.

An example is provided by Data Interchange Format (DIF). This is a utility origianlly designed by Daniel Bricklin (the inventor of Visicalc) to permit the exchange of data between programs specifically from mainframe files to his spreadsheet. The DIF format is expressed in terms of:

- Table (file);
- Vectors (fields);
- Tuples (records);
- Main data; *and*
- End of data.

---

* See also D.N. Chorafas and S. Legg "The Engineering Database", Butterworths, London and Boston, 1988

A primary advantage of a standard data interchange format is the elimination of costly, time-consuming re-entry of information which is already in use by one program. If two applications cannot access the same IE, but require the same information, that information must somehow be entered twice - which violates the efficiency principle: one entry but *many uses* of information.

An Office Document Architecture

The communications requirements developing during the coming years will increasingly focus on document interchange. Between 1984 and 1987 there were two standards: one by the Consultative Committee for International Telecommunications and Telephony (CCITT); the other by IBM. With IBM now supporting X.400 things might be changing.

Like all CCITT standards, X.400 is described in a series of recommendations. The processes that these elaborate fit within the upper-most layer of ISO/OSI (International Standards Organisation/Open Systems Interconnection). They are grouped into two layers: UAE (User Agent Entity) and MTAE (Message Transfer Agent Entity).

IBM's DCA (Document Content Architecture) defines data formats that are compatible among dissimilar office systems. The DIA sublayer corresponds to the UAE of CCITT, but it is not compatible with it.

Within either X.400 *or* DIA/DCA, the protocol compatibility which exists allows office systems (whose design is based on the corresponding architecture) to communicate with other offices in a transparent manner. Since there is no similitude of formats between DCA and UAE, interfaces are necessary for applications living in these two different worlds.

IBM's DIA (Document Interchange Architecture) defines functions for interchanging documents and other information between separate office systems connected through the network. It permits its users to:

- Request document distribution and processing functions;
- Address recipients; and
- Retrieve documents from a library without having to know anything about the physical organisation of the network.

These services permit a user to enter a single request to distribute a document to multiple recipients, confirm delivery, and so on.

This interest in *document protocols* is justified by the fact that productivity can be significantly increased by eliminating the discrete islands of automation which result when:

- Departments implement computers independently of one another; and
- Cannot exchange information electronically.

Without doubt, a full integration perspective permits a higher level of automation than can be possible with departmental barriers. As Figure 14.3 suggests, the latter lead toward a lower level of sophistication.

An organisation needs efficient document protocols permitting an efficient internal and external exchange of information. Examples of internal exchange in a banking

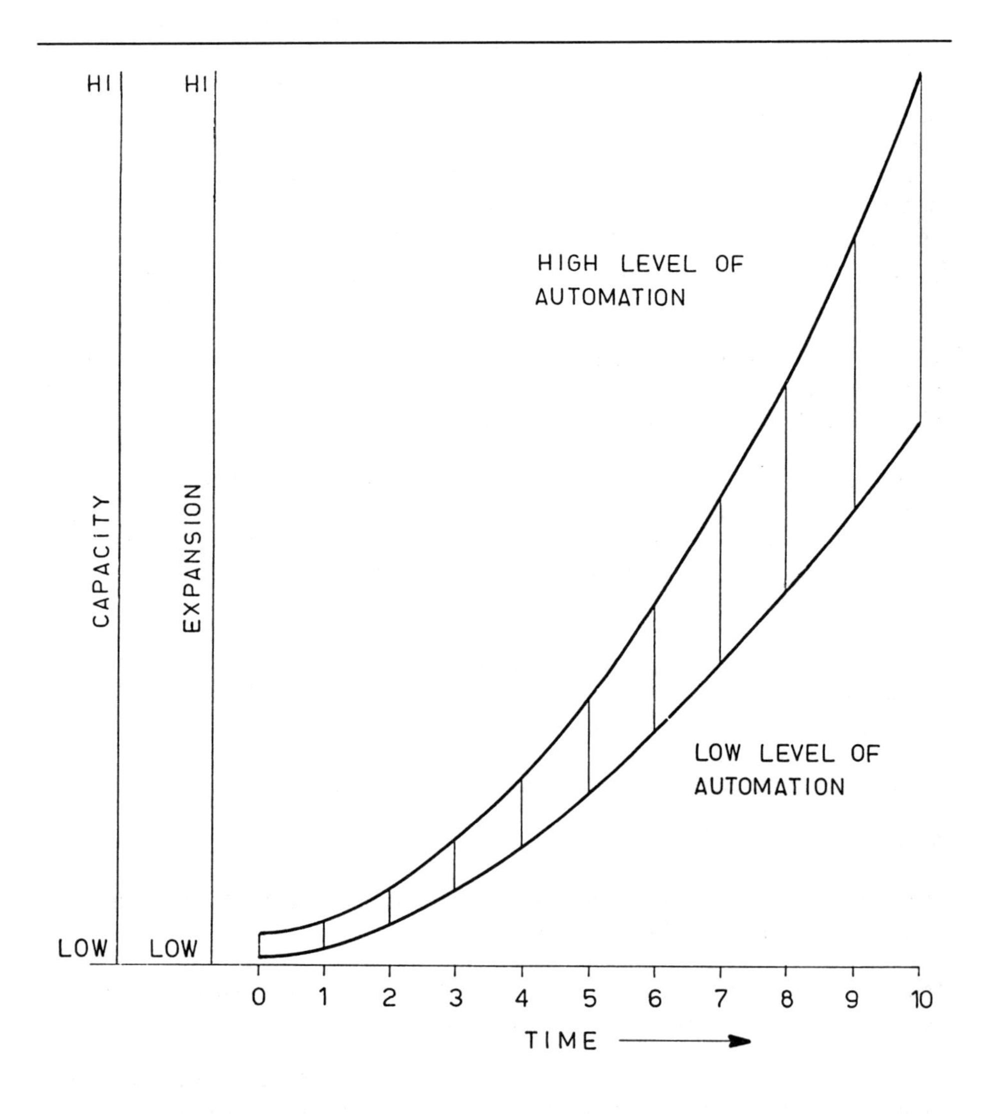

Figure 14.3

environment can be found among investment and commercial banking, securities, forex, finance, engineering, graphic arts and publishing departments.

Using Lotus Graphwriter II and Lotus Freelance Plus, for instance, the finance department may put together an analysis chart, then send it to the graphic arts department for visual enhancement and the creation of a slide to supplement the report. Once the graph is enhanced, it too can be sent to the electronic publishing department through the same protocol. In the publishing department, the images from the two departments have to be integrated with other documents.

As these requirements expand, new protocols are needed to handle them. Since 1987, ISO's Office Document Architecture (ODA) has helped to create a new generation of document exchange automation. Its aim is to provide interoperability between different proprietary word processing and desktop publishing systems, thus assuring a foundation for true multimedia communications.

We should be keen to both integrate text, data, voice, image and graphics into a multimedia system and to make possible the interchange of fully processable documents. Some of the leading suppliers have chosen ODA as the basic architecture for compound electronic document exchange; others are using ODA as a basic architecture for their office automation product - to be used while employing Message Handling Systems (MHS) and File Transfer Access and Management (FTAM).

ODA is an ISO-recommended standard for interchanging "what you see is what you get" (WYSIWYG) integrated text and graphics documents *between dissmilar systems* in an Open Systems Interconnetion environment.

- It prescribes an independent model of the content, logical and layout structures in typical documents; *and*

- Provides for processable documents interchange.

By observing the ODA protocol, documents with features such as page numbering, graphics, footnotes, graphs, paragraph format and so on, can be edited and reformatted. They can make round trips between systems without losing information which the users did not intentionally delete.

ODA includes three models: document processing, structural and descriptive. It features document classes, layout and logical structure, overview attributes and styles. The content architecture addresses itself to:

- Character Sets;

- Raster Graphics; *and*

- Geometric Graphics.

There is a document profile and interchange format, as well as interchange mechanisms and document specification language(s).

The ODA aim is to provide a high level protocol basis for text and graphics documents to be sent using Message-handling System and File Transfer, Access and Management - the OSI Application Layer service. Increasing availability of X.400 MHS implementations and rapidly growing interest in FTAM illustrates the significance of these services, particularly when attribute qualifiers and defaulting mechanisms are specified as well as when functional requirements, technical specifications and attribute applicability are included.

Integrating Voice with Text and Data

In an intelligent network, the integration capability and range of information facilities to be provided should permit text, data, graphics, image and voice to be transmitted, switched, stored and presented anywhere within the organisation. This must be done in a manner enhancing mental productivity, promoting client service, and ensuring a high quality product.

As I have had the opportunity to emphasize on a number of occasions, eventually all services of an information society should be accessed remotely. Operating through an integrated information network, endusers and computer professionals are no longer two distinct populations. For any practical purpose they are one.

Independently of the specific professional orientation of each member of this integrated population of users, the need is increasingly felt for bringing together voice with text and data. We have to go in this direction and proceed as fast as the best available technology permits, in spite of the difficulties in terms of integration underlined in the preceding chapter.

Integration in an information-wide sense comes in stages. Since the early 1980s data processing (DP) and word processing (WP) have integrated. The new challenge is that of integration at the multimedia level, including voice and image. This is practically synonymous to the creation of large scale information systems for knowledge processing.

Advantage should be taken of the fact that new and interesting types of voice products have been introduced to the market. ''Data'' and ''voice'' people have begun working together to plan their future integrated information systems - which should span a whole range of services:

- transmission;
- switching; *and*
- workstations.

Attention should be paid to voice input/output because, to the human user, the voice is the most natural and efficient means of communications. If voice could be employed as a means of exchange between man and machine, it would be the most desirable communication method, as it imposes no pressure on change of habits on the part of man.

If man and machine are able to communicate with each other over a public telephone line, information can easily be input and output from an individual home or a distant location, using the basic telephone installed in most homes as a terminal. In realisation of this fact, recognition and audio response systems combine:

- voice response functions, *with*

- speaker-independent voice recognition offering a solution which has for some time attracted management's attention.

Such systems are already in service at banks in various countries. They are now providing information such as payment guides, balance queries and other customer data directly to clients by voice via a telephone set. New equipment integrates various functions, such as speech recognition, audio response, pushbutton telephone signal receiver and so on, into one unit.

Artificial intelligence-enriched solutions, agile software support and physical compactness are currently aiming at expansion of the application range as a further step forward. A good reference is: PSI-X, an AI-enriched PBX under development by NTT in Japan, based on ICOT's PSI engine.

The main problems in the speaker-independent isolated-word recognition are voice pattern variations due to different speakers and speaking rates. To overcome these problems, classification functions and time normalisation have been adopted. The recognition process has four stages:

1. Speech analysis;

2. Time normalisation;

3. Template matching; *and*

4. Classification.

An important parameter for specifying automatic speech recognition (ASR) technologies is vocabulary size. In a limited-word ASR system which recognises up to several hundred words, the basic recognition unit can be the whole word, and the recognition process is quite simple. To the contrary, unlimited-word ASR requires sophisticated phoneme-based processing.

The limited-word ASR can be divided into the speaker-dependent and speaker-independent types. Each of them is further distinguished into discrete word and connected word recognition. A generalised two-level dynamic programming/matching algorihtm incorporates a finite state automation syntax.

Already several types of speech recognition and synthesis systems have been developed and commercialised. Some of these systems have been widely used in real world applications. Still, there are many problems yet to be solved to establish a really useful man-machine interface through spoken language. Such problems are closely related with natural language processing technologies such as lexical, syntactic, and semantic analysis.

There is a different manner of presenting these concepts. Voice technology can be expressed at three levels of sophistication: the easiest is single word, and it has been applied since 1974. Then comes full sentence without meaning. The highest layer is full sentence with meaning - the true natural language dialogue.

Single word voice recognition applications have been typically voice interfaces to computer menus. The menu selection is spoken rather than pointed to or keyed in. However, single word applications have not generated much interest because they add relatively little value. They may have, however, found a market on the industrial floor and in banking, for queries about current accounts.

Single word or discrete speech recognition is more error-free than connected speech recognition because the processor does not have to guess at word boundaries. Hence, larger vocabularies and richer domains are achievable with discrete speech. The cost is lower, but the approach presents user inconvenience in the sense that he cannot behave in his natural manner, although he uses his natural language.

Still, in applications where speech is the preferred mode of input, these disadvantages are outweighed by the advantages. Also, users are able to learn to speak machine recognisable discrete speech more easily than they learn to speak machine recognisable connected speech.

Typically, in such a system, the voice recogniser drives an error-correcting parser which sends output to a language semantics processing system. The latter displays the result to the user. A voice response system can be added for answerback, prompting and error messages.

Connected speech can be approached at two different levels of sophistication. An example of full sentence without meaning implementation is voice-activated typewriters. The user dictates an entire letter into the computer. These applications take advantage of the larger vocabularies that recent voice systems can recognise. They do not require full natural language facilities because there is no need for the computer to understand the text. It only has to format it.

Ongoing research in speech technologies utilises artificial intelligence and knowledge-base systems, in connection with natural language processing. It looks at natural languages as the focal point of developments for the years to come.

One of the most promising fields with speech technology is automatic language translation. This is a subject which we will consider in section 6 in connection with the increasingly sophisticated communications requirements of a knowledge society.

There is a similarity between natural language manipulation and computing. When we wish to compute we have to devise:

1. Computational rules; *and*

2. Representation.

Linguistics have similar semantics requirements. The wider approach is to view words as pointers to concepts. For instance, *Credit*, in a banking sense, means: credit proper, credit procedures, and credit risks. This notion underlines communications and leads to phrase semantics.

Some of the basic ideas behind this presentation come from compositional semantics, which have had an impact on computation only since 1981/82. Still, there is no real theoretical guidance on how to handle language translation in a systematic way. There is, however, a recognised need for representation in phrase semantics and the thought that first order logic can do the work.

The Japanese project (DUALS) aims at discourse understanding through logic-based systems through AI. It is also layered and involves:

1. *Semantic structure*, using situation theory, assignments, individuals, relations, locations, conditions, events, parameters, numbering, layout, etc. can be re-considered.

2. *Syntax analyses*, including grammatical rules, executed by a parser;

3. *Anaphora processing*. This is an algorithm identifying pronouns and zero-pronouns;

4. *Plan-goal-based discourse structures*, including metarules for linguistic discourse; and a:

5. *Sequence generating module*. Its goal is the generation of surface sentences from internal meaning structures using grammar rules.

Anaphora is *situation semantics* (framing). It is done in the context of natural language processing. Anaphora precedes the natural language (NL) sentence, and is important in defining the meaning of relative pronouns.

The competent handling of natural language and speech constructs will play an increasingly important role in the years to come. NL solutions will dominate business applications of communications and computers, serving databasing requirements. They will be helped through breakthroughs in artificial intelligence.

As we can see in conclusion, integration perspectives are in a process of full evolution. The aims and nature of activities change with time, progressing toward higher levels of design reference as we acquire implementation experience at the relatively lower levels.

Let us keep in mind the fact that no manufacturing company or financial institution anywhere in the world has patent rights on telecommunications facilities, intelligent workstations, expert systems, supercomputers or realspace solutions. At the same time, the cost of staying in touch with high technology is high. That is why we should use, in the most profitable manner, every dollar that we invest - always adjusting to the advancing state of the art. If investments do not give us a competitive advantage, they are not worth making.

Natural Language Solutions in Communications

An important component of the skilful use of computers as communications links is a natural, user-friendly interface allowing people without extensive computer skills to employ the system. In addition, the computer must be able to explain its reasoning and justify its conclusions, so that users can evaluate the validity of results - which is assisted through expert systems.

As we said in Chapter 7, natural language recognition represents one of the most significant advances in the history of computers. Today, for the first time, spoken requests for information can be accepted as an input. A sequence of words handled by voice recognition technology, as well as their meaning, can be understood and the user's questions answered by the natural-language system.

The combination of voice recognition and natural language is likely to have the same impact on enduser computing that the addition of sound had on the motion picture industry. A still greater impact will come from *language translation.* In its background is the premise that the ability of computers to understand and interpret natural language will increase their:

- utility;
- accessibility; *and*
- flexibility.

Therefore, the Japanese 5GC project has steadily aimed to develop natural language solutions. So do other computers and communications projects around the world. These projects have found that a logic programming framework is the most suitable for implementing natural language processing. The basic mechanism of logic programming is unification, including semantics pragmatics and inference, leading to an AI orientation.

While automatic translation between languages has been a goal for nearly two decades, only now do we start seeing basic breakthroughs. Technological advances have been most helpful in this domain, and the same is true of our ability to closely follow and understand the cycle of communications.

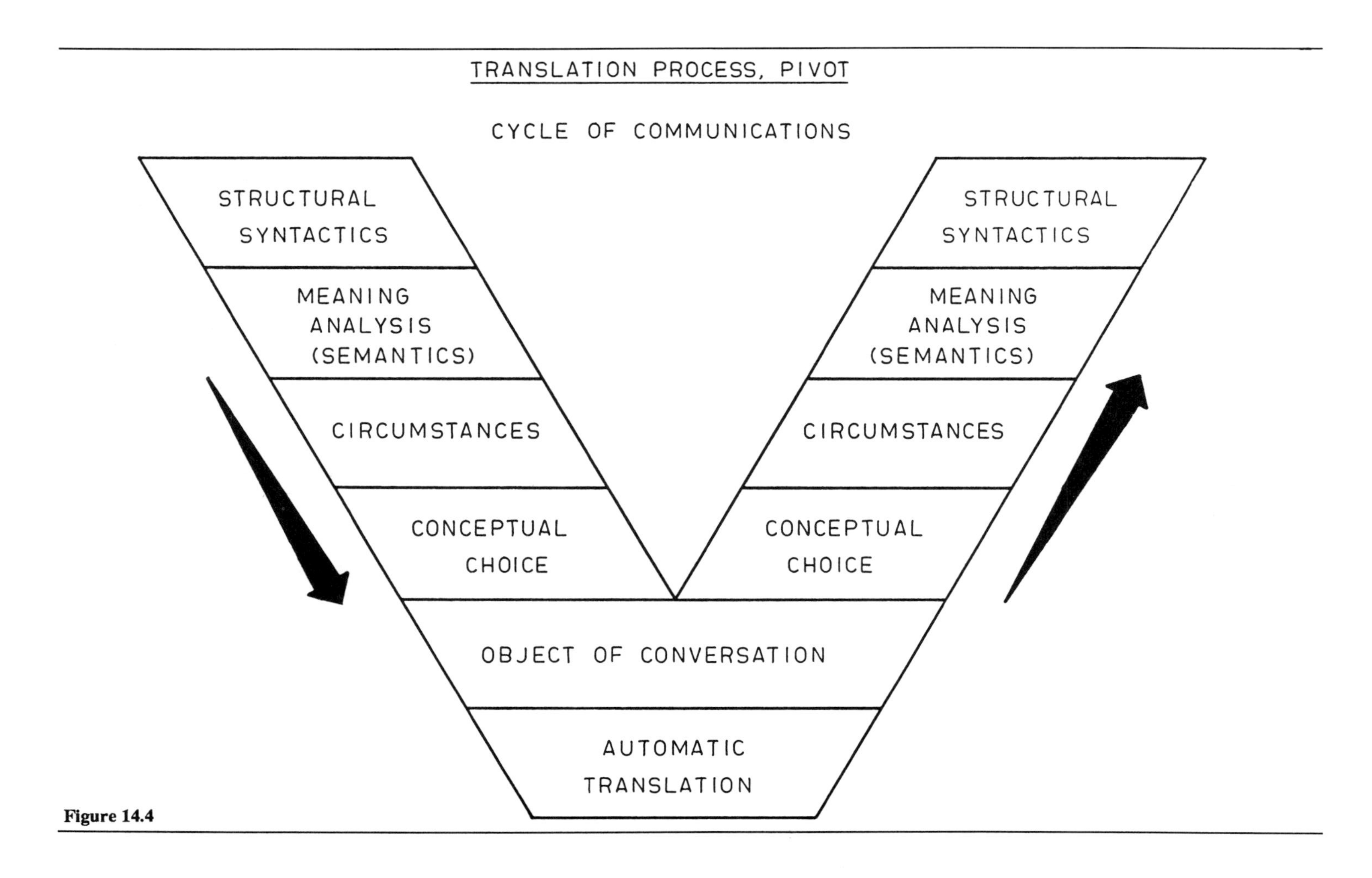
TRANSLATION PROCESS, PIVOT
CYCLE OF COMMUNICATIONS
STRUCTURAL SYNTACTICS
MEANING ANALYSIS (SEMANTICS)
CIRCUMSTANCES
CONCEPTUAL CHOICE
STRUCTURAL SYNTACTICS
MEANING ANALYSIS (SEMANTICS)
CIRCUMSTANCES
CONCEPTUAL CHOICE
OBJECT OF CONVERSATION
AUTOMATIC TRANSLATION

Figure 14.4

Automatic language translation is an integral part of this effort, with the Japanese finding themselves in the lead. Figure 14.4 outlines the translation process adopted by the Pivot project. The full cycle of communications is described, but the reader's attention is brought to the fact that the topmost issue is one of cultural understanding.

As it is easy to appreciate, a layered approach has been adopted. The translation process goes through structural syntactics, meaning analysis, and identifiers regarding circumstances. Bringing under perspective contextual aspects is a major contribution of the Japanese approach. The AI construct:

- Starts working at a source language through structural syntactics;
- Proceeds with meaning analysis;
- Accounts for circumstances and culture;
- Leads into conceptual choice; *and*
- Only then attacks the object of conversation, subjecting it to automatic translation.

Let me repeat this reference, as it constitutes a new departure. The language translation process focuses on conceptual choices and the object of the conversation, prior to proceeding with automatic translation as such. Circumstances, conceptual choices and the object of conversation are cultural ingredients that the Japanese researchers have ingeniously employed. They help overcome the constraints that automatic translation encountered in the past. But they also call for AI assistance.

The same procedure is followed to reach the object language. Building a lexicon is like defining data descriptions for a database management system. It is the simplest part. Much more is necessary to achieve commendable results.

Two conclusions can be derived from this short description. First, the need for powerful equipment to help in linguistic communications. Hence, 5th Generation Computing. Second, the equally basic need for a methodology able to handle the representation mechanism:

- What are the types of things we can talk about?
- What can we say about them?
- How can we understand somebody elses communication?

Other issues characterising linguistic logic are properties and relations. Properties are one point of view of relations. Objects are properties and stand in notation to one another.

Current research on automatic language translation uses propositional connectivity to build more complex structures. It also employs intentional operators, which have to do with the meaning rather than the expression.

The message to be retained from this discussion is that significant work is now in process to realise an automatic language translation mechanism. Some results are already available. Others are forthcoming. NEC has a realtime implementation able to understand English and respond in Japanese, or German and respond in French. It runs on a mainframe. With the CHI engine it will reside in a small-size, high-capacity logical box.

Significant benefits can be derived from combinations of different technologies. For example, language translation and electronic mail. Synergy may be of value in multiple-task situations, where a user could carry on a telephone conversation and request data or send a message at the same time. To determine specific applications, we have to see how the solutions develop and at what cost.

Benefits never come as a matter of course. For every system, specific studies are necessary to respond to queries such as:

What training would be required to teach users to speak in machine recognisable sentences?

- What additional learning would occur while individuals use the system and start to ask for more functionality?

- What word error rates would be delivered by the voice recogniser or automatic translator?

- To what extent would these errors be correctable by a machine?

The whole man-machine interactivity has to be studied. How fast would users speak individual sentences? How fast could tasks be completed? Also, what is the expected response? How would the enduser judge his ability to do useful work through the automatic translation mechanism put at his disposal?

The highest sophistication of voice technology can be realised in full synchronism with a growing range of developments. Some of them are basic for extracting and analysing information from computer databases. Others permit the user to hold a meaningful conversation with the computer - or with another user through the computer. At the bottom line, they automate the process of asking questions and receiving answers.

Natural language technology brings the computer squarely into the communications process. It permits the machine to understand the meaning of the request, and to answer the questions. Sophisticated voice recognition technology propels the natural language approach to a new level of interaction. Many more developments will be required before the whole process settles.

What Can We Expect in the Coming Years?

While handling full sentences with meaning is not yet an available technology, research is now being conducted in this direction at several American, European and Japanese universities and at electronics companies. One such project, launched by Japan's Advanced Telecommunications Research Institute International, received $107 million, for a 7-year budget, from the Japanese government, Nippon Telegraph and Telephone, and a handful of other corporate giants are also at work in this field.

IBM is one sponsor of similar efforts at Carnegie-Mellon University. The goal is a system that will produce text out of the speech sounds of one language, analyse and translate it in context and reconvert the translated signals into speech.

This and similar projects employ already-achieved breakthroughs in speech recognition and speech synthesis. The latter is on its way to becoming a mature technology where further advances focus on economic justification rather than feasibility: reducing the price of parts and finding more applications for the system.

Still, in spite of ongoing developments, *speech recognition* is still in its early stages. While the size of vocabularies is large, and advances in memory density are impressive, we still need logical constructs as described in the preceding paragraphs.

Speech synthesis has been a simpler undertaking. Semiconductor technology can synthesise speech by sending a combination of voiced and unvoiced sources through a filter representing the five different formats. A lookup table is required to set the filters for each of the different sounds or phonemes.

Differentiation between male and female voice is created by setting the female formants at higher frequencies. Analog circuitry, or even switched capacitor technology, can be used to set up the filters. Another approach is to implement the filters digitally with a signal processor and then convert the digital signal to analog with a digital/analog converter.

The human ear is very perceptive, so precision circuits are required for high quality sound, making the parts expensive, but advances in design rules have brought prices down to consumer application levels. Typically, a microcontroller would be used to convert text to speech by breaking up words into phonemes and sending the right data into the filters and then out to a speaker.

We have seen the evidence regarding why speech recognition calls for more complex solutions. The most widely used voice recognition systems are based on the *pattern matching* paradigm. Initially, the user registers one or more samples of every vocabulary word in the machine. At later times, recognition is done by comparing the time-spectral characteristics of unknown words with the known samples.

This is the sense of *device training*. The unknown word is assumed to be identical to the stored prototype which is closest to it by some distance measure. But, if the unknown word has no near matches, the recogniser may reject it by refusing to make a selection. Connected speech on such machines is typically done by a dynamic programming algorithm that guesses word boundaries at one level and matches words at the other level. We have also spoken of a complete sentence understanding.

New systems are presently designed to eliminate the need for learning the speaker's voice beforehand. Input can be made from an ordinary telephone set, and a nationwide system can be established through the use of the public telephone network.

High technology recognition methods effectively absorb voice pattern fluctuations, resulting in a high recognition accuracy of more than 95 percent. High-speed computers make realtime recognition possible. By incorporating a time-division matrix switch for the voice input part, the system accommodates a number of voice-output channels for greater overall efficiency.

One day, callers may simply need to hook their telephones up to personal computers, plug in voice recognition and synthesising units, and converse in a foreign language. They will not care about the grammar of their own language and those they don't speak. Neither will they care about vocabularies. From the interpretation of the *cultural environment* to the enrichment of the lexicon, the job will be done by computers.

15

Macroengineering, CIM and Robotics

Introduction

Our society is fascinated by two things: fast development and gigantic structures. Both call for the ability to continue significant investments in research and, in a number of cases, both involve unknown risks. We have seen examples with biotechnology, but there are also many others.

Experts in the relatively new discipline of risk analysis caution that assessing the collective risk of the world's potential calamities is not an easy matter. Experimenting on the individual probabilities provides at least a first approximation, and computer-based studies can be relatively accurate. But, there is no certainty as there is no experience in pre-evaluating the very large aggregates that we construct, and in assessing their long-term lifecycle with its associated dependability.

One way of evaluating future risk is through the use of probabilities. Norman C. Rasmussen, Professor of Nuclear Engineering at the Massachusetts Institute of Technology, put his thoughts in the following manner: ''The probability of a dam failure is about 1 in 10,000 a year. There are about 10,000 dams. And they fail about one a year''. The numbers change as plants age - and that, too, must be taken into consideration. We have also spoken of Chernobyl, and of the educated guesses regarding the risk of failure which have proved to be wrong.

The reason why such estimates cannot be considered as being awfully reliable is that there is, at present, a great inconsistency in the *assessment* of differing *risks*. This is an art which, in today's interdependent world, involves balancing very small risks against very large consequences, and the means for doing so is not under control. Steady vigilance is the answer.

The risk from hydroelectric and nuclear plants is not the only one we are faced with in this age of gigantism. The present chapter provides the necessary evidence to back up this statement. As technology moves forward at full speed, and we are eager to put every breakthrough immediately into practice - which is essentially commendable - we fail to appreciate that we are opening our flanks to hazards. However, everyday life gives us reasons to consider this issue.

Risk management has its own focal points for study and research:

- What are the management approaches which will permit us to keep new and unknown processes under control?
- What sort of very specific and detailed technical safeguards do we provide and which sort of supercontrol is exercised?
- What are the long range directions that are required to maintain and improve upon our civilisation without unwanted risk?
- What design tools are being used to ensure correct procedures at every milestone of a new project?
- How is prototyping handled so that current processes and systems are not adversely affected until they are definitely replaced?
- How can laboratories and companies obtain qualified designers and management personnel? How should the latter control the form? How is testing being handled by the laboratory? the manufacturers? the social organisation as a whole?
- How are the consumer interface problems, that inevitably occur, being resolved?

While hazards may be easier conceived in fields such as biotechnology, nuclear engineering and very large dams, the mass production of other goods can also create severe problems - unless we act with insight and foresight. One of the better examples is automobiles.

Based on an inner city, which in Europe quite often dates back to the Middle Ages, but which in America has, at most, two hundred years history, the sprawling metropolis feels the pinch of traffic jams and congestions. Rome annually adds 120,000 cars to its motor vehicle inventory. About 20,000 cars are retired yearly; the balance increases the weight of traffic growth and congestions. The large metropolis phenomenon is reshaping the cities in general and our living space in particular - polluting it in the process.

This is one of the reasons (though not necessarily the only one) why we are so much interested in developing *efficient, intelligent networks.* One of the aims is: ''Communicate, don't commute''. This can be helped by networks which provide, not only plain telephony, but which also permit the introduction of new services capable of reducing communications costs and improving management efficiency. We have spoken of this fact.

Large corporations are actively reviewing their networks, and planning for the new generation. They want advanced services, economy, high grade service, network reliability, and the right kinds of terminals. In short, they would like access to an intelligent network as soon as possible.

An intelligent network will definitely be a large scale system. It will operate under a distributed architecture that allows for enhanced network management capabilities and an infusion of new services. It will feature a significant number of AI constructs and aim to offer its users a leading position in the market in which they work.

There is no clear definition yet of what an intelligent network may feature. A first approximation devised in France tends to include:

- Fully digital transmission;
- Integrated multimedia capabilities;
- Toll-free calling;
- Alternate billing services;
- Emergency response;
- Private virtual networking; *and*
- Wide-Area centrex.

Welcome as these services may be, they are only the bottom line. Design should be done so that, once the infrastructure of the intelligent network is in place, it will be possible to integrate other, AI-enriched services with little problem. A fully digital network makes it easy to move quickly into intelligent networking, but the architectural perspectives are all important.

The macroengineering aspect is vital. Answers would not come easy and they cannot be left to the level of cheap political arguments. In a number of systems projected for the 1990s and well into the 21st Century, factual and documented answers require a close collaboration between design engineers, city planners, far-sighted sociologists, imaginative architects and macroengineering experts. We must plan beyond Year 2000 if we wish our constructs to have a good lifecycle.

Macroengineering Projects

Macroengineering projects have several common characteristics. The three topmost are size, complexity and cost. Such projects typically include: large telecommunications networks; interconnected multimainframe computer centres; energy plants and associated distribution networks; planned cities; regional development enterprises; large transportation projects; industrial complexes; and space programs.

Typically, macroengineering projects require very substantial capital investment - from hundreds of millions to billions of dollars, and are distinguished by four factors:

1. They are beyond the capacity of a single organisation to implement.

Hence, they require collaborative efforts. This is not only true of funding but also of technology. The sheer size of a macroengineering project makes imperative: skilful resource planning; tough development timetables; and impressive knowhow - from design to control and implementation.

2. They utilise to their limits advanced technology and scientific breakthroughs.

This includes adaptations for the utilisation of new means and processes, but above all it calls for imaginative research undertakings, aimed to push the frontiers of knowledge well beyond where they stand today. The Star Wars project is an example.

3. They involve considerable complexity, and complexity can mean different things: conceptual, descriptive and computational.

Even a one-line algorithm may be complex in terms of its expression, but can execute very quickly. Then we say that its descriptive complexity is high but its computational complexity is low. This is not true of macroengineering. There are also interesting analogies between complexity issues in science and in economics - where conceptual, descriptive and computational complexity tend to merge.

4. They have far-reaching economic, sociological, cultural, and environmental impact.

Such impact extends further than the project's sponsors even imagine, and faster than the sponsors think it ever will. It includes other organisations: bankers, export credit agencies, insurers, suppliers, shippers, marketing outlets, nations, governmental ministries, universities and public interest groups.

A project is *macro* if we know very little about it. As we enter into the merits of the subject, it settles down, although it may still be a complex system. When we get to know, not only its components, but also how to integrate them effectively, it becomes a package, and we can start marketing it as such. With experience covering all its aspects, the formerly complex system finally settles into a simple product and it is often offered off-the-shelf.

An example of projects with far-reaching economic, sociological and cultural implications is 5th Generation Computing and artificial intelligence. Shortly after these problems are solved, technology will be facing a second and third challenge rolled into one. For instance, a computer that could execute computations at a rate of 1 trillion operations per second by the end of the century.

Precisely because the implications and aftermaths of the high technology age are so great, and yet still so unknown, it is imperative that a country, company or person remains at the very forefront of technology. More is at stake than national prestige. The issue is one of ensuring both *economic momentum* and *cultural survival.*

This is the other side of macroengineering projects. A knowledge of both technical and historical forces pushes towards a sense of proportion. In a way, macroengineering can be compared to gardening. You have to be patient; you have to understand the soil and the plants; you cannot pull things up every week to see whether the roots have grown. You have not only to project in the right way, but also to inspect regularly.

An alien subject to macroengineering is the often obstinate indifference to anything except immediate issues and trivial results. Another alien concept is the conviction, in the teeth of the evidence, that any problem can be fixed: "That there is nothing that money cannot buy". We do not solve complex problems by throwing money at them.

While technology, particularly high technology, needs comfortable budgets to work with, money will not buy everything. The difference is made up of ingenuity, skill and dedication. The knowhow necessary to see projects through; the insight to understand the hidden aspects; the foresight to project well into the future.

One Logical Network

As with all macroengineering projects, specialising on systems integration means tackling large jobs. Systems integration involves purchasing, and putting together, all of the necessary pieces of, say, a computers and communications aggregate; caring both in the process of selection and in the course of subsequent design, implementation and maintenance.

An example of a macroengineering project with system integration objectives is the *one logical network* that EDS is building for General Motors. The goal is:

- To integrate the many physical networks in existence;
- To expand GM's communications capacity; *and*
- To bend the curve of cost increases.

To reach this goal, EDS has been tearing out much of AT&T's existing equipment, lines, telephones and Centrex systems. It is replacing them with modern devices, including high capacity digital private branch exchanges (PBX). Through microwaves and satellites, EDS plans to reduce the cost of GM's communications by about 20 percent to 30 percent per transmission.

The Information Network System (INS) in Japan is another example of macroengineering focusing both on communications and on system integration. It has been elaborated by Nippon Telegraph and Telephone Corporation (NTT), and is now being tested. It foreshadows a society in the not-too-distant future - and, most particularly, its technological requirements. The programme is expected to furnish new telecommunications networks and services in Japan in the next decade.

The components of the network system are not in themselves new. What is original about NTT's plan is the idea of linking the entire telecommunications network to one trunk line. In other words, telephone, facsimile, data and image communication signals will be merged into a single digital network and then linked across Japan through integrated optical-fibre cables.

As a result, various communication services would work as one aggregate. Transmission and calls will be quicker, and in the long run more economical, NTT claims. An official estimated that the programme would lower the cost of a telephone call by 30 percent, when implementation becomes effective.

Presently, the INS program (which has been tested in two Tokyo suburbs) is still too costly to be commercially immediately viable. But the future promises a service that could be applicable in many fields.

A telephone line linked to a facsimile can transmit messages simultaneously to several destinations, at two seconds per page. Another, for colour prints, sends a page in six seconds. The sketch phone could be an ideal communications vehicle for the deaf, says NTT. If the person you call is out, you can leave messages at the other end.

The video teleconference system will allow businessmen to hold meetings with colleagues in other cities. Hospitals could benefit from the reflection-circuit service with which they can monitor patients, even from a distance. A message-communication service will act as a sort of message board for subscribers.

Macroengineering projects of this type first define a system architecture. It is then the duty of the system architect to choose a system network solution. These two layers constitute the communications part of system integration, as Figure 15.1 suggests.

Building, implementing and maintaining complete standalone systems through their traditional life cycles does not respond to the requirements of the 1990s. Integration should be pursued through steady, directed effort. A rational approach is to build and experiment with prototypes. Prototyping permits us to validate system requirements. It also makes it feasible to verify the rules characterising data about data, meaning the *metadata.*

For instance, metadata associated with a macro-engineering project could be project identification, scope, return on investment, scheduled start date, scheduled completion date, design reviews and approvals. Metadata associated with a capital investment account include account number, maximum approved expenditure level, subordinate expense accounts, personal responsibility and accountability, expense accumulation, and plan/actual budgetary comparisons.

Such information may exist in discrete and incompatible databases. The latter may not necessarily be stored online or, if they are, they may belong to separate networks which are not interconnected. The data may be coded in formats, and code may be difficult to exploit or completely untranslatable. It is not enough to access a repository of information, it is also necessary to be able to exploit the IE which it contains.

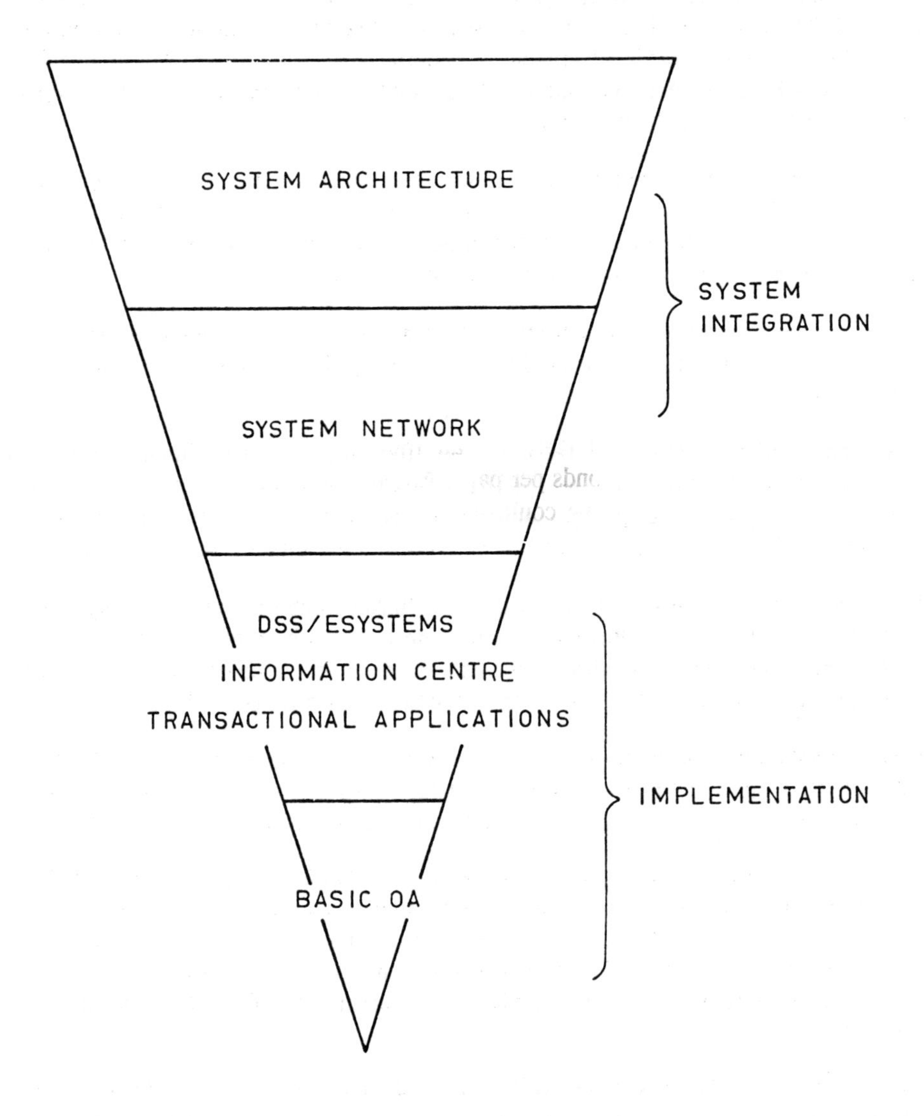

Figure 15.1

One integrated network structure will not merely facilitate communications among distributed databases, and between databases and workstations. It will make possible any-to-any interconnection, at any time for any purpose. It will do so in an ad hoc, non-preplanned manner, and will ensure that the whole process is transparent to the enduser - no matter how complex the access may be. This is the sense of *one logical network.*

Largely based on communications disciplines and database systems, integration requires:

1. Concepts;

2. Definitions;

3. Instrumental Design;

4. Competent Implementation; *and*

5. Flexibility for renewal and expansion.

Solutions must be any-to-any, and any-to-any connectivity is one of the most challenging jobs. Yet it is necessary, because we are not smart enough to see the future, and all the evolving requirements. At the same time, our network is in steady evolution.

Strategically, the integration of information systems is one of the most difficult tasks, the more so when combining corporate resources. Prior to the merger of Northwestern Bell, Mountain Bell and Pacific Northwest Bell into *US West*, management wisely put priority on the integration of the information systems of the three firms. Synchronisation proved to be a demanding task, in spite of common AT&T origins.

Key players in the systems integration market specialise in building highly-efficient private communications networks. Typically, these are large organisations. For instance, the Fortune-1000-sized companies with geographically distributed sites, as well as govern-ments. For such organisations, one logical network becomes the necessary infrastructure for continuing success in the marketplace.

Banks and insurance companies are prime candidates in this systems integration effort. Any labour-intensive industry which has been, for at least 10 to 15 years, actively seeking to automate its operations, is a fertile area where the integration technology can be of help - and the same is true of the new concepts which dominate networking.

The Thrust for Greater Competitiveness

Innovation, rigourous cost control, imaginative product design and marketing, and an overall willingness to take risks, are a necessity rather than a choice for companies which wish to remain competitive in order to survive. Precisely the reverse is true of those organisations living in the past, which are scared of preparing for the future. Corporations effectively responding to present-day challenges resort to a wide variety of tactics:

1. *Full automation* - usually a wise move, but also dangerous if the technological and organisational leap is too large and the human resources cannot follow.

We have paid considerable attention to this fact. What should also be underlined is that it is not enough to make now and then automation breakthroughs, and then rest on the laurals: In 1976, a high mark was text processing; but, in 1979, competitiveness meant using spreadsheets; three years later micro-database management systems (DBMS) were a "must" on workstations. In 1985 interest focused on PC-to-mainframe links. Today it is supercomputers and end-to-end, any-to-any networks.

2. *Cutting labour costs.* This is a key to survival for many firms in traditionally high-wage industries. A basic move is cutting payrolls, from stage hands to middle management.

Results would not be obtained as a matter of course. Technology planning is very important. We must carefully look after computers and communications developments and judge how we can use them to our advantage. Corporate chief technology officers (CTO) are not necessarily strategic planners, but they can contribute much to the strategic plan - although the company also needs focused strategic planning for business purposes:

3. *Streamlining* the entire corporation by selling off extraneous businesses and employing excess cash to buy back shares and/or boost dividends. The food, chemical, pharmaceutical and media industries are notable examples.

Computers and communications are only a tool, not a goal. The goal is survival. Streamlining must weed out products, people, investments and operations which are not strategic. Strategic products are those on which our organisation depends for its continuing prosperity during the next decade.

4. *Slashing overheads* by shutting down obsolete or excess capacity, and abandoning the principle of vertical integration by *outsourcing* parts that can more cheaply be purchased than produced within the company.

Strategic considerations should dominate. Merging with other companies can be a path to prosperity, or at least survival. Many firms in the airline, oil, oil service, computer, hospital management, banking and advertising industries, among others, have found the value of such a policy, with larger units benefiting from the effect of mass. Taken in combination, these trends have transformed the competitive line-up of most industries.

Just as one swallow does not make a summer, a radical change in one company may help make that company mean and lean, but does not turn around a country's complex economy. The whole infrastructure must change. No wonder that users of West Germany's telecommunications system grumble that the Bundespost, Europe's biggest *service enterprise* (with annual sales of more than DM 50 billion ($32 billion) and 547,000 employees), is slow to innovate - and this hurts their business. They also complain that little has been done to improve its service, and that little has cost too much.

Awakening is rude, sometimes. In the Bundespost's case, even the government's own council of economic advisers, citing the privatisation of British Telecom and the break-up of American Telephone & Telegraph, has called for its overhaul. However, these wishes are being frustrated by an unholy alliance that includes conservative politicians to whom change is anathema, trade unionists worried about job losses, and West German producers of telecommunications equipment. The latter fear that a reorganisation would upset their cosy links with the Bundespost.

However, if an organisation is to survive, the thrust of restructuring should not be stopped because conflicts of interest raise barriers. The evidence is more than enough to show that large chunks of the manufacturing industry, for instance, can survive only if they shift from being labour-intensive to being knowledge-intensive.

- Machine operators getting high wages for doing unskilled, repetitive work must be replaced by intelligent robots.

- Knowledge-workers may get even higher wages but these will be for designing, controlling and servicing processes and products, or for managing information, not for automatable work.

We pay people for their brains, not for their hands. And they should be using their brains.

The shift in emphasis also fits in with demographics. In every developed country more and more young people, and especially young males, stay in school beyond the secondary level, and consequently are no longer available for blue-collar jobs, even for well-paid ones. A knowledge society needs an educated, skilled workforce that cares about what it is producing. This becomes clearer as new production technologies are introduced.

Getting the kind of labour force necessary for *flexible automation* requires management and unions to abandon the old adversarial attitudes and to strive for cooperation. If unions resist such change, they will guarantee a continuing loss of jobs and production to foreign competitors. That is what has been happening over the last 15 years in America. Manufacturers, for their part, can flee any country. However, they should be keener in making their domestic operations pay better by investing more in training, and encouraging worker participation in obtaining high quality results.

Computer Integrated Manufacturing

The human factors in factory automation are as crucial as the financial and technological factors. Basic to an effective Computer Integrated Manufacturing (CIM) strategy is a strong commitment to intercompany communication. It is essential to have a spirit of cooperation and understanding from the highest level of management down, synchronised with readiness to accept new technologies.

Computer Integrated Manufacturing brings with it new approaches, both internal and external, to enterprise communications. It also introduces a great deal of change - much of it cultural. People must experience change before they believe it, but change made through experience is lasting.

Defined in simple terms, CIM is the application of Information Technology (IT) in order to *solve business problems* and gain advantage over competition. As such, it involves a great many corporate departments, not just manufacturing, as shown in Figure 15.2:

- Marketing;
- Engineering;
- Research and Development;
- Finance; *and*
- Administration.

are an integral part of CIM, both in a conceptual and in an implementation sense.

A basic objective in computer integrated manufacturing is to close the gap which exists today between:

- The Factory; *and*
- Headquarters.

That is, between the planning and control (P+C) of manufacturing operations and the business system within which it runs. This is just as true of process control as of any other production system: automobiles, home appliances, and computers themselves, among other examples.

The existing information gap between the business system and factory P+C is at the origin of:

- higher than necessary costs;
- delays;
- errors; *and*
- incompatibilities.

CIM basically ensures integrated plant information, but also sees to it that such information enters the business system in a homogeneous, compatible and comprehensive way.

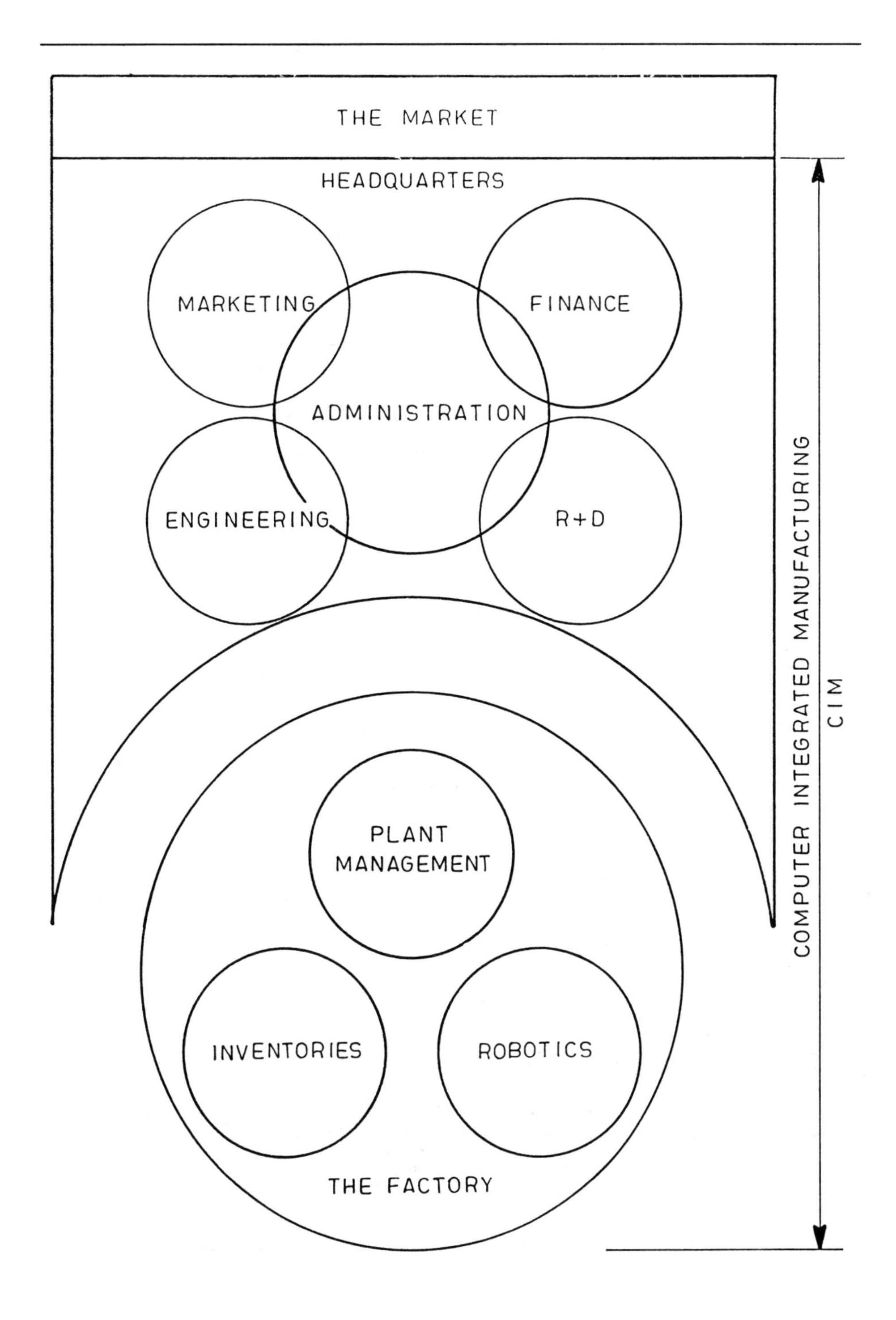
THE MARKET
HEADQUARTERS
MARKETING
FINANCE
ADMINISTRATION
ENGINEERING
R+D
PLANT MANAGEMENT
INVENTORIES
ROBOTICS
THE FACTORY
COMPUTER INTEGRATED MANUFACTURING
CIM

Figure 15.2

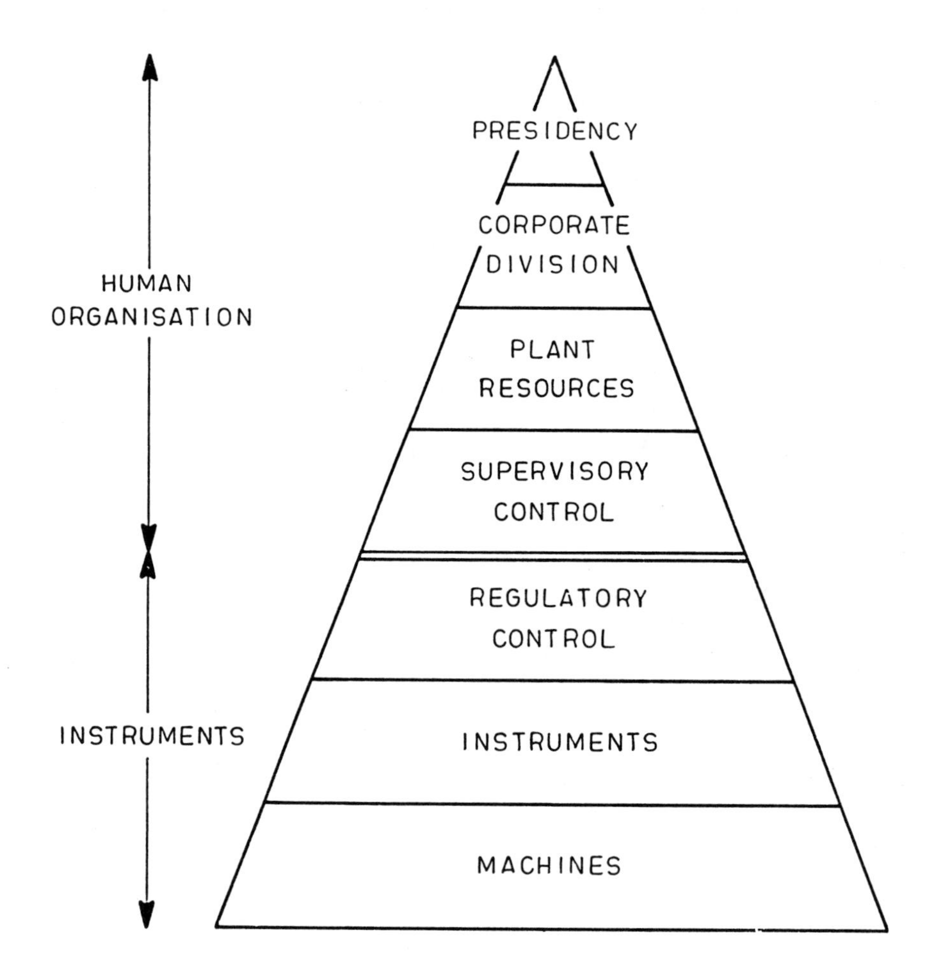

Figure 15.3

Figure 15.3 demonstrates the structure of production systems. They are typically built in layers. Such structure has little to do with computers. It has been the outgrowth of an improved organisation. The problem is that, until recently, computers have not been used as they should to:

- eliminate intermediate structural layers;
- flatten the organisation; *and*
- close the gap between the plant and business management.

The process to which CIM addresses itself specifically is *resource management.* It includes production tracking but also involves a number of important interdependent areas:

- Production Planning and Control (PPC);
- Just-in-Time Inventories;
- Elimination of Waste;
- Maintenance Management;
- Energy and Tools Administration;
- Quality Control;
- Strengthened Accountability; *and*
- Coordination with Marketing.

Some of these issues are long-standing pillars of production management. Others are recent areas of focus where CIM is instrumental. For instance: Just-in-time inventories, steady quality assurance, and elimination of waste. But CIM performs no miracles. We need organisational consistency to successfully implement these types of applications.

In other words, the technology is at *our* service, but is *our* organisation ready for it? The principal issues to be faced with CIM are:

- Organisational;
- Cultural;
- Architectural;
- Response Planning; *and*
- Coping with the rate of Change.

When it comes to CIM implementation we see that variations do exist by industry - but also that all industries face common requirements. Successful companies:

- Focus on cost;
- Rethink and retrofit their information technology; *and*
- Update and restructure their organisational architecture.

To be successful, not only must we cope with, but also be somewhat ahead of, change in information technology, which grows every year and whose rate of change will keep on accelerating, at least until the late 1990s. Therefore, we must focus on training and upgrading our human resources to cope successfully with such change.

Because of the fast developing environment, our personnel loses its skills. Current projections talk of spending 15 percent to 20 percent of our budget on training just to keep up with change. Machines alone will be of no help in:

- Exploiting the existing possibilities for better quality;
- Providing online cost accounting capabilities;
- Having a better integrated PPC;
- Assuring product tracking;
- Swamping investments in inventories;
- Improving upon management control; *and*
- Truly upgrading product quality.

Among many industries the inability to effectively handle these problems is cultural rather than technical. However, competition is moving that way.

Organisational rethinking and restructuring is that much more important because of globality in doing business. Global markets drive for homogeneity, yet many multinational companies today solve problems by country rather than providing solutions for the multinational market to which they appeal. Their management fails to realise that the critical success factors today are:

- Improved margins;
- Increased marketshare;
- Sharply reduced costs;
- Better product quality;
- Shorter time from conception to design, production, and marketing;
- Shorter time for transition from production and marketing to effective distribution;
- Continued after sales service; *and*
- Faster billing and receivables.

In fact, one of the top benefits we want from CIM is an understanding of the real cost of manufacturing. Another key benefit we are seeking is integrated coordination. The tough issues that avant-garde companies have overcome are those of:

1. Getting started;

2. Developing the necessary vision;

3. Replacing the antiquated accounting system;

4. Establishing a basis for cost justification;

5. Staying with the plan they have established; *and*

6. Providing for necessary cultural change.

Such change may be painful, as it upsets the status quo. But, there is no lasting result without it. It is just like a baby being born - and this is part of the human side of CIM.

The emphasis on new accounting principles and systems should not escape attention. We need new cost accounting concepts and procedures to work well with CIM. The old cost accounting is based on direct labour and direct materials (DL+DM). But, computers and robotics eliminated most DL. They weeded it out of the factory. Still we did not change our accounting principles. The result is that incorrect accounting procedures cause errant decisions by management.

Digital Equipment Corporation provide a good example of this last point. Its current accounting system indicated that a certain terminal should be bought rather than made by the company itself - but the indicator missed the point. As bids were invited for submission, a DEC plant manager supplied a bid to make the terminal at his plant, and won. The old cost accounting system did not reflect the level of technology, and what could be done with it by a management alert to the profitability concept.

In another case, a division of Monsanto Chemicals needed to significantly increase productivity of an ageing fibre plant to remain viable. The old cost accounting would not indicate the weakest points, but a thorough systems study was able to achieve significant productivity improvements through a combination of:

- Human resources planning;
- Total quality control;
- Just-in-time inventories; *and*
- Computer integrated manufacturing.

The lesson is that we must always be looking for competitive advantages and ways to exploit them. A great deal of the challenge is open communications, precisely as underlined in the opening paragraphs. We must approach the problem with a new frame of mind, keep high technology under perspective, conceive new approaches, and be able to communicate them to all cognisant people.

CIM and the Communications Prerequisites

We said that, while the design perspectives are important and, in several ways, challenging, CIM implementation means communications. It also means change. It affects the psychology of people and also crucial factors such as:

- organisation and structure;
- required new skills;
- attitudes and awareness; *and*
- performance measurements.

This means that management must be keen in human resources development. Human resources development should both update people and teach them how to manage change. It should do so while the computer integrated manufacturing programme is young and no definite attitudes have yet become ingrained - the classical "do" and "don't" of an organisation.

While technology takes the spotlight, the ability to manage people - the most overlooked resource in an organisation - is quite crucial. Although the automated factory is supposed to run without people, this is only true of the less-skilled labour. The knowledge resources are still essential.

The master plan that we develop should look at computer integrated manufacturing as a technology which aims to unify the projecting, planning and operating activities of the production processes, as well as logistics (administrative management) functions, with the development work characterising products and processes. There are two types of projects in this connection:

1. Projects of large dimensions which must be precisely defined in their details; *and*

2. Smaller projects, less exactly defined and generally oriented to prototyping.

The foundation on which CIM is constructed is typically represented by various forms of automation already existing in the factory. These have been optimised for a particular application, but they have not been integrated. CIM provides the opportunity of doing so.

Based on concepts and trends which I obtained during my research in 1988, Figure 15.4 presents a layered approach to computer integrated manufacturing. It emphasises three levels:

1. Global information architecture;

2. Computer-based linkages; *and*

3. Existing level of automation.

Layer No. 2 acts as the integrator of the discrete island characterizing the lower level, but it also requires a *metalayer* able to provide the global perspective. CIM addresses itself to integration, as described in the upper two layers of this figure.

The goal of computer-based linkages is that of inter-connecting islands of automation, so that they can effectively communicate among themselves. That is where computers

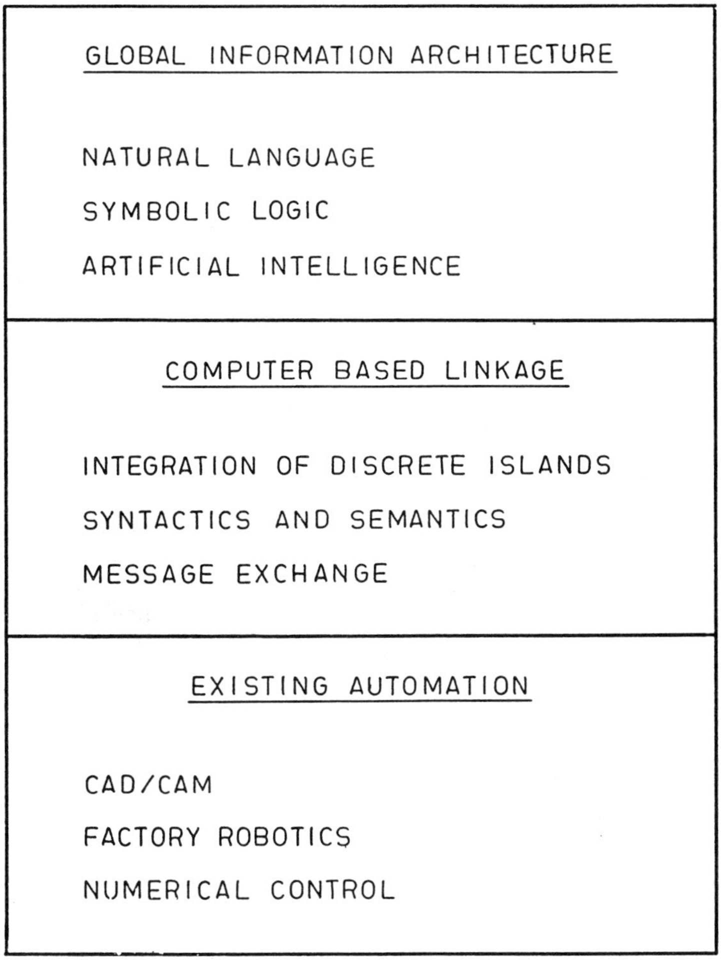

GLOBAL INFORMATION ARCHITECTURE

NATURAL LANGUAGE
SYMBOLIC LOGIC
ARTIFICIAL INTELLIGENCE

COMPUTER BASED LINKAGE

INTEGRATION OF DISCRETE ISLANDS
SYNTACTICS AND SEMANTICS
MESSAGE EXCHANGE

EXISTING AUTOMATION

CAD/CAM
FACTORY ROBOTICS
NUMERICAL CONTROL

Figure 15.4 A Layered Approach to Computer Integrated Manufacturing

play a key role. The proper semantics, syntactics and pragmatics guarantee that the same information will be interpreted by the various automated machines and control units, in precisely the same manner.

The top layer of CIM consists of the global information architecture. At this conceptual level numerous aspects such as perception, cognition, psychology, language, symbolic logic and artificial intelligence come into play. The implementation of CIM strategies requires input from all major functions within the organisation. It also calls for logical constructs which fit a particular CIM strategy.

Artificial intelligence solutions integrate with and assist, rather than compete with, the human element. This approach should be education-intense and involve: cross-functional training for better overall company understanding; active retraining programmes for inividuals to acquire new skills in automation; the teaching of accountability for integrated automation; as well as the notion of personal responsibility.

Once a strategy is established and enters the implementation process, it must be constantly monitored and fine-tuned. Technology continues to change, and people can adapt to that change. People will be constantly learning while performing their new skills.

The system must allow for feedback from everyone in the organisation so that accurate, timely information can be evaluated, improvements made and rewards for performance advancement implemented. A long, hard look must be taken at all activities which enter into the processes to be automated.

This is where the macroengineering skills are necessary. The job is not easy. If we examine the current span of industrial activities we discover that tasks such as order handling are mostly automated. This reference is much less true of activities such as sales, inventory management and production control - particularly in terms of analysis. Production functions have rarely been subjected to highly-intellectual scrutiny such as has been experienced in basic and applied research.

Tying together human factors, financial factors, strategic thinking and logic programming can lead to a successful use of today's technology. Many companies begin to investigate or implement CIM because of the priorities and competitive pressures that the marketplace is placing on the manufacturer today:

- Consistent quality;
- Dependable delivery;
- High-performance products;
- Low prices, hence low costs;
- Rapid design changes;

- Faster new-product introduction; *and*

- Reliable after-sales service.

are issues closely connected to perceived competitive importance. But, the obstacles are many and so are the challenges. Therefore, a CIM program is one of the most intensive and extensive in which a corporation can engage. It is also among the largest capital investment areas facing industry.

The decision to invest in CIM is evolving into a policy investment rather than just a capital investment. Corporations are now concerned with the rate at which they should make their investments, thus initiating a change in standard business practices. With the implementation of CIM, it will be necessary for marketing, accounting, engineering, data processing and manufacturing personnel to become familiar with one another's area of expertise. This knowledge will allow the design of a total system which accommodates the entire organisation's information requirements.

Controlling the Complexity of System Solutions

Our high technology systems can manufacture while unattended, rapidly detect defects, correct machine tolerances without human intervention, and provide total quality control. Taking advantage of these capabilities should lead to greater productivity, lower inventories and substantially lower costs. But, it does not always work that way.

A major part of the difficulty in providing integrated system solutions lies in the fact that organisations include heterogeneous equipment, incompatible operating systems, organisational schemes which are both archaic and modern, installations in multiple locations and ill-defined requirements for disaster backup - among other negatives.

At the same time, system goals call for online security assurance, valid solutions for timely response, the need for common access to all resources and increasing stress on reliability and availability.

An automated environment has inherent in it the need for fast applications development. It also presents requirements for fairly homogenous - hence common solutions. Such solutions must properly support:

- *Uninterrupted production processes*, ideally with zero inventory on the production floor, and minimal inventory in raw materials and ready goods;

- *High Product Quality* despite potential problems in tooling, fabrication and assembly procedures. Detecting and solving problems before they cause serious downstream damage;

- *Shorter project time spans.* As time is money, reducing the time required to perform a task is a valid basis for avoiding undue overhead expense, and also for providing increasing productivity; *and*

- *Assurance of an integrated database* compatible with high-level automation. Descriptive qualitative and quantitative data residing in the computer's memory are easily accessible for updating, changing, and generally providing feedback from design.

An artificial intelligence enriched production activity should go from the planning of the productive sequences and the scheduling of resources to the administration of material deliveries and the coordination with the suppliers' manufacturing plans. This, too, is a macroengineering project.

Such tasks have many elements which are known, and other, which are new. Simple products, manufactured at a high volume, as well as chemical processes, have been automated for many years. Products manufactured in a single copy have little to do with automation, apart from the CAD perspective and the order control requirements. But, between these two references lies the majority of manufacturing activities.

Integrating computer technology and manufacturing is not an optional matter for a company that intends to remain in business. If a company's objective is to prevail over both foreign and domestic competitors, CIM is the strategy of choice. Yet, few managements presently give a commitment to implementing CIM, and low levels of achievement are giving the approach a bad name.

Real commitment implies four things:

1. The company uses CIM as a system to manage the business;

2. The system works in all, or virtually all, areas of the business;

3. The obtained results are truly outstanding; *and*

4. The personnel has been upgraded and brought fully into the picture.

The Japanese are active in all four fields. They are particularly outstanding in the human aspect. That is the story of the *quality circles* originally developed in manufacturing industry and now migrating into banking and other service industry sectors. This is fundamental in creating a good environment where people enjoy coming to work.

The results of getting employees to accept the cultural change necessary for CIM implementation add impressively to a company's performance. Customer service can improve while inventory turnover increases. From scheduling to capacity planning, setups, and routing accuracy, shop efficiency rises.

Because the computer-run system is flexible and reprogrammable, management can vary standard production to keep step with changing market needs and add more options. The small batch environment can become economic, providing a high level of flexibility. Having streamlined their operations through CIM, Japanese car manufacturers, for instance, offer the client the facility to design his car through an interactive, customer-activated approach.

The premise is to provide computer aided design (CAD) workstations at sales showrooms. Rather than going through catalogues of parts, prices and colour samples, the client will make his choices through CAD, literally designing his car from standard components. He then can drive his car through simulators under conditions of his choice - and see the results on screen.

From sales to design and production, CAD and CIM differ from more traditional approaches in several important ways. First, the flexibility inherent in software allows industry to respond rapidly to changing competitive situations. This leads to a synergetic integration of systems and markets, whose output is enriched through computer graphics.

Second, as the implementation grows, costs associated with CIM decrease. This is attributed to upward compatibility of hardware and improving price/performance. They help reduce the effective cost of downstream installations. The benefits of CIM further grow as software performance improves.

Third, manufacturers are better able to accommodate the needs of their customers and suppliers because of the increased flexibility in the manufacturing facility. Improvements in communications resulting from the implementation of a CIM program allow a company to respond to opportunities in a more timely manner and to benefit the clients in the process.

Knowledge Engineering and the Challenge of Robotics

A sense of gigantism does not exist only in physical projects of a large dimension. It is also present in immense logical undertakings which either stress the limits of imagination (such as the origin of this world) or which require logical solutions well beyond those we know from the past and to which there is reference. Knowledge engineering for mobile robots is an example.

Robots first moved into the world of reality as remote manipulators in nuclear power plants. They were used in manufacturing by the early 1980s, particularly aiming at liberating humans from tiresome or hazardous labour. Currently, science is leaning towards the development of a new robot generation with greatly expanded abilities, able to perform inspection and maintenance of plants and to engage in ocean engineering in environments hostile to man.

As the preceding section underlined, computer controlled manufacturing requires machines and humans to cooperate in the same workplace. In turn, this makes mobility an important factor in new generation robots.

The role of *knowledge engineering* in a robotics setting starts with this reference. The typical real life environment exhibits continuous variation for the mobile robot. Hence, it needs intelligence to understand it and make plans autonomously of central resources, although these resources may have established the robot's master plan.

As Kamejima, Funabashi, and Ichikawa underlined in a study which they conducted, * robots are much more useful if they can steer themselves around the workplace. New technological breakthroughs in the implementation of machine intelligence for environment understanding and planning, permit the use of knowledge-based approaches, including:

- hierarchical modelling of surroundings;
- object recognition; *and*
- identification of environment status

Experimental studies carried out in Japan with mobile robots have demonstrated the practicality of this technology in the near future. They give evidence that mobile robots can manoeuvre autonomously using an environment model in which various levels of a priori knowledge about the surroundings are stored. Such knowledge is matched with data observed through sensors and artificial vision constructs.

The robot moves by processing sensory data based on its stored map.**The latter shows place locations and the shape and structure of objects. A node map combines passages and places at which motion primitives must be altered. Such a map is used to determine the path from the current node to a pre-assigned goal, optimising the tradeoff between elapsed time and the energy consumption required for each passage.

- Robots utilise the map describing their surroundings from various points of view;
- The node map specifies the connection of nodes and manoeuvre cost for passages;
- 3-dimensional segments associated with object contours are classified according to passage direction and are stored at each node; *and*
- An image structure list contains a set of parameters to identify the status of the environment.

This is the logical part of the structure necessary to create and implement a mobile robot environment. A good example of the physical structure is a three-ton robot

---

* Sponsored by ICOT, the Japanese New Computer Generation project.

* * An approach which will be greatly enriched by integrating into it smart skin effects, as covered in a preceding chapter.

which came to life in a workshop at Ohio State University, in late 1985. Each one of its legs is powerful enough to crush a car but gentle enough to manipulate a fragile box.

The builders of Ohio State's Adaptive Suspension Vehicle (ASV) which has so far taken two decades of research and development, believe it is almost ready. The project began as a research exercise for developing the theory of robotics, and in the early years, financial support came mainly from the National Science Foundation. But, because a large walking robot has obvious military potential, the Defense Department has been the main supporter of the project since 1980.

Controlled partly by a battery of computers and partly by a human driver, the robot will manoeuvre through forests and desert sand, up and down steep hills, across ditches up to 2.7 metres (9 feet) wide and over obstacles. Scanning its path with infrared laser vision, the 5 metre-long machine plans each step it takes, looks for footholds, avoids holes, and makes the best of whatever terrain it has to cover.

The moving robot uses acoustic echoes to gauge the proximity of neighbouring objects.

- Pressure sensors in its footpads inform its computer of the nature of the ground it is covering.

- A gyroscope balance sensor serves the same purpose as an animal's vestibular system.

The robot is fairly sure-footed. At first, a human driver will control the higher functions of the machine. The driver can override the computer's control of individual leg movements, but since these operations are so complex and difficult, he would rarely choose to take them on.

In future years, the researchers intend to field a version of the walker so improved that it would perform automatically, with no driver aboard. The radar-like data sent to onboard computers by the scanning infrared laser beam will enable the machine, not only to steer and manoeuvre itself, but to carry out a programmed sequence of actions. Here the Japanese research on knowledge-based reasoning seems to integrate nicely.

Moving robots enriched with knowledge engineering will be part of a global network of production facilities which will itself depend on computers for integration. Throughout business and industry, the direction is towards global networking with communications supported anywhere, with anyone (man or machine) all the time. Solutions point towards two drives. First, the ability to integrate into one logical network many physical networks in operation today. We have spoken of such a requirement in Section 3. Second, the proper inter-connection among wide area networks (WAN) and local area networks (LAN) with peer-to-peer capability.

Communications networks are a robotics structure by excellence. While autonomous intelligent vehicles are considered to be the real testbed of 6th Generation Computers, an AI-enriched, photonics-based network is just as challenging an achievement. Its impact can be wide-ranging; the level of automation which it features underpins future developments in competitiveness; the sophistication of the expert systems which it includes makes feasible a greater degree of integration than may otherwise be possible.

Organisational Prerequisites for an Advanced Integration Capability

Within a business environment, system integration should be making a company's collection of computers, software, PBX and other communications devices effectively work together. But, to integrate all facilities, we first have to automate the:

- planning premises;
- day-to-day office work;
- engineering design chores;
- manufacturing plants; *and*
- a considerable number of decision issues.

We said that interconnecting an organisation's distributed computer operations requires a large and sophisticated private telecommunications network. It must be engineered to cut costs by raising productivity and improving inventory control. And it must be designed to connect a number of intelligent terminals, which would tend to be equal to the employment level.

The systems solution we will adopt must reflect the fact that users must increase their sophistication to be able to make decisions about new telecommunications technologies. They must be in a position to choose network services to fit their unique needs. They must also be smart enough to use the network in a competent manner, capitalising on the control, flexibility and service that it provides. The networks themselves must be as widely available as possible, sharing the benefits of the Information Age with as many users as is feasible.

Such networks cannot be designed by hand. They are so large-size that they can be tackled only through computer aided approaches. A good example is an expert system for networking designed by Bolt, Beranek and Newman (BBN). Steadily projected factors estimated by the expert system are:

● Costs;

● Busiest node identification; *and*

● Link utilisation.

Results are mapped into graphics. The expert system developers observed what an expert designer is typically doing, and used it as basic command structure. "No matter what artificial intelligence is, it is a very useful system", said one of the BBN executives.

Activities involved in network design and implementation are shown in Figure 15.5. They include network development, implementation and administration. Each divides into further detail, and each can, and should, be enhanced through artificial intelligence.

Major part pf the expert systems support should be diagnostics intimately connected to network control centre (NCC) operations. NCC duties include: network-wide supervisory activities; testing (line, modem, terminal, software, text and data); quality control database (playbacks and analyses); providing for fail-soft operations; restart and recovery; assuring statistics for dimensioning studies; as well as journaling and security.

At every place where a network is installed, an NCC facility is necessary. Network facilities will be increasingly implemented in manufacturing as the industry seeks to integrate discrete islands of implementation. We have reviewed many of them but, by way of conclusion, we can now classify them into five major categories:

1. *Computer Aided Design (CAD)*

CAD systems typically run on specialised graphics-oriented engines. Most often they are not tied with the rest of the manufacturing system. But they should be. Only what is integrated and operates online is of importance.

2. *Manufacturing Requirements Planning (MRP)*

MRP is complex, tracking orders, supplies, suppliers and the production floor. The problem is that there are few, if any, automated links between CAD and MRP. Computer Integrated Manufacturing aims to correct this imbalance.

3. *The Automated Shop Floor (ASF)*

Today, a typical ASF has computer controlled tools and robots - which cannot communicate. As a result, a study carried out not long ago demonstrates that some 40 percent of investment is watered down, to pay for interfaces in sprawling incompatible environments. The need for a thorough IT study on system integration cannot be dramatised in a better way.

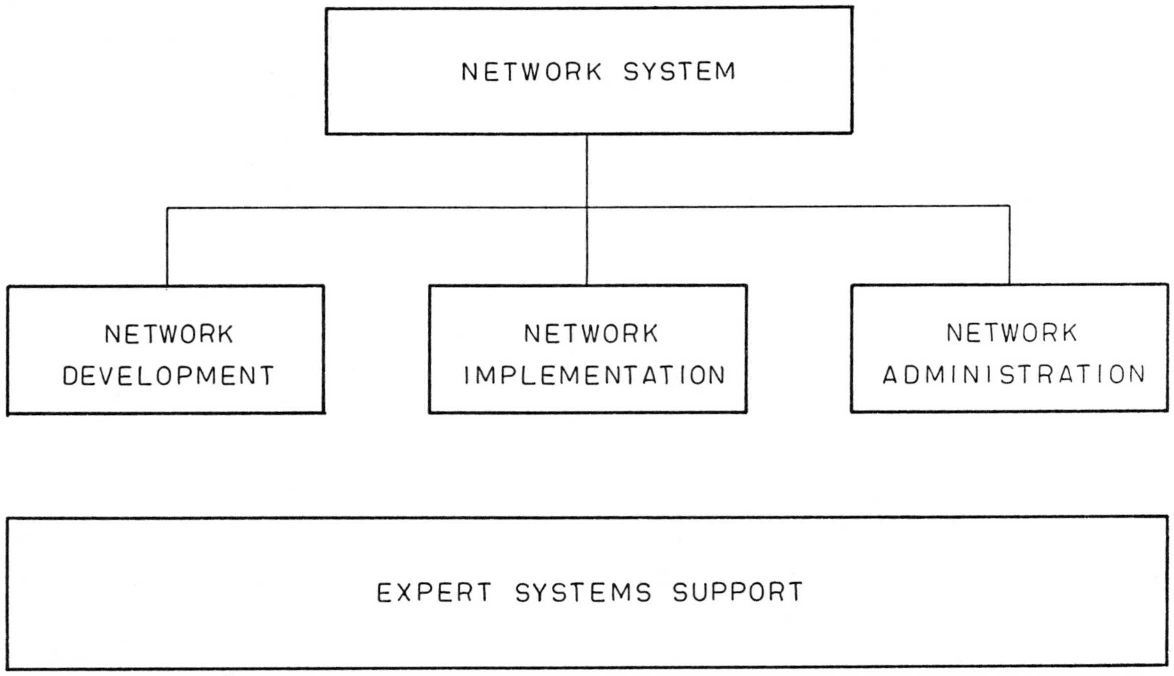

NETWORK SYSTEM

NETWORK DEVELOPMENT

NETWORK IMPLEMENTATION

NETWORK ADMINISTRATION

EXPERT SYSTEMS SUPPORT

Figure 15.5

4. *Word processing at headquarters, factories and sales offices*

An estimated 75 percent of all paper-based information is for internal consumption. Compound electronic documents go beyond data processing/word processing (DP/WP) integration and point to the future direction of a comprehensive, all-inclusive environment.

5. *Integrated Mainframe Support*

Every organisation has a variety of incompatible mainframes and/or operating systems under which they run. Infocentre implementation magnifies the need for transparency. Networking must be done both for the remote sites and inhouse - but without integration it will provide suboptimal results.

A number of trends are emerging that offer opportunities for meaningful developments. At the same time, these trends are creating additional challenges for engineering, marketing and maintenance of large scale systems.

One of the least appreciated facts is that the area of software engineering is just beginning to grow out of its infancy. Such growth is spurred in part by the maturity and dissemination of basic programming practices, in part from the new tools at our disposition, and in part by the fact that IT management starts living up to the challenge. But some pillars are still missing in the structure of the new approaches. For instance: metrics, and the ability to develop higher quality, more maintainable software systems than in the past.

At the same time, system demands for much larger, more complex software aggregates continue to expand. Size and complexity of projects, such as autonomous vehicles, pattern recognition in weapons systems, SDI or the advanced technology fighter dwarf what has been done through software so far. As a result, software engineering finds it increasingly difficult to meet requirements, even with modern programming practices.

There is also the hurdle of extremely large inventories of programs which have become absolutely unmanageable. According to some estimates, the US Air Force has today 17 billion statements in inventory to be maintained. Even at 20 percent per year maintenance cost, this means 3.4 billion statements as if to be written new every year.

Ways and means which have been used for 35 years to tackle the software problem can no longer be of help. Emulating approaches in more traditional engineering disciplines, professionals try to identify and resolve implementation risks earlier in the development cycle. They also aim to develop designs of software systems based on the systematic use of standard, interchangeable parts which can undergo thorough tests.

Of significant importance is prototyping, as the sophistication of the effort increases in a function of time, and the same is true of the requirements to be met. As Figure 15.6 suggests:

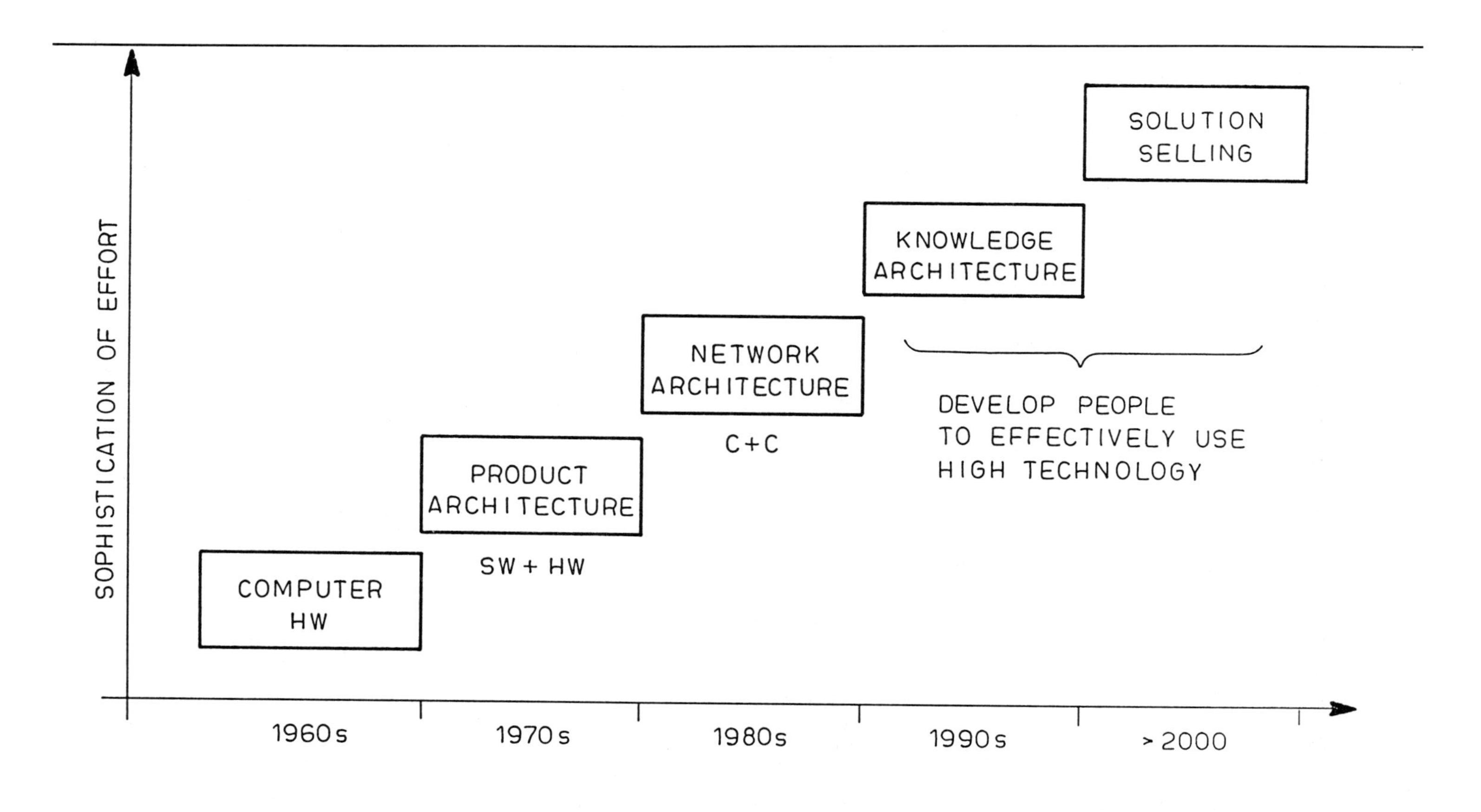
SOPHISTICATION OF EFFORT
SOLUTION SELLING
KNOWLEDGE ARCHITECTURE
NETWORK ARCHITECTURE
C+C
DEVELOP PEOPLE TO EFFECTIVELY USE HIGH TECHNOLOGY
PRODUCT ARCHITECTURE
SW + HW
COMPUTER HW
1960s
1970s
1980s
1990s
> 2000

Figure 15.6

- The focal point of the 1960s was computer hardware, turning the ALU-based machine of the 1950s into a memory-based system;

- The 1970s saw an emphasis on product architecture, which valued software higher than hardware;

- During the 1980s, a great deal of interest has been concentrated on network architectures, integrating communications and computers into one aggregate; *and*

- The 1990s will see large investments in Knowledge Engineering, with the payoff increasingly depending on AI - and this from software development to enduser-oriented implementation.

The design of integrated knowledge-based systems will bring forward new engineering practices. The use of modules and standard exchange parts for most system design, for instance, will allow system architects and designers to concentrate design efforts on those areas that have novel requirements or constraints. Typically, these may be a small fraction, 10 to 15 percent, of the modules of finally assembled systems - but they will be the most important ones.

As a matter of principle, to act as a true standard part, the AI/SW/HW construct should be characterised by a blackbox-type specification. Its details should be opaque to the user, the external specification being sufficient to uniquely characterise it. This will make it possible to catalogue parts based on an identification system keyed to these specifications -and to increasingly use off-the-shelf parts as a way to more efficient design.

Such an approach will not only permit us to control the sprawling complexity of software systems and help avoid redundant effort. It will also make possible *solution selling* which, throughout the 1990s, will characterise communications and computer installations designed and marketed with the aim of obtaining measurable results.

INDEX

C

D

J

K

L

M

N

O

P

Q

R

S

T

U

V

W

X

Y

Z